S0-BET-598

Multilingual Dictionary of Analytical Terms

International Union of Pure and Applied Chemistry
Analytical Chemistry Division

Multilingual Dictionary of Analytical Terms

English, French, German, Spanish, Russian, Chinese and Japanese

PREPARED FOR PUBLICATION BY

ROBERT A. CHALMERS

Department of Chemistry, University of Aberdeen, Scotland

ERRATUM

Chalmers: Multilingual Dictionary of Analytical Terms

p. ix The name of the Chinese translator 9 lines from the foot of the page should be Prof. Zhang Yonghua

OXFORD

Blackwell Scientific Publications

LONDON EDINBURGH BOSTON

MELBOURNE PARIS BERLIN VIENNA

Published by
Blackwell Scientific Publications
Editorial Offices:
Osney Mead, Oxford OX2 0EL
25 John Street, London WC1N 2BL
23 Ainslie Place, Edinburgh EH3 6AJ
238 Main Street, Cambridge
Massachusetts 02142, USA
54 University Street, Carlton
Victoria 3053, Australia

Other Editorial Offices:
Librairie Arnette SA
1, rue de Lille
75007 Paris
France

Blackwell Wissenschafts-Verlag GmbH
Düsseldorfer Str. 38
D-10707 Berlin
Germany

Blackwell MZV
Feldgasse 13
A-1238 Wien
Austria

First published 1994

Printed and bound in Great Britain
at the Alden Press Limited, Oxford
and Northampton

A catalogue record for this title is available from the British Library

ISBN 0-86542-859-X

Library of Congress
Cataloging-in-Publication Data

Multilingual dictionary of analytical terms:
English, French, German, Spanish, Russian, Chinese and Japanese/prepared for publication by Robert A. Chalmers.
p. cm.
Includes bibliographical references and index.
ISBN 0-86542-859-X
1. Chemistry, Analytic—Dictionaries—Polyglot.
2. Dictionaries, Polyglot.
I. Chalmers, Robert Alexander.
QD71.5.M85 1994
543'.003—dc20 94-11359
CIP

DISTRIBUTORS
Marston Book Services Ltd
PO Box 87
Oxford OX2 0DT
(*Orders*: Tel: 0865 791155
Fax: 0865 791927
Telex: 837515)

Australia
Blackwell Scientific Publications Pty Ltd
54 University Street
Carlton, Victoria 3053
(*Orders*: Tel: 03 347-5552)

Distributed in the USA
and North America by
CRC Press, Inc.
2000 Corporate Blvd, NW
Boca Raton
Florida 33431

Contents

PREFACE, vi

ACKNOWLEDGEMENTS, ix

ENGLISH, 1

CROSS-REFERENCE TABLE, 27

FRENCH, 71

GERMAN, 97

SPANISH, 121

RUSSIAN, 145

CHINESE, 179

JAPANESE, 233

APPENDIX: CORRECTIONS TO INDEXES, 273

Preface

The idea of a multilingual dictionary for analytical chemistry was conceived at the 1977 IUPAC General Assembly in Warsaw, but in the two years of preparatory work it became evident that compilation of a full-scale dictionary would involve too great an expense. It was therefore decided at the 1983 Lyngby Assembly that the subject index of the 'Orange Book' (the *Compendium of Analytical Nomenclature*) should be translated into several languages so that the English definitions of the terms concerned would be more readily and unambiguously available to analytical chemists of other nationalities. Translation of the index into Chinese, French, German, Japanese, Russian and Spanish was organized, but by the time it was completed the second edition of the Orange Book was in preparation and it seemed more useful to amalgamate the two indexes.

When this task was approached, various problems arose. The first edition of the Orange Book had 1716 entries in its index, and the second edition had almost 3000. There was considerable duplication of material because of extensive cross-indexing and the appearance of some items in both editions, but even with elimination of this there remained more than 2500 entries to be translated. The most economical way of presenting the information seemed to be listing the English terms alphabetically, along with a serial number and a reference to the Orange Book edition and page number where the definition or use of the term could be found, then adding individual alphabetically arranged and serially numbered lists of the corresponding terms in the other languages (each term being accompanied by the appropriate English term number). Finally there would be a complete cross-reference listing of the English numbers and the corresponding numbers for the other languages. In this way a given term could be translated from one language into another through the cross-reference listing.

Another problem was that of word order and the choice of a key word in English. Most of the terms indexed appear in the text as main or sub-headings, or are underlined or italicized, but some occur only as part of the text and are not distinguished in some way. The terms also generally occur as groups of words in the order normally used in speech or writing (although there are exceptions to this in some headings, because of the punctuation used), so it seemed more logical to list them alphabetically according to the first word of the term, and in the order used in the Orange Books (which explains why T-tubes occurs as the first item in the 't' section of the index). The same approach was used for the other languages, but more than one version may be given for the same English term, as explained later. Where no standard equivalent could be found, an equivalent phrase was invented.

The original plan was to include the IUPAC-approved definitions of the terms, but some terms are not specifically defined and are used only in context. To prepare definitions of such terms and obtain IUPAC approval of them would have been extremely time-consuming, and in any case definitions of these terms are readily available in technical dictionaries and encyclopaedias. The approved definitions are available in the Orange Books or the recommended nomenclature reports published in *Pure and Applied Chemistry*, and also in the IUPAC 'Gold Book' (V. Gold, K.L.

Loening, A.D. McNaught and P. Sehmi, *Compendium of Chemical Terminology*). An exception was made in response to a request from the compiler to include some terms that were commonly used in the older literature but might be unfamiliar to younger chemists, and also some 'modern' terms less familiar to much older chemists. These 'unofficial' terms are indicated by an asterisk, and are accompanied by a definition (or a reference if they happen to have been later approved by IUPAC). The definitions given are sometimes paraphrases of published definitions, but sometimes concocted by the compiler (to whom any complaints should be addressed). Some of these terms may seem trivial to the reader, but have been added to match analogous terms in the Orange Books. The compiler has also taken the editorial liberty of adding comments (sometimes etymological, sometimes correcting misprints in the Orange Books, and occasionally offering an alternative or an expansion for an approved definition) on the IUPAC terms.

In the English listing the Orange Book edition and page number are given (for example, I/5, II/9 would indicate page 5 in the first edition and page 9 in the second). This was done because the compiler thought it easier to find a word on a page than to track down the multiply-divided decimal section numbers given in the two indexes. He also used his editorial prerogative to decide on spelling and punctuation (British spelling was used in the first edition, American in the second, and the punctuation originally used sometimes created problems). He apologizes for the fact that the English entries do not always agree exactly with the original index entries; this is sometimes the result of a misprint in the original index or text, but is sometimes a typing error on his own part.

In the French, German and Spanish lists, the serial number for that language is followed by the corresponding English number. If the same word is used as an equivalent for more than one English term, it is repeated. If alternatives are offered, they may be given in parentheses or separated by a comma.

In the Russian, Chinese and Japanese, the serial number is followed by the appropriate term, then in Chinese and Japanese there is a phonetic rendering of the pronunciation, and last there is the English number. This mode of presentation is adopted because terms are not themselves repeated if they are used for more than one English term, but all the English term numbers are given, and it would look rather untidy if these numbers appeared on the left of the written terms. In the Chinese, characters in parentheses are intended to give the context of the term, and characters in square brackets are alternatives for the same number of characters immediately preceding them. Alternative phrases or terms may be separated by a comma, or given as separate terms. The Chinese terms are arranged in phonetic alphabetical order (according to the Pinyin system, but the diacritical marks indicating the different tonal values have been omitted). Because the same ideogram may have more than one tonal value it may seem to some readers that the phonetic alphabetical order has not always been strictly followed, but this is not the case. In the Japanese section the terms are arranged in the order of the Japanese alphabet, and are divided by appropriate headings.

It will be found that a few terms in the original indexes have been omitted. These were terms such as 'abbreviations', or 'choice of coordinates for electrochemical plots', which seemed to the compiler not to need translation within the proposed scope of the dictionary. A few terms that were either not translated or were accidentally omitted from the English listing, have now been translated and are placed at the end of each language section. They are given numbers prefixed by A, and the corresponding English numbers are given; the cross-referencing is done either within or at the end of the cross-reference table, as appropriate. This table is placed between the English section and the other language sections, for convenience in use; the columns are arranged in the order of appearance of the languages, namely English, French, German, Spanish, Russian, Chinese, Japanese, and are headed by the initial letter of the language concerned.

In principle it would have been possible to include defined terms from IUPAC nomenclature reports that were not available until after compilation of the second edition of the Orange Book was completed, and from other official sources, but it would never be possible to produce a completely up to date dictionary in this way. It would be simpler to produce supplementary material as need or demand arose. One way would be to publish translations along with the reports, but this would create problems of its own.

The accuracy of the translations has been checked by back-translation as far as lay within the power of the compiler (with considerable help from a number of colleagues), who hopes to be forgiven for any errors he introduced or failed to detect.

The dictionary could not have been achieved without the enormous amount of hard and careful work put in by the translators and their typists, for which IUPAC and the compiler are extremely grateful.

The compiler is particularly indebted to Dr. Mary Masson for her invaluable tuition in the mysteries of word-processing and for her assistance in preparing the camera-ready copy, and also to his wife for her encouragement and the use of the kitchen table.

Acknowledgements

The first version of this dictionary, based on the first edition of the Orange Book, was organized by Prof. E. Pungor (Budapest) but for reasons given in the Preface was never brought into printed form. The translators were as follows:

French: Prof. J. Robin
German: Prof. G. Henrion (assisted by Dr. F. Scholz, Dr. G. Michael, Dr. R. Schmidt, Dr. P. Heiningen, Mrs. B. Boeden)
Spanish: Prof. R. Gallego Andreu
Russian: Prof. Yu.A. Zolotov (assisted by Dr. V. Izvekov)
Chinese: Prof. H. Kao
Japanese: Prof. Shizuo Fujiwara (the index was prepared by Dr. Akira Yamasaki).
A numerical cross-reference index for that version was prepared by Dr. Éva Fekete (Budapest).

Valuable though their work was, it was not directly used in preparing the present version, mainly because it was easier to compile the new English listing and then have that translated anew. Their work was very useful, however, in checking the new translations. The new translation team, jointly organized by Prof. Pungor and the compiler, was as follows:

French: Prof. M. Gross (Strasbourg)
German: Prof. G. Ackermann (Dresden)
Spanish: Antonio Losada (Córdoba)
Russian: Prof. Yu.A. Zolotov and Dr. E.Ya. Neiman (Moscow)
Chinese: Prof. Zhang Honghua and Prof. Wang Erkang (Changchun)
Japanese: Profs. Y. Hasegawa (Nagoya), K. Hirokawa (Sendai), K. Hozumi (Kyoto), H. Kawaguchi (Nagoya), T. Kawashima (Tuskuba), K. Izutsu (Matsumoto), I. Mita (Tokyo), T. Nasu (Sapporo), M. Senda (Fukui), Y. Sugitani (Hiratsuka), M. Taga (Sapporo), M. Tsuchiya (Yokohama), A. Tsuji (Tokyo) and K. Tomura (Tokyo).

In addition, the compiler was given linguistic and other help and advice by several colleagues and friends, but special thanks are due to Dr. S.-C. Tam, Mr. Lam Wang, Dr. S. Garcia-Martin and Miss S. Taudte, and in particular to Dr. A.N. Davies and Dr. Helgard Staat (Dortmund) for checking and expanding the German listing.

English

0001 abnormal glow discharge: II/149, 150
0002 absolute counting: II/209
0003 absolute full energy peak efficiency: II/209
0004 absolute photopeak efficiency: II/209
0005 absolute temperature: I/143
0006 absorbance: I/141
0007 absorbed dose: II/209
0008 absorbed electrons: II/247
0009 absorber: II/210
0010 absorption: II/121, 187
0011 absorption coefficient: II/122, 210
0012 absorption edge jump ratio: II/175, 176
0013 absorption edge wavelength: II/174, 176
0014 absorption energy: II/210 (indexing error for energy absorption)
0015 absorption factor: I/101, 141
0016 absorption of gamma radiation: II/231
0017 absorption of particles: II/225
0018 absorption path-length: I/141
0019 absorption profile: II/122
0020 absorption transition probability: II/121
0021 absorptivity*: absorption coefficient
0022 accuracy: I/9, 116, II/6
0023 acid–base indicator: I/29, II/48
0024 acid–base interaction: II/189
0025 acid–base titration: I/41, II/47
0026 acidimetric titration: I/41, II/47
0027 acidimetry: I/35, II/47
0028 acidity constant: I/57, II/13
0029 acidity functions: I/51, II/31
0030 acoustical shock-wave sifter system: II/126
0031 activatable material: II/225
0032 activation: II/210
0033 activation analysis: II/210
0034 activation cross-section: II/218
0035 active medium: II/141
0036 active solid: I/65, II/98
0037 activity: II/9, 211
0038 activity coefficient: I/46, II/7, 18, 25
0039 activity concentration: II/211
0040 additives: I/107, 131, II/159
0041 additivity*: the principle that the analytical signal from a mixture is the sum of the signals that would be given by the individual components if measured separately
0042 adjusted retention volume: I/67, 83, II/104
0043 adsorbate: I/20, II/85
0044 adsorbent: I/20, II/85
0045 adsorption: I/20, II/85
0046 adsorption chromatography: I/75, II/94
0047 adsorption current: II/52
0048 adsorption indicator: I/39, II/48
0049 aerosol: I/121, II/126
0050 aerosol carrier gas: II/136
0051 agglomeration: I/19, II/126
0052 aggregate: I/19
0053 aggregation: I/19
0054 aging (ageing): I/21
0055 air-peak: I/67
0056 alkalimetric titration: I/41, II/47
0057 alkalimetry: I/35, II/47
0058 alkalize*: make alkaline
0059 α-cleavage: I/34, II/207
0060 α-coefficient*: the side-reaction coefficient, defined as the ratio of the conditional to the equilibrium concentration of a reaction component (A. Ringbom, *Complexation in Analytical Chemistry*, Interscience, 1963)
0061 alpha decay: II/211
0062 α parameter: I/142
0063 alpha particle: II/211
0064 alternating current amplitude: II/53
0065 alternating current arc: II/123
0066 alternating current chronoamperometry: I/157, II/75
0067 alternating current polarography: I/160, II/79
0068 alternating voltage: II/64
0069 alternating voltage amplitude: II/64
0070 alternating voltage chronopotentiometry: I/161, II/80
0071 alternating voltage polarography: I/161
0072 aminopolycarboxylic acid: I/35
0073 amperometric end-point detection: I/36
0074 amperometric titration: I/154, II/47, 73
0075 amperometry: I/154, II/73
0076 amphiprotic solvent: I/50, II/30, 31, 32
0077 amplification reaction*: *Pure Appl. Chem.*, 1982, **54**, 2554
0078 analogue to digital converter (ADC): II/211
0079 analysis element: I/101, 102, II/111
0080 analyte: I/130, II/111
0081 analyte addition technique: II/172
0082 analyte atoms: II/147
0083 analyte ions: II/147
0084 analytical addition technique: II/139
0085 analytical balance: I/13, II/37
0086 analytical calibration: I/131
0087 analytical calibration functions: I/116
0088 analytical curve: I/103, 116, 131, II/158
0089 analytical curve technique: I/132, II/139
0090 analytical electrode: II/125
0091 analytical electron microscopy: II/126

0092 analytical evaluation curves: I/116
0093 analytical functions: I/116
0094 analytical gap: II/125, 129
0095 analytical radiochemistry: II/211
0096 analytical result: I/131
0097 analytical samples: I/101 (see also *Pure Appl. Chem.*, 1990, **62**, 1200, 1206)
0098 analytical standards*: reference or certified materials
0099 analytical systems*: combinations of reactions, equipment and techniques
0100 angle of divergence: II/141
0101 angular dispersion: I/101, II/158
0102 angular nebulizer: I/122, II/165
0103 anion effect: I/136, II/170
0104 anion-exchange: I/89
0105 anion-exchanger: I/88
0106 anionic complex*: a complex carrying a negative charge
0107 anisotropy: II/198
0108 annihilate: II/212
0109 annihilation: II/211
0110 annihilation radiation: II/211
0111 annular anode: II/150
0112 anode dark space: II/147
0113 anode gap: II/153
0114 anode glow: II/147
0115 anodic stripping chronoamperometry with linear potential sweep: II/66
0116 anodic stripping controlled-potential coulometry: I/150, II/66
0117 anodic stripping voltammetry: I/150, II/66
0118 anodic vaporization: II/125
0119 anti-Compton gamma ray: II/211 (indexed unnecessarily)
0120 anti-Compton gamma ray spectrometer: II/227 (cross-referenced from II/211 via II/243)
0121 antiparticle: II/212
0122 anti-Stokes fluorescence: I/140
0123 aperture stop diameter: I/100, II/156, 157
0124 apex: II/51, 53
0125 apparent concentration: I/134
0126 apparent pH*: the pH recorded for a solution when the solvent is not purely aqueous but the pH-meter has been standardized with a purely aqueous solution
0127 appearance energy: I/33, II/205
0128 appearance potential: I/33, II/205
0129 appearance temperature*: *Pure Appl. Chem.*, 1992, **64**, 257
0130 applied potential: II/61, 64
0131 arc: II/123
0132 arc atmosphere: II/128
0133 arc column: II/124
0134 arc gap: II/125
0135 arc lamp, II/150
0136 arc-like discharge: II/133
0137 arc line: I/107
0138 artificial radioactivity: II/212
0139 aspirate: I/121, II/163
0140 aspiration*: drawing up by means of suction; aspirating
0141 aspiration rate: I/123
0142 aspirator*: apparatus for aspiration (indexed in I but not used in text)
0143 Aston dark space: II/147
0144 asymmetry potential: II/27
0145 atmospheric pressure microwave-induced plasma: II/140
0146 atomic absorption*: absorption of electromagnetic radiation of suitable wavelength by free atoms
0147 atomic absorption spectrometry*: measurement of an analyte by means of atomic absorption
0148 atomic bombardment: II/145
0149 atomic emission*: emission of photons by thermally excited free atoms
0150 atomic emission spectrometry*: measurement of an analyte by means of atomic emission
0151 atomic fluorescence: I/139
0152 atomic fluorescence spectrometry*: determination of an element by measuring its atomic fluorescence
0153 atomic line: I/107, II/118
0154 atomic mass: II/212
0155 atomic mass units: II/212
0156 atomic number: II/212
0157 atomic spectra: II/128
0158 atomic spectral lines: II/118
0159 atomic spectroscopy: II/111
0160 atomic weight: II/212
0161 atomization: I/122, II/118, 143, 165
0162 atomization efficiency: I/123, 124, II/139
0163 atomized fraction: I/124, II/169
0164 atomizer: I/122, II/165
0165 attenuation: II/212
0166 attenuation coefficient: II/212
0167 Auger effect: II/212
0168 Auger electron: II/212
0169 Auger electron spectroscopy: II/249
0170 Auger electron yield: II/252
0171 Auger peak energy: II/252
0172 Auger spectroscopy: II/246
0173 Auger yield: II/212
0174 auto-ionization: I/33, II/206
0175 automate: I/23
0176 automation: I/23
0177 automatization: I/23
0178 automatize: I/23
0179 autoradiograph: II/212
0180 autoxidation*: spontaneous oxidation (usually self-catalysed) in the presence of air or oxygen, without combustion
0181 auxiliary discharge: II/144
0182 auxiliary electrode: I/151, II/59, 65
0183 auxiliary spark gap: II/132
0184 average: I/9, II/4
0185 average flow-rate: II/63
0186 average gas pressure: II/117
0187 average life: II/213
0188 average ligand number: I/57, II/13
0189 axial electrode: II/139
0190 back-extract (noun): I/62
0191 back-extract (verb): I/62
0192 back-extraction: I/62
0193 background: I/103, II/159
0194 background correction: I/115, II/139

0195 background mass spectrum: II/207
0196 background radiation: II/213
0197 back-scatter: II/213
0198 back-scattered electrons: II/247, 248
0199 back-scattering spectroscopy (BSS): II/250
0200 back-titrate*; titrate the excess of a titrant already added
0201 back-titration: I/36
0202 Baker–Sampson–Seidel transformation: I/109, II/161
0203 balance*: an analytical balance
0204 band spectra II/118
0205 bandwidth: II/154
0206 barn: II/213
0207 base*: a substance with ability to react with acids to form salts; frequently used as a synonym for alkali
0208 base electrolyte: II/60
0209 baseline: I/66, 77, II/43, 96, 101
0210 baseline technique in atomic spectroscopy: indexed in I but not used in text; see *Pure Appl. Chem.*, 1976, **45**, 117
0211 base peak: I/33, II/206
0212 basicity*: tendency of a compound to act as a proton acceptor
0213 Bates–Guggenheim convention: II/19, 21
0214 bathochromic shift*: *Pure Appl. Chem.*, 1983, **55**, 1292
0215 becquerel: II/213
0216 bed volume: I/89, II/100
0217 bed volume capacity: I/90
0218 β-cleavage: I/34, II/207
0219 beta decay: II/213
0220 beta particle: II/213
0221 biamperometric end-point detection: I/37
0222 biamperometric titration: I/154
0223 biamperometry: I/154
0224 bias: I/10, 134
0225 biased linear pulse amplifier: II/211, 213
0226 bifunctional ion-exchanger: I/89
0227 binary collisions: II/148
0228 binding*: bond formation, bonding; indexed in II but not defined in text
0229 binding energy: II/174, 245
0230 binuclear complexes*: complexes containing two metal atoms
0231 bipotentiometric titration: I/153
0232 bipotentiometry: I/153
0233 black body: I/79, II/154
0234 black body radiation: II/119
0235 blank*: an analysis performed under the same conditions as the test analysis, but in the absence of the sample
0236 blank background: I/131
0237 blank correction: I/115
0238 blank measure: I/131, II/199
0239 blank scatter: I/133
0240 blank solution: I/131
0241 blank titration: I/36
0242 blaze angle: I/100, II/157
0243 blaze wavelength: I/100, II/157
0244 bleachable substance: II/143
0245 block: II/43
0246 blow-in procedures: II/126
0247 blue shift*: *Pure Appl. Chem.*, 1983, **55**, 1292
0248 Boersma-type arrangement: II/45
0249 boiling point*: the temperature at which the vapour pressure of a liquid is equal to atmospheric pressure
0250 boiling temperature: II/141
0251 bolometer: II/193
0252 boosted hollow cathode: II/152
0253 boosted magnetic field glow discharge: II/153
0254 boosted output glow discharge: II/153
0255 bracketing procedure: I/132, II/24, 139 (also indexed as bracketing technique)
0256 Bragg angle: II/176
0257 Bragg equation: II/176
0258 branching decay: II'213
0259 branching fraction: II/213
0260 branching ratio: II/213, 214
0261 breakdown jitters: II/133 (incorrectly indexed as breakdown litters)
0262 breakdown time: II/132
0263 breakdown voltage: II/132
0264 breakthrough capacity: I/90
0265 breakthrough volume*: the volume of eluent that has passed through an ion-exchange column when a solute transported from the start of the column first appears in the detector
0266 bremsstrahlung: II/213
0267 bridge solution: II/28
0268 bridge solution of a double-junction reference electrode: II/28
0269 broadening: II/121, 122
0270 brush discharge: II/139
0271 buffer (noun)*: a solute with the ability to mimimize certain concentrational changes that would otherwise occur as the result of a chemical reaction
0272 buffer (verb)*: add a buffering agent
0273 buffer addition techniques: I/137, II/141, 172
0274 buffer capacity: I/36
0275 buffered arcs: II/128
0276 buffer index: I/36
0277 bulk concentration: II/52
0278 bulk sample*: *Pure Appl. Chem.*, 1990, **61**, 1198
0279 Bunsen burner: II/166
0280 burning-off effect: II/125
0281 burning spot: II/134
0282 burning velocity: I/126
0283 burning voltage: II/123
0284 burn-up: II/214
0285 burn-up fraction: II/214
0286 by-pass injector: I/64, 79, II/98, 99
0287 calibration curve: I/168
0288 calorimetry*: measurement of the quantity of heat involved in a process
0289 Calvet-type arrangement: II/45
0290 capacitively coupled plasma: II/139
0291 capacitive microwave plasma: II/140
0292 capacitor voltage: II/129
0293 capacity (of a balance): I/12, II/35
0294 capacity current: II/53
0295 capacity factor: II/107
0296 capillary arc: II/127

0297 capillary electrodes: II/126
0298 capillary tube: II/151
0299 capture: II/214
0300 capture cross-section: II/214
0301 capture gamma radiation: II/214
0302 carbon cup atomizer: I/125
0303 carbon filament atomizer: I/125
0304 carbon rod furnace: I/125
0305 carrier: II/214
0306 carrier distillation arc: II/128
0307 carrier electrode: II/126
0308 carrier-free: II/124
0309 carrier gas: I/65, 79, 127
0310 catalytic current: II/53, 56
0311 cataphoresis: II/118
0312 cathode dark space: II/147
0313 cathode fall region: II/148
0314 cathode fall voltage: II/123
0315 cathode glow: II/147
0316 cathode layer arc: II/125
0317 cathode process sign convention: I/166, 167
0318 cathode ray polarography: I/158
0319 cathodic sputtering: I/125, II/145, 148
0320 cathodic stripping: I/150
0321 cathodic vaporization: II/125
0322 cation effect: I/136, II/170
0323 cation exchange: I/89
0324 cation exchanger: I/88
0325 cementation*: precipitation of a metal from solution by means of a metal that is lower in the electromotive series
0326 Cerenkov detector: II/215 (Cherenkov is a more accurate transliteration of the Russian name)
0327 Cerenkov effect: II/215
0328 Cerenkov radiation: II/215
0329 certified reference material: II/17, 19, 21
0330 chain fission yield: II/215
0331 chamber saturation: I/79, II/99
0332 chamber-type nebulizer: I/122, II/165
0333 channelling: II/149
0334 characteristic concentration: I/131
0335 characteristic mass*: *Pure Appl. Chem.*, 1992, **64**, 258
0336 characteristic potential: II/51
0337 characteristic X-radiation: II/215
0338 charged particle activation*: inducing radioactivity by bombardment with charged particles
0339 charged particles: II/147
0340 charge exchange: II/148
0341 charge-step polarography: I/160
0342 charge transfer: II/148
0343 charging circuit, II/130
0344 charging current: II/130
0345 charging resistor: II/130
0346 charging time constant: II/132
0347 chelatometric titration: I/41, II/47
0348 chemical actinometer: II/193
0349 chemical aging: I/21, II/86
0350 chemical dosimeter: II/215
0351 chemical equilibrium: I/143
0352 chemical equivalent: I/185
0353 chemical flames: II/112
0354 chemical interference: I/135
0355 chemical ionization: I/31, II/204
0356 chemical isotope exchange: II/125
0357 chemical shift: II/252
0358 chemical yield: II/215
0359 chemi-ionization: I/31, II/204
0360 chemiluminescence: II/184
0361 chemiluminescence analysis: II/199
0362 chemiluminescent indicator: I/39
0363 chemisorption: II/252
0364 chopper: I/128
0365 chromatogram: I/66, 76, II/95, 101
0366 chromatograph (noun): I/77, II/95
0367 chromatograph (verb): I/77, II/95
0368 chromatography: I/74, II/92
0369 chronoamperometry: I/155, II/73, 76
0370 chronocoulometric constant: II/51
0371 chronocoulometry: I/155, II/73
0372 chronopotentiometric constant: II/51
0373 chronopotentiometric end-point: I/37
0374 chronopotentiometric titration: II/47
0375 chronopotentiometry: I/154, II/75
0376 circular polarization: II/122
0377 clean bench*: a bench on which the number of dust or contaminant particles per unit area is kept below a specified limit
0378 clean room*: a room in which the number of dust or contaminant particles per unit volume is kept below a specified limit
0379 clean surface*: a surface on which the number of dust or contaminant particles per unit area is kept below a specified limit
0380 clean up: II/149
0381 climbing of the arc: II/125
0382 coagulation: I/19, II/85
0383 coaxial waveguide: II/140
0384 coefficient of variation: II/5
0385 coherent electromagnetic radiation: II/112
0386 coherent radiation: II/141
0387 coherent sources: II/112
0388 coherent system of units: I/96
0389 coincidence circuit: II/216
0390 coincidence counting*: use of two or more particle detectors for registration of occurrence of counts in a given time interval
0391 coincidence resolving time: II/216
0392 co-ions: I/88
0393 cold hollow cathode: II/151
0394 cold neutrons: II/232
0395 collection: I/19, II/84
0396 collector: I/19, II/84
0397 collimation: /216
0398 collimator: II/216
0399 collimator resolution: II/178
0400 collision*: impact of two or more bodies, resulting in abrupt change of motion
0401 collisional broadening: II/122
0402 collisional chains: II/149
0403 collisional cross-section: II/120
0404 collisional de-excitation: II/120
0405 collisional half-width: I/142
0406 collisional processes: II/120
0407 collisional shift: II/122
0408 column: I/78, II/97
0409 column chromatography: I/75, II/94
0410 column performance: I/69, 85, II/107

0411 column temperature: I/80
0412 column volume: I/80, 89, II/100
0413 combination electrode: I/170
0414 combustion*: oxidative decomposition by means of heat
0415 comparator*: a device for colorimetric analysis by comparing the colour of a mixture of sample and reagent, with that of the sample and a series of graded coloured glass discs
0416 comparison solution: I/36
0417 complete evaporation: II/125
0418 complexan*: synonymous with complexone
0419 complex anion*: an anion containing a central metal atom and negatively charged ligands
0420 complexation*: formation of a complex of a metal atom
0421 complex formation equilibria: II/11
0422 compleximetric (complexometric) titration: I/41, II/47
0423 complexometry : II/47
0424 complexone: I/35
0425 component: II/118
0426 compromise conditions: II/138
0427 Compton angle*: the angle between the directions of travel of the scattered photon and electron in Compton scattering
0428 Compton effect: II/216
0429 Compton electron: II/216
0430 Compton scattering: II/216
0431 concentration: I/102, 114, 131, II/111
0432 concentration constants*: equilibrium constants expressed in terms of concentrations
0433 concentration distribution ratio: I/84, 91, II/88
0434 concentration equilibrium constant*: an equilibrium constant expressed in terms of concentrations
0435 concentration index: I/103, II/159
0436 concentration ratio: I/102
0437 concentric nebulizer: II/165
0438 concomitants: I/130, II/118
0439 condensed arc: II/127
0440 conductivity*: ability to transfer heat or electricity
0441 conductometric titration: I/152, II/47, 68
0442 conductometry: I/152, II/68
0443 confiner plate: II/133
0444 constant: I/94
0445 constant current: II/127
0446 constant temperature arc: II/128
0447 constant voltage: II/127
0448 constituent: I/16, 101
0449 constituent classification: I/16
0450 constituent content: I/15
0451 constricted glow discharge:II/150
0452 constriction of arc plasma:II/124
0453 contamination by entrapment:I/20
0454 contamination of a precipitate: I/19
0455 contamination phenomena in precipitation: I/18
0456 continuous flow analysis*: analysis in which a sample is mixed with or injected into a continuous flow of reagent
0457 continuous procedure: II/126
0458 continuous radiation: II/119
0459 continuous supply: II/126
0460 continuous wave (CW): II/141
0461 continuous wave operation (CW operation): II/142
0462 continuum: I/137, II/119
0463 control band*: the region between the upper and lower control limits in a control chart for a process
0464 control gap: II/131
0465 controlled alternating current arc: II/123
0466 controlled atmosphere: II/128
0467 controlled-current coulometry: I/154, II/72
0468 controlled-current potentiometric titration: I/153, II/72
0469 controlled current potentiometry: I/153, II/72
0470 controlled duration spark: II/132
0471 controlled flow nebulizer: I/122, II/165
0472 controlled high-voltage spark generator: II/132
0473 controlled-potential coulometric titration: I/155, II/74
0474 controlled-potential coulometry: I/155, II/74
0475 controlled-potential electrogravimetry: I/156, II/74
0476 controlled-potential electroseparation: I/156
0477 controlled-potential polarography: I/150
0478 controlled waveform spark: II/132
0479 control spark gap: II/131
0480 control titration: I/36
0481 convection: II/118
0482 convective chronoamperometry: I/155
0483 convective chronocoulometry: I/155
0484 conversion electron: II/126
0485 conversion factor: I/105
0486 convolution-integral linear-sweep voltammetry: I/151, II/66
0487 cooled hollow-cathode lamp: I/125
0488 cooling curves: I/25
0489 cooling rate curves: I/25
0490 coprecipitation: I/20
0491 corrected decay times of fluorescence: II/197
0492 corrected decay times of phosphorescence: II/197
0493 corrected emission spectrum: II/196
0494 corrected excitation spectrum: II/197
0495 corrected intensity: II/179
0496 corrected luminescence emission polarization spectrum: II/198
0497 corrected luminescence excitation polarization spectrum: II/198
0498 corrected quantum yield of luminescence: II/198
0499 corrected retention time*: *Pure Appl. Chem.*, 1993, **65**, 842
0500 corrected retention volume: I/67< II/104
0501 correction of direct weighings: I/13, II/37
0502 cosmic radiation: II/147
0503 coulometric titration: I/41, 154, II/47, 72
0504 coulometry*: analysis based on measurement of amount of electric charge involved in a reaction
0505 count: II/217
0506 counter electrode: I/151, II/59, 65, 125
0507 counter-ions: I/88

0508 counter tube: II/217
0509 counting efficiency: II/217
0510 counting geometry: II/227
0511 counting loss: II/217
0512 counting precision: II/183
0513 counting rate: II/217
0514 count rates: II/208, 239
0515 coupled simultaneous techniques: II/39 (called multiple techniques on I/97)
0516 coupling efficiency: II/135
0517 coupling unit: II/136
0518 crater: II/143
0519 crater depth: II/143
0520 crater diameter: II/143
0521 crater shape: II/143
0522 critically damped discharge: II/131
0523 Crookes dark space: II/147
0524 crossed electric and magnetic fields: I/30
0525 cross-section: II/217, 218
0526 cross-section for ionization: II/173, 176
0527 crystal characteristics: II/178
0528 crystal-controlled oscillator: II/136
0529 crystal diffraction: II/176
0530 crystal diffraction spectrometer: II/218, 243
0531 crystal lattice: II/149
0532 crystalline ion-selective electrode: I/171
0533 cumulative distribution*: the total number of results not greater than a value *x*, plotted as a function of *x*
0534 cumulative fission yield: II/218
0535 cumulative formation constant: I/56, II/11
0536 cumulative protonation constant: I/57, II/13
0537 cumulative stability constant: I/56, II/11
0538 cumulative sum*: the sum of deviations from the mean
0539 cup electrode: II/126
0540 curie: II/218
0541 current*: electric current
0542 current-carrying arc plasma: I/124, 125, II/123
0543 current-cessation chronopotentiometry: I/156
0544 current-free plasma: I/124
0545 current-reversal chronopotentiometry: I/156
0546 current-scanning polarography: I/156
0547 current–time curve: I/123
0548 curve corrector: I/130
0549 cyanogen molecular bands: II/128
0550 cyclic chronopotentiometry: I/156, II/75
0551 cyclic current-step chronopotentiometry: I/156, II/75
0552 cyclic triangular wave polarography: I/158
0553 cyclic triangular wave voltammetry: I/159, II/77
0554 cyclotron: II/218
0555 cyclotron resonance mass spectrometry: I/29
0556 cylinder collector: II/204
0557 cylindrical shape: II/146
0558 damping constant: I/142
0559 dark current: I/129
0560 data*: items of information; often used as synonymous with results
0561 daughter products: II/218
0562 deactivation process: II/185
0563 dead-stop end-point: I/37
0564 dead time: II/179, 181, 219
0565 dead time correction: II/219
0566 dead volume: I/81, II/101
0567 Debye–Hückel equation: II/33
0568 decay: II/219
0569 decay chain: II/219
0570 decay constant: II/219
0571 decay curve: II/219
0572 decay scheme: II/219
0573 degree of dissocation: I/144
0574 degree of ionization: I/144
0575 degree of titration*: ratio of the number of moles of titrant added, to the number required to react with the amount of titrand present
0576 deionized water*: water that has been freed from ions other than hydrogen and hydroxide ions by passage through a mixed bed ion-exchange column
0577 delayed fluorescence: II/185
0578 delayed neutrons: II/23
0579 Delves cup*: a nickel capsule for blood samples in atomic-absorption spectrometry (H.T. Delves, *Analyst*, 1970, **95**, 439)
0580 demasking*: liberation of a metal from a complex that has been formed for the purpose of preventing interference by the metal
0581 demineralized water*: a more accurate term for deionized water
0582 demodulation polarography: I/162, II/81
0583 densitometer: II/160
0584 depolarizer: II/58
0585 depth profile: II/252
0586 depth resolution: II/253
0587 derivative: I/24, 150, II/65
0588 derivative chronopotentiometry: I/154
0589 derivative dilatometry: I/15, 27
0590 derivative end-point*: an end-point determined by plotting dS/dV against V, where S is the signal and V is the volume of titrant added
0591 derivative polarography: I/158
0592 derivative potentiometric titration: I/153, II/65
0593 derivative pulse polarography: I/159
0594 derivative spectrophotometry*: use of the first or higher derivative of absorbance with respect to wavelength, plotted against wavelength
0595 derivative thermal analysis: I/25
0596 derivative thermogravimetry: I/25, 26
0597 derivative voltammetry: I/157
0598 derivatographic analysis: I/25
0599 derivatography: I/25
0600 designated volume: I/36
0601 desolvated fraction: I/123
0602 desolvation: I/121, II/165
0603 detection: I/79
0604 detection efficiency: II/219
0605 detection limit: I/117, 133, 168
0606 detector: I/65, II/99
0607 detector dead time: II/179, 181
0608 detector efficiency: II/220
0609 detector resolution: II/179
0610 develop: II/95
0611 deviation: I/9, II/4
0612 devolatilizer: I/108, II/159

0613 diatomic*: describes a molecule consisting of two atoms
0614 dielectric constant*: ratio of the permittivity of a substance to that of a vacuum
0615 dielectrometric titration: I/152, II/69
0616 dielectrometry: I/152, II/69
0617 differential: I/24, II/65
0618 differential amperometry: I/154
0619 differential chromatogram: I/78, II/95
0620 differential chromatography*: chromatography with use of a differential detector
0621 differential detector: I/65, 80, II/99
0622 differential dilatometry: I/25
0623 differential polarography: II/158
0624 differential potentiometric titration: I/153, II/65, 71
0625 differential potentiometry: I/153, II/71
0626 differential pulse polarography: I/160, II/79
0627 differential pumping: II/150
0628 differential scanning calorimetry: I/25, 27, II/44
0629 differential thermal analysis: I/25, 27, II/42
0630 differential thermal analysis in an isothermal environment: II/44
0631 differential thermogram: I/24
0632 differential thermogravimetric analysis: I/25
0633 differential thermogravimetry: I/25
0634 differential voltammetry: I/157
0635 diffracted electrons: II/248
0636 diffused junction semiconductor detector: II/220, 242
0637 diffusion: II/118
0638 diffusion coefficient: I/92
0639 diffusion current: II/54
0640 diffusion current constant: II/59
0641 digital to analogue converter (DAC)*: device for conversion of digital data into analogue form
0642 dilatometry: I/25, 27
0643 diluent: I/62, 107, II/89, 91
0644 dilution test: I/134
0645 diode-array detector*: a contiguous array of diodes, for simultaneous recording of part or all of a spectrum
0646 direct current: II/54
0647 direct current arc: II/113, 117, 123
0648 direct current biased oscillating discharge: II/131
0649 direct current conductometry: I/152
0650 direct current polarography: I/157
0651 direct fission yield: II/225
0652 direct-injection burner: I/122, II/165
0653 direct-line fluorescence: I/140
0654 direct probe: II/207
0655 discharge: II/147
0656 discharge atmosphere: II/132, 134
0657 discharge circuit: II/131
0658 discharge current: II/129
0659 discharge gas: II/145
0660 discharge polarography: I/160
0661 discharges in vacuum: II/134 (incorrectly indexed as discharge vacuum)
0662 discontinuous procedures: II/125
0663 discontinuous simultaneous techniques: II/39
0664 discriminator: II/220
0665 disintegration constant: II/219
0666 disintegration rate: II/220
0667 dispersion: I/101, II/158
0668 dispersion in flow-injection analysis*: ratio of the analyte concentration in the sample solution to the analyte concentration recorded as the sample plug (or bolus) passes through the detector
0669 dispersion of a material: I/101, II/158
0670 displacement chromatography: I/74, II/92
0671 dissociation: I/143, II/118
0672 dissociation constant: I/145
0673 dissociation interferences: I/136, II/171
0674 dissociation potential: I/144
0675 dissolution: I/18, II/83
0676 distillation: II/118
0677 distribuend: II/91
0678 distribution*: division of an analyte between two phases; division into classes
0679 distribution coefficient: I/61, 91, II/90
0680 distribution constant: I/84, II/88
0681 distribution equilibria*: the various individual equilibria involved in distribution of an analyte between two phases
0682 distribution equilibrium constant: II/87
0683 distribution laws: I/20, II/85
0684 distribution ratio: I/59, 61, II/88
0685 Doerner and Hoskins equation: I/20, II/86
0686 Doppler broadening: II/121
0687 Doppler half-width of spectral line: I/142
0688 Doppler shift: II/121
0689 dose (noun): II/215, 220
0690 dose equivalent: II/221
0691 double arc: II/128
0692 double-beam spectrometer: II/191
0693 double-beam system: I/128
0694 double escape peak: II/222
0695 double-focusing mass spectrometer: I/28
0696 double-layer current: II/54
0697 double potential-step chronoamperometry: I/159, II/78
0698 double potential-step chronocoulometry: I/160, II/78
0699 double (spectral) beam spectrometer: II/191
0700 double (synchronous) beam spectrometer: II/191
0701 double-tone polarography: I/162, II/82
0702 drift: I/169
0703 drilled electrode: II/126
0704 drop generator: I/122, II/165
0705 drop (generator) nebulizer: I/122, II/165
0706 dropping electrode coulometry: I/155
0707 dropping mercury electrode: I/149
0708 drop time: II/59
0709 dry aerosol: I/121, II/165
0710 Duane–Hunt short wavelength limit: II/173
0711 duty cycle: II/124
0712 dye lasers: II/143
0713 dynamic differential calorimetry: I/25
0714 dynamic field mass spectrometer: I/30, II/202
0715 dynamic system: II/146
0716 dynamic thermogravimetric analysis: I/25
0717 dynamic thermomechanometry: II/45
0718 effect: I/36, II/170

0719 effective cadmium cut-off energy: II/221
0720 effective dose equivalent: II/221
0721 effectively infinite thickness: II/182
0722 effective theoretical plate number: I/85, II/108
0723 effective thermal cross-section: II/218
0724 efficiency*: ratio of observed to theoretical performance
0725 effluent gas analysis: I/25
0726 effluent gas detection: I/25
0727 eigen-frequencies: II/142
0728 Einstein coefficient for spontaneous emission: I/140
0729 Einstein transmission probability: I/140
0730 ejection energy: II/147
0731 elastic collision: II/120
0732 elastic scattering: II/122, 231
0733 electrical arc: II/123
0734 electrical conductivity: II/125
0735 electrical discharge: II/129
0736 electrical excitation: II/141
0737 electrical parameter: II/127
0738 electrical source parameter: II/127
0739 electrical sparks: II/129
0740 electric current: II/52
0741 electroactive substance: II/58, 59
0742 electrodeless discharge lamp: I/127
0743 electrodeless excitation: II/144
0744 electrodeless resonant cavity plasma: II/140
0745 electrode resonant cavity plasma: II/140
0746 electrode–solution interface: II/51
0747 electro-erosion: II/127 (incorrectly indexed as electrode-erosion)
0748 electrography: I/155, II/74
0749 electrogravimetry: I/155, II/74
0750 electrolyte: II/60
0751 electromagnetic radiation: II/111
0752 electromotive force: II/7
0753 electron capture: II/149, 221
0754 electron capture detector*: an ionization detector in which free electrons are captured by suitable solutes
0755 electron diffraction: II/248
0756 electron energy: I/30, II/ 203 (incorrectly indexed as electron voltage in II)
0757 electron energy loss spectroscopy: II/249
0758 electronic energy transfer processes: II/187
0759 electron impact ionization: I/30, II/203
0760 electron kinetic energy: II/245
0761 electron microscopy: II/248
0762 electron multiplication: II/148
0763 electron pressure: II/134
0764 electron probe microanalysis: (EPMA): II/247
0765 electron probe X-ray microanalysis (EXPMA): II/247
0766 electron spectra: II/247
0767 electron spectroscopy for chemical analysis (ESCA): II/245, 250
0768 electron temperature: II/119
0769 electro-optical shutters: II/143
0770 electrophoresis*: cataphoresis
0771 electrophoretogram (electrophorogram)*: a visual display of the separation obtained by electrophoresis
0772 electroseparation: I/155
0773 electroseparation at controlled potential: I/156
0774 electrospray*: an ionization technique in mass spectrometry
0775 element*: a substance which cannot be made by chemical union or decomposed by chemical change
0776 elemental analysis*: determination of the elemental composition of a substance
0777 elementary particle: II/234
0778 elliptical polarization: II/122
0779 eluate: I/79, II/99
0780 eluent: I/79, II/97
0781 elute: I/77, II/95
0782 elution band: I/77, II/96
0783 elution chromatography: I/74, II/92
0784 elution curve: I/77, II/95
0785 elution peak*: equivalent to elution band
0786 elution volume*: the volume of eluent required to give complete elution of an analyte
0787 emanation thermal analysis: II/42
0788 emission monochromator: II/191
0789 emission profile: II/122
0790 emission spectra: II/196
0791 emulsion calibration curve: I/109, II/161
0792 emulsion calibration function: I/109, II/161
0793 endotherm: II/43
0794 endothermic peak: II/43
0795 end-point: I/36, II/47
0796 energy dispersion: II/180
0797 energy-dispersive X-ray fluorescence analysis: II/222
0798 energy flux density: II/222
0799 energy levels: II/118
0800 energy resolution: II/222
0801 energy scale: II/245
0802 energy threshold: II/222
0803 energy transfer: II/118
0804 energy transition: II/118
0805 energy yield of luminescence: II/198
0806 enhanced phosphorescence analysis: II/188
0807 enhancement: I/134, II/175
0808 enrichment: II/222
0809 enrichment factor: I/62, II/222
0810 enthalpimetric end-point; I/38
0811 enthalpimetric titration*: titration based on measurement of heat change in a reaction
0812 entrance slit height: I/129
0813 entrance slit width: I/129
0814 enzyme substrate electrode: I/172
0815 epicadmium neutrons: II/222
0816 epithermal neutrons: II/222
0817 equilibrium constant: I/54, 59, II/9, 10, 13
0818 equivalence point: I/38, 179, II/47
0819 equivalent: I/181
0820 error: I/10, II/6
0821 escape depth: II/252
0822 E-type delayed fluorescence: II/185
0823 evaporation: II/117
0824 evaporation equilibrium: II/134
0825 evolved gas analysis: I/25, 27, II/40, 42
0826 evolved gas detection: I/25, 27, II/40, 41
0827 excimer luminescence: II/188
0828 excitation: I/143, II/117, 223
0829 excitation–emission spectra: II/197

0830 excitation energy: I/105, 144, II/138, 223
0831 excitation interferences: I/136, II/171
0832 excitation levels: II/138
0833 excitation mechanism: II/138
0834 excitation mode: /184
0835 excitation monochromator: II/190
0836 excitation potential: I/106, 144
0837 excitation region: II/139
0838 excitation source: II/144, 190
0839 excitation spectra: II/197
0840 excitation temperature: II/119
0841 excited fluorescence spectrum: II/197
0842 excited phosphorescence spectrum: II/197
0843 excited states: II/184, 228, 240
0844 exciting energy: II/142
0845 exoenergetic condition: II/117
0846 exotherm: II/43
0847 exothermic peak: II/43
0848 exothermic reactions: II/112
0849 experimental surface: II/251
0850 exponential decay: II/223
0851 exposure: I/98, 102, II/223, 240
0852 external heavy atom effects: II/190
0853 externally ignited spark discharge: II/133
0854 external standard: I/111
0855 extract (noun): I/62, II/91
0856 extract (verb): I/77, II/95
0857 extractability: I/61
0858 extractant: I/62, II/89, 91
0859 extraction*: transfer into a liquid phase from a liquid phase immiscible with it or from a solid phase
0860 extraction coefficient: I/61, II/90
0861 extraction constant: I/60, II/89
0862 extraction indicator: I/39
0863 extrapolated onset: II/44
0864 extrapolated range: II/223
0865 factor weight: I/38 (also means a sample weight that is a decimal multiple or submultiple of a gravimetric conversion factor)
0866 Fano factor: II/180, 181
0867 faradaic current: II/54
0868 faradaic demodulation current: II/54
0869 faradaic rectification current: II/55
0870 Faraday constant*: the charge of one mole of electrons
0871 Faraday cup collector: I/31, II/204
0872 Faraday dark space: II/147
0873 fast atom bombardment (FAB)*: ionization by bombardment with a beam of accelerated atoms
0874 fast neutrons: II/223
0875 feed-back system: I/23
0876 fertile material: II/223
0877 fertile nuclides: II/223
0878 field ionization: I/30, II/203
0879 filament chromatography: I/76 (listed in text, but not defined; has now been dropped from IUPAC nomenclature for chromatography; its meaning is obscure)
0880 fill gas: I/127, II/145
0881 filling solution: II/27, 28
0882 film badge: II/224
0883 filter: II/224
0884 filter off*: separate a solid from a liquid by collection on a filter
0885 fissile: II/224
0886 fissile nuclide: II/224
0887 fission: II/224
0888 fissionable: II/225
0889 fission fragments: II/224
0890 fission neutrons: II/224
0891 fission products: II/224
0892 fission yield: II/224, 225
0893 fixed frequency: II/136
0894 fixed ions: I/89
0895 flame: I/118, II/165
0896 flame background: I/131, II/169
0897 flame-geometry interference: I/136, II/171
0898 flame spectrometry: I/118, II/163
0899 flame temperature: I/126
0900 flash-back, I/122, II/166
0901 flash fluorimetry: II/197
0902 flash lamp: II/190
0903 floating frequency: II/135
0904 flocculation: I/19, II/85
0905 flow-injection analysis (FIA)*: a technique in which a plug (or bolus) of sample is injected into an unsegmented continuous flow of reagent
0906 flow-programmed chromatography: I/76, II/92
0907 flow-rate: I/82, 126, II/63, 102
0908 fluorescence: I/139, II/184
0909 fluorescence analysis: II/199
0910 fluorescence efficiency: I/142
0911 fluorescence power efficiency: I/142
0912 fluorescence quantum efficiency: I/142, II/198
0913 fluorescence spectrometer: II/190
0914 fluorescence total quantum efficiency: I/142
0915 fluorescence yield: II/174
0916 fluorimetric end-point: I/37
0917 fluorometric titration: II/47
0918 flushed: II/146
0919 flux: I/111
0920 flux density: II/225
0921 flux depression: II/225
0922 flux integration: I/112 (indexed as integration of flux)
0923 flux monitor: II/225
0924 flux perturbation: II/225
0925 flux through monochromator: I/113
0926 focal length: I/100, II/157
0927 focal spot: II/141
0928 focal spot diameter: II/143
0929 focusing lens: II/141
0930 focusing mirror: II/141
0931 fog: I/103, II/159
0932 foil detector: II/225
0933 formality: I/39
0934 formation constant: I/55, II/11
0935 Fourier transform mass spectrometer: II/208
0936 fractional distillation: II/125
0937 fragment ion: I/32, II/205
0938 free–bound transition: II/119
0939 free-burning arc: II/123
0940 free electrons: II/149
0941 free–free transition: II/119

0942 free-running operation: II/142
0943 frequency: II/60, 136
0944 frequency multiplication: II/136
0945 frequency of modulation: I/129
0946 frontal chromatography: I/74
0947 fronting: I/77
0948 front surface (geometry): II/199
0949 fuel: I/122, II/165
0950 fuel cycle: II/225
0951 fuel element: II/226
0952 fuel-rich flame: I/123, II/166
0953 full energy peak: II/226
0954 full energy peak efficiency: II/226
0955 full width at half maximum (FWHM): II/226
0956 fusion: II/135
0957 gamma cascade: II/226
0958 gamma quantum: II/226
0959 gamma radiation: II/226
0960 gamma radiation capture: II/227
0961 gamma ray: II/226
0962 gamma ray spectrometer: II/227
0963 gamma ray spectrum: II/227
0964 gap: II/153
0965 gap resistor (incorrectly indexed as gas resistor): II/130, 132
0966 gas amplification: II/179
0967 gas chromatography: I/64, II/93
0968 gaseous discharges: II/112
0969 gaseous form: II/145
0970 gas hold-up volume: I/66, 81, II/100
0971 gas lasers: II/143
0972 gas–liquid chromatography: I/64, 74, II/93
0973 gas-permeable membrane: I/172
0974 gas proportional detector: II/179
0975 gas-restricting orifice: II/146
0976 gas–solid chromatography: I/64, 74, II/93
0977 gas-stabilized arcs: II/124
0978 gas temperature: II/119
0979 gated photodetectors: II/193
0980 Geiger–Müller counter tube: II/227
0981 Geiger–Müller region: II/227
0982 Geiger–Müller threshold: II/227
0983 Geissler lamps: II/151
0984 gel filtration: I/75, II/94
0985 gel-permeation chromatography: I/75, II/94
0986 geometrical quantities: II/156
0987 geometric attenuation: II/227
0988 geometry factor: II/227
0989 Gerlach homologous lines: I/103
0990 gettering: II/149
0991 giant pulse: II/142
0992 glass electrode: II/26
0993 glass electrode error: II/27
0994 globule arc: II/125
0995 glow discharge: II/147, 149, 150
0996 gradient elution: I/76, II/93
0997 gradient layer: I/79, II/98
0998 gradient packing: I/79, II/98
0999 graphite-cup atomizer: I/125
1000 graphite-rod furnace: I/125
1001 graphite spark: II/135
1002 graphite-tube furnace: I/125
1003 gravity-fed nebulizer: I/122, II/165
1004 gravity-fed powder sifter system: II/126
1005 gray: II/227
1006 ground state: II/121, 227
1007 growth curve of activity: II/227
1008 G-value: II/227
1009 half-intensity width of absorption line: I/129
1010 half-intensity width of emission line: I/129
1011 half life: II/228
1012 half-peak potential: II/51, 61
1013 half-thickness: II/228
1014 half-value layer: II/228
1015 half-value thickness: II/228
1016 half-wave potential: II/51, 62
1017 Hammett acidity function*: a function describing the strength of strong acids
1018 Hammett indicator*: an indicator suitable for determination of a Hammett acidity function value
1019 harmonic overlap: II/179
1020 heat-flux differential scanning calorimetry: II/45
1021 heating-curve determination: II/40, 42
1022 heating curves: I/25, 26
1023 heating rate: II/39
1024 heavy atom effect: II/190
1025 height equivalent to an effective theoretical plate: I/86, II/108
1026 height equivalent to a theoretical plate: I/85, II/108
1027 height of observation: I/129
1028 heterochromatic photometry: I/111
1029 heterogeneous membrane electrode: I/171
1030 high-current arc: II/123
1031 higher harmonic alternating current polarography: I/161
1032 higher harmonic alternating current polarography with phase-sensitive rectification: I/161
1033 high-frequency (conductometric) end-point: I/37
1034 high-frequency conductometric titration: I/152, II/68
1035 high-frequency conductometry: I/152, II/68
1036 high-frequency shorting capacitor: II/127
1037 high-frequency step-up transformer: II/133
1038 high-frequency titration: II/47
1039 high-intensity hollow cathode: II/152
1040 high-level faradaic rectification: I/162, II/82
1041 high-pressure arc: II/151
1042 high-pressure xenon lamp: I/127
1043 high reflectivity: II/141
1044 high repetition rate presparking: II/134
1045 high repetition rate sparks: II/132
1046 high-resolution energy-loss spectroscopy (HRELS): II/249
1047 high-voltage sparks: II/132
1048 Hittorf dark space: II/147
1049 hold-back carrier: II/228
1050 hold-up volume: I/81
1051 hollow cathode discharge: I/125, II/151
1052 hollow cathode lamp: I/127, II/151
1053 hollow cathode sources: II/151
1054 hollow electrode: II/126
1055 homogeneous distribution coefficient: I/20, II/86
1056 homogeneous magnetic field: II/124
1057 homogeneous membrane electrode: I/171

1058 homologous line: I/103
1059 horizontal electrodes: II/126
1060 host–guest chemistry*: intercalation chemistry
1061 host material: II/142
1062 hot atom: II/228
1063 hot cell: II/228
1064 hot hollow cathode: I/125, II/152
1065 Hurter and Driffield (H & D) curve: I/110 (correct in text but indexed as Hunter and Driffield)
1066 hydrodynamic voltammetry: I/157, II/76
1067 hydrogen gas electrode: II/27
1068 hyperchromic shift*: *Pure Appl. Chem.*, 1983, **55**, 1292
1069 hypochromic shift*: *Pure Appl. Chem.*, 1983, **55**, 1292
1070 hypsochromic shift*: *Pure Appl. Chem.*, 1983, **55**, 1292
1071 hysteresis: I/169
1072 ignited alternating current arc: II/128
1073 ignition: II/124
1074 ignition circuit: II/133
1075 ignition pulse: II/133
1076 ignition spark: II/124, 133
1077 ignitor circuit: II/127
1078 Ilkovič equation sign convention: I/167
1079 image devices: II/193
1080 immunoassay*: analysis by means of antigen–antibody interaction
1081 impedance converter: II/136
1082 impinging photon: II/117
1083 incandescent bodies: II/112
1084 incandescent radiation: II/119
1085 incident power: II/136
1086 incident radiation: II/217
1087 inclusion: II/85
1088 incoherent radiation pulse: II/142
1089 incremental-charge polarography: I/160, II/78
1090 indicator: I/39, II/48
1091 indicator blank: I/40
1092 indicator correction: I/40
1093 indicator electrode: I/151, II/59, 66
1094 indifferent electrolyte: II/60
1095 indirect titration: I/42
1096 induced radioactivity: II/228
1097 induced reactions*: an induced reaction is one having a rate which can be increased by another reaction taking place in the same solution, both reactions having one reactant in common
1098 induction coil: II/135
1099 inductively coupled plasmas: I/125
1100 inductor: II/135
1101 inelastic collision: II/120
1102 inelastic scattering: II/122, 228
1103 inert discharge atmosphere: II/132
1104 initial current: II/129
1105 initial temperature: I/80, II/41
1106 initiate (an arc): II/150
1107 injection gas: II/136
1108 injection temperature: I/80, II/100
1109 injection tube: II/136
1110 injector: II/136
1111 inlet system*: a means of introduction of a sample into a mass spectrometer ionization chamber
1112 inner bremsstrahlung: II/228
1113 inner zone: I/122, II/165
1114 in situ microanalysis: II/247
1115 in situ micro X-ray diffraction: II/249
1116 instantaneous current: II/55
1117 instantaneous flow-rate*: the flow-rate at any given point in a gas chromatographic column
1118 instantaneous measurements: II/138
1119 instrumental activation analysis: II/228
1120 instrumental indication: II/36
1121 instrumental neutron activation analysis*: instrumental activation analysis with neutrons as the activation agent
1122 instrumentation: I/23
1123 integral absorption: I/141
1124 integral chromatograms: I/67, 78, II/95
1125 integral detector: I/65, 80, II/99
1126 integral reflection coefficient: II/178, 181
1127 integral standard*: indexing error in II – should have been internal standard, which was omitted from the index
1128 integrating sphere: I/135
1129 intensity: I/102
1130 intensity bridge: I/111, II/163
1131 intensity bridge ratio: I/111, II/163
1132 intensity calibrating device: I/110, II/162
1133 intensity-modulated beam: II/193
1134 intensity-modulated hollow cathode lamp: II/152
1135 intensity of radiation: I/102, 140, II/228
1136 intensity of spectral line: I/141
1137 intensity relative to base peak: II/206
1138 interaction time: II/134
1139 interatomic collisions: II/147
1140 interchromophoric radiationless transition: II/187
1141 interconal zone: I/122, II/166
1142 inter-element effects: II/118
1143 interface: I/18, II/251
1144 interference: I/134, II/169
1145 interference curves: I/134
1146 interferent: I/134
1147 interfering lines: I/103, 135, II/159, 169
1148 interfering substance: I/169
1149 intermediate neutron: II/232
1150 intermediate plasma gas: II/137
1151 intermediate tube: II/137
1152 intermolecular radiationlesss transition: II/187
1153 internal absorbance: I/141
1154 internal conversion: II/187, 228
1155 internal conversion coefficient: II/229
1156 internal electrogravimetry: I/155
1157 internal electrolysis*: an electrochemical reaction occurring spontaneously in a short-circuited cell
1158 internal filling solution: II/27
1159 internal filling solution of a glass electrode: II/27
1160 internal oscillation amplitude: II/142
1161 internal reference electrode of a glass electrode: II/27

1162 internal reference line: I/102, II/139
1163 internal standard: I/78, 111, II/97 (not indexed in II)
1164 interrupted arcs: II/124
1165 interstitial fraction: I/80, II/100
1166 interstitial velocity: I/82, II/103, 109
1167 interstitial velocity at outlet pressure: I/82, II/102
1168 interstitial volume: I/65, 80, II/99
1169 intersystem crossing: II/187
1170 interzonal region: I/122, II/166
1171 intrachromophoric radiationless transition: II/187
1172 intrinsic detector efficiency: II/220
1173 intrinsic efficiency: II/229
1174 intrinsic full energy peak efficiency: II/229
1175 intrinsic photopeak efficiency: II/229
1176 inverse cooling rate curves: I/25
1177 inverse derivative potentiometric titration: I/153
1178 iodimetric titrations: I/42, II/47
1179 iodimetry: I/42, II/47
1180 iodometric titrations: I/42
1181 iodometry: I/42
1182 ion: II/118, 229
1183 ion bombardment: II/145
1184 ion chromatography*: *Pure Appl. Chem.*, 1993, **65**, 826
1185 ion current*: the current carried by the ions detected in the detector of a mass spectrometer
1186 ion cyclotron resonance mass spectrometer: II/202
1187 ion-exchange: I/88
1188 ion-exchange chromatography: I/75, II/94
1189 ion-exchange isotherm: I/91
1190 ion-exchange membrane: I/92
1191 ion-exchanger: I/88
1192 ionic activity coefficient: I/46, II/7
1193 ionic line: I/107, 138
1194 ionic medium: I/55, II/10
1195 ionic path-length: II/150
1196 ionic spectra: II/128
1197 ionic spectral lines: II/118
1198 ionic strength: I/173
1199 ionic strength adjustment buffer: I/170
1200 ionization: I/143, II/117, 118
1201 ionization buffers: I/137, II/172
1202 ionization by sputtering: I/31, II/204
1203 ionization chamber: II/229
1204 ionization coefficient: II/148
1205 ionization efficiency curve: I/33, II/206
1206 ionization energy: I/106, 144, II/229, 245
1207 ionization interferences: I/136, II/171
1208 ionization needle: II/133
1209 ionization potential: I/144, II/123, 245
1210 ionization probability: II/148
1211 ionization temperature: II/119
1212 ionizing particle: II/229
1213 ionizing radiation: II/229
1214 ionizing voltage: I/30, II/203
1215 ion microscopy: II/249
1216 ion–molecule reaction: I/33, II/206
1217 ionogenic groups: I/88
1218 ion probe microanalysis (IPMA): II/249
1219 ion scattering spectrometry (ISS): II/250
1220 ion-selective electrodes: I/168
1221 ion source*: the ionization chamber of a mass spectrometer
1222 ion-specific electrodes: I/168
1223 irradiance: I/98, II/155
1224 irradiation: II/229
1225 irregular radiation pulses: II/142
1226 isobaric mass-change (weight-change) determination: I/25, 26, II/40, 41
1227 isoformation: II/135
1228 isomeric states: II/229
1229 isomeric transition: II/229
1230 isopotential point: I/171
1231 isotherm: I/91
1232 isothermal weight-change determination: I/26
1233 isotones: II/230
1234 isotope: II/230
1235 isotope dilution: II/230
1236 isotope dilution analysis: II/230
1237 isotope exchange: II/230
1238 isotopes of atoms: II/230
1239 isotopic abundance: II/230
1240 isotopic analysis: II/152
1241 isotopic carrier: II/215
1242 isotopic ion: I/32
1243 isotopic separation: II/230
1244 isotopic tracer: II/244
1245 Kalousek polarography: I/159
1246 Kerr cell: II/143
1247 kinetic analysis*: analysis based on measurement of reaction rates
1248 kinetic current: II/55
1249 Kossel technique: II/249
1250 label (noun)*: an isotopic marker atom
1251 label (verb)*: to mark a compound with an isotopic atom
1252 laminar flame: I/122, II/165
1253 laminar flow*: non-turbulent flow
1254 laminar sheath: II/128
1255 lamp: II/145
1256 laser: II/141
1257 laser action: II/141
1258 laser active substance: II/142
1259 laser analysis: II/145
1260 laser atomization and excitation: II/144
1261 laser atomizers: II/143
1262 laser beam: II/141
1263 laser beam ionization: I/31
1264 laser erosion: II/127
1265 laser local analysis: II/144
1266 laser micro emission spectroscopy (LAMES): II/250
1267 laser micro mass spectrometry (LAMMS): II/250
1268 laser microprobe analysis: II/144
1269 laser mirror: II/141
1270 laser output: II/143
1271 laser ouput energy: II/142
1272 laser plume: II/143
1273 laser-produced vapour cloud: II/141
1274 laser pulse: II/141
1275 laser Raman microanalysis (LRMA): II/249
1276 laser resonator: II/141
1277 laser shots: II/145

1278 laser source: II/141
1279 laser spikes: II/142
1280 laser (lasing) threshold: II/142
1281 lateral diffusion interference: I/136, II/171
1282 law of mass action: II/119
1283 layer equilibration: I/79, II/99
1284 length of effective prism base: I/100, II/157
1285 level of titration: I/40
1286 ligand*: a charged or neutral species capable of formimg complexes with metal ions
1287 ligand-bridged complexes*: complexes in which two or more metal atoms are linked by ligands
1288 ligand number: I/57
1289 light modulation: I/128
1290 limited volume: II/126
1291 limiting adsorption current: II/52
1292 limiting catalytic current: II/53
1293 limiting current: II/55
1294 limiting diffusion current: II/54
1295 limiting kinetic current: II/55
1296 limiting migration current: II/56
1297 limit of detection: I/117, 133, 168, II/5
1298 limit of determination*: the lowest concentration or amount of a substance that can be determined by the procedure used
1299 limit of quantification*: synonymous with limit of determination
1300 line: I/106, 137
1301 linear acceleration mass spectrometer: II/202
1302 linear attenuation coefficient: II/175
1303 linear decrement: II/132
1304 linear dispersion: I/101, II/158
1305 linear electron accelerator: II/230
1306 linear electro-optical effect: II/143
1307 linear energy transfer: II/230
1308 linear energy-transfer dependent factor: II/236
1309 linear flow-rate: I/82
1310 linear photoelectric absorption coefficient: II/175
1311 linear polarization of luminescence: II/198
1312 linear polarizer: II/193
1313 linear potential sweep: II/76
1314 linear pulse amplifier: II/211
1315 linear scattering coefficient: II/175
1316 linear sweep voltammetry: I/157
1317 linear titration curve: II/49
1318 line-broadening parameter: I/142
1319 line profile function: II/121
1320 line shape: II/121
1321 line shift: II/121
1322 line width: I/100
1323 liquid chromatography: II/93
1324 liquid extraction: I/60, II/87
1325 liquid–gel chromatography: I/75, II/93
1326 liquid junction: II/27
1327 liquid junction potential: II/27
1328 liquid lasers: II/143
1329 liquid–liquid chromatography: I/75, II/93
1330 liquid–liquid distribution: I/60, II/87
1331 liquid–liquid equilibria: I/58
1332 liquid–liquid extraction: I/60, II/88
1333 liquid–liquid junction: II/7
1334 liquid phase: I/65
1335 liquid sample: II/126, 182
1336 liquid sample injection: II/126
1337 liquid scintillator counter: II/231
1338 liquid scintillator detector: II/231
1339 liquid–solid chromatography: I/75, II/93
1340 liquid stationary phase: II/98
1341 liquid volume: I/65, II/99
1342 load: I/12, II/35
1343 load resistor: II/127
1344 local analysis: I/125, II/145
1345 local efficiency of atomization: I/124, II/169
1346 local fraction atomized: I/124, II/169
1347 local fraction desolvated: I/123, II/168
1348 local fraction volatilized: I/124, II/168
1349 local gas temperature: II/120
1350 local thermal equilibrium: II/120
1351 logarithmic decrement: II/132
1352 logarithmic distribution coefficient: I/20
1353 logarithmic titration curve: II/49
1354 Lomakin–Scheibe equation: I/103
1355 long tube device: I/124
1356 Lorentz broadening*: asymmetric broadening (and shift of maximum) of an atomic absorption profile as a result of atomic collisions in the observation zone
1357 low-current arc: II/123
1358 low-energy collisions: II/173
1359 low-energy electron diffraction (LEED): II/248
1360 low loss: II/141
1361 low-pressure arc lamps: II/150
1362 low-pressure discharge lamps: I/127
1363 low-pressure electrical discharge: II/145
1364 low-pressure microwave-induced plasma: II/140
1365 low-voltage sparks: II/133
1366 luminescence: II/231
1367 luminescence parameters: II/184
1368 luminescence quenching: II/188
1369 luminescence spectrometer: II/190
1370 luminescence spectrometry*: measurement of luminescence with a spectrometer
1371 luminescence spectroscopy: II/111
1372 luminous efficacy: I/99, II/156
1373 luminous efficiency: I/99, II/156
1374 luminous flux: I/99
1375 luminous layer: II/147
1376 luminous quantity: I/99
1377 lump circuit: II/135
1378 macroscopic cross-section: II/218
1379 magnetically stabilized arcs: II/124
1380 magnetic deflection: I/29, II/201
1381 magnetic field: II/124
1382 magnetic field glow-discharge: II/153
1383 magnetic Penning effect: II/148
1384 magnetron: II/140
1385 mains supply voltage: II/133
1386 major constituents: I/16, 101
1387 marker: I/78
1388 masking*: conversion of an interferent metal ion into a non-interfering form by formation of a stable complex or by change in oxidation state
1389 masking agent: I/40
1390 mass attenuation coefficient: II/175
1391 mass/charge ratio: II/207

1392 mass distribution*: distribution between phases, described in terms of amount in each phase
1393 mass distribution ratio: I/84, II/107
1394 mass number: II/207
1395 mass spectrometer: I/28, II/204
1396 mass spectrometry: I/28, II/201
1397 mass spectrum: I/31, II/204
1398 mass-transfer controlled electrolysis rate constant: II/60
1399 matching capacitor: II/136
1400 matrix: I/101
1401 matrix effects: I/101, 136, II/118, 170, 189
1402 matrix matching: II/141
1403 matrix modifiers*: substances added to minimize matrix effects
1404 Mattauch–Herzog geometry: I/29
1405 Maxwell–Boltzmann law: II/119
1406 McLafferty rearrangement: I/34, II/207
1407 mean: I/9, 117, II/4
1408 mean activity coefficient: I/47, II/7
1409 mean interstitial velocity of carrier gas: I/82, II/103
1410 mean ionic activity coefficient: II/25
1411 mean life: II/213
1412 mean linear range: II/239
1413 mean mass range: II/239
1414 measured fluorescence excitation: II/197
1415 measured intensity: II/179
1416 measured phosphorescence excitation: II/197
1417 measured quantum yield of luminescence: II/197
1418 measurement: II/4
1419 mechanical entrapment: I/20, II/85
1420 mechanical shutters: II/143
1421 mechanism: I/22
1422 mechanization: I/22
1423 mechanize: I/22
1424 medium: I/54, II/9
1425 medium-current arc: II/123
1426 medium power (spikes): II/142
1427 medium-voltage spark: II/133
1428 medium-voltage spark generator: II/133
1429 Méker burner: II/166
1430 membrane: I/169
1431 mercury pool electrode: I/151
1432 meso: I/16
1433 meso sample: I/16
1434 mesotrace analysis: I/17
1435 metal filament atomizer: I/125
1436 metallochromic indicator: I/39
1437 metallofluorescent indicator: I/39
1438 metal vapour arc: II/150
1439 metal vapor arc lamp: II/150
1440 metastable (adjective)*: of transient stability
1441 metastable (noun)*: commonly used synonym for metastable ion
1442 metastable decomposition: I/32, II/205
1443 metastable ion peak: I/32, II/205
1444 metastable peak: I/32, II/205
1445 metastable state: II/120
1446 microanalysis: I/16
1447 microbalance*: microchemical balance
1448 microchemical balance: I/14, II/37
1449 microcomponent; I/19
1450 microcomputer*: a microprocessor plus input/output devices and other elements needed for functioning as a computer
1451 microcoulometry: I/155
1452 microphotometer: I/108, II/160
1453 microprocessor*: a computer processing unit made on one or more integrated-circuit chips
1454 microsample: II/145
1455 microscopic cross-section: II/217
1456 microtrace: I/16
1457 microwave: II/140
1458 microwave-induced plasma: II/140
1459 microwave plasmas: II/140
1460 Mie scattering: II/122
1461 migration: II/118
1462 migration current: II/55
1463 millicoulometry: I/155
1464 milligram equivalent of readability: I/13, 14, II/36
1465 minimal line width: I/100
1466 minor constituent: I/16
1467 mirror configuration: II/141
1468 mist: I/121, II/163
1469 mixed complexes: I/56, II/12
1470 mixed crystal: I/20, II/85
1471 mixed indicator: I/40, II/48
1472 mixed-ligand complexes*: mixed complexes
1473 mixed-metal complexes*: complexes containing more than one metal
1474 mixed-solution method*: for determination of potentiometric selectivity coefficients by use of a solution containing both the primary ion and the interfering ion
1475 mobile carrier electrode: I/171
1476 mobile phase: I/65, 78, II/97
1477 moderation: II/231
1478 moderator: II/231
1479 modes (laser): II/142
1480 modified active solid: I/79, II/98
1481 modified Nernst equation: I/172
1482 modifier: II/89
1483 modulation frequency: I/129
1484 modulation of linear polarized radiation: II/193
1485 modulation polarography: I/162, II/82
1486 molal activity*: activity expressed on the molal scale
1487 molal activity coefficient: I/51
1488 molality: I/178
1489 molar absorptivity*: molar absorption coefficient
1490 molar activity*: activity expressed on the molar scale
1491 molar activity coefficient*: activity coefficient expressed on the molar scale
1492 molar ionization energy: II/245
1493 molarity: I/178
1494 molecular absorption spectrometry*: measurement of molecular absorption by use of a spectrometer
1495 molecular absorption spectroscopy: II/111
1496 molecular anion: I/32
1497 molecular bands: II/128
1498 molecular cation: I/32

1499 molecular emission spectrometry*: use of a spectrometer to measure molecular emission
1500 molecular emission spectroscopy: II/111
1501 molecular ion: I/32, II/205
1502 molecular luminescence spectrometry*: determination by measurement of molecular luminescence with a spectrometer
1503 molecular methods: II/184
1504 molecular radiation: II/117, 118
1505 molecular spectrometry*: determination by means of molecular spectroscopic techniques
1506 molecular spectroscopy: II/111
1507 molecule*: a group of atoms bound by chemical forces
1508 momentum transfer: II/148
1509 mono-alternance: II/132
1510 monochromatic radiation: II/250
1511 monochromator: I/99, 128, II/156
1512 monofunctional ion-exchanger: I/89
1513 monolayer: II/251, 252
1514 monolayer capacity: II/252
1515 mononuclear binary complexes: I/56, II/11
1516 mononuclear ternary complexes*: complexes containing a single metal atom but three different ligands
1517 Mössbauer effect: II/231
1518 multichannel analyser: II/181
1519 multichannel pulse-height analyser: II/231
1520 multilayer: II/251, 252
1521 multipass method (system): I/128
1522 multiplication methods*: *Pure Appl. Chem.*, 1982, **54**, 2554
1523 multipurpose spark sources: II/133
1524 multislot-burners: II/166
1525 multisweep polarography: I/158
1526 mutual interference: I/134
1527 nanotrace: I/16
1528 natural abundance*: the abundance of an element as found in nature
1529 natural broadening: II/121
1530 natural impedances: II/132
1531 natural isotopic abundance: II/231
1532 natural radiation: II/237
1533 natural radioactivity: II/231
1534 nebulization: I/121, II/163
1535 nebulization efficiency: I/123, II/167
1536 nebulizer: I/122 II/165
1537 nebulizer–flame systems: I/123, II/163
1538 needle valve: II/146
1539 negative glow: II/147
1540 negative ion: II/118
1541 negative-ion mass spectrum: I/32, II/204
1542 negative zones: II/147
1543 negatron: II/221 (indexed and in text as negaton)
1544 neodymium glass: II/142
1545 nephelometric end-point: I/37
1546 nephelometric titration: II/47
1547 nephelometry*: analysis of a suspension by measurement of light scattered by it (usually at right angles to the incident beam)
1548 Nernst equation*: describes the relation between the working electrode potential and the activities (or concentrations) of the species involved in the electrode reaction
1549 nernstian response: I/170
1550 net faradaic current: II/54
1551 net retention time*: *Pure Appl. Chem.*, 1993, **65**, 842
1552 net retention volume: I/68, 83, II/104
1553 neutral*: neither acidic nor alkaline
1554 neutron*: neutral particle with approximately the same mass as the proton
1555 neutron activation: II/231
1556 neutron activation analysis*: activation analysis in which activity is induced by exposure to a flux of neutrons
1557 neutron capture: II/223
1558 neutron density: II/231
1559 neutron flux*: neutron flux density
1560 neutron flux density: II/225
1561 neutron multiplication: II/231 (indexed incorrectly as multiplication neutron, and neutron, multiplication, and left unpunctuated in text
1562 neutron number: II/230
1563 neutron radiation: II/217
1564 neutron temperature: II/232
1565 Nier–Johnson geometry: I/29, II/201
1566 noble gases: II/145
1567 no-load indication: I/12, II/36
1568 nominal linear flow: I/82, II/102
1569 non-additive current: II/53
1570 non-aqueous titration: I/42
1571 non-coherent electromagnetic radiation: II/112
1572 non-coherent optical sources: II/112
1573 non-conducting powder: II/126
1574 non-crystalline electrode: I/171
1575 non-current-carrying plasmas: II/124
1576 non-destructive activation analysis: II/210
1577 non-discrete continuous radiation: II/112
1578 non-faradaic admittance: I/152
1579 non-homogeneous magnetic fields: II/124
1580 non-homogeneous material: II/128
1581 non-specific interferences: II/170
1582 non-spectral interference: I/135
1583 normal eye: I/99, II/156
1584 normal glow discharge: II/149
1585 normality: I/176
1586 normal solution: I/175, 181
1587 norm temperature: II/120 (indexed as normal temperature)
1588 nuclear activation analysis: II/233
1589 nuclear chemistry: II/233
1590 nuclear decay: II/219
1591 nuclear fission: II/224
1592 nuclear fission reaction: II/239
1593 nuclear fuel: II/226
1594 nuclear fusion: II/226 (the definition is wrong: it is the process in which nuclei undergo nuclear fusion reactions)
1595 nuclear fusion reaction: II/226
1596 nuclear isobars: II/229
1597 nuclear isomers: II/230
1598 nuclear particle: II/233
1599 nuclear reaction: II/243
1600 nuclear reactor: II/133
1601 nuclear system: II/213
1602 nuclear track: II/244

1603 nuclear track detector: I/244
1604 nuclear transformation: II/233
1605 nuclear transition: II/233
1606 nucleation: I/19, II/84
1607 nucleon: II/233
1608 nucleon number: II/233
1609 nuclide: II/233
1610 nuclidic mass: II/234
1611 observation: II/4
1612 observation height: I/129
1613 observation zone of plasma: II/138
1614 occlusion: I/20, II/85
1615 one-colour indicator: I/39, II/48
1616 one-electrode microwave plasma: II/140
1617 one-step excitation source: II/144
1618 on–off ratio: II/124
1619 open-ended hollow cathode: II/151
1620 opening out*: bringing a solid sample into solution by treatment with acid or alkali, with preliminary fusion with a suitable flux if necessary
1621 open-tube chromatography: I/75, II/94
1622 open tubular column: I/78, II/97
1623 operational pH-cell: I/44, II/26
1624 operational pH scale: I/44, 49, II/29
1625 operational pH standards: I/45, II/21, 22
1626 optical conductance: I/101, 111
1627 optical density: I/110
1628 optical emission spectrometry*: use of a spectrometer for measurement of optical emission
1629 optical fibres*: glass (or similar) fibres used as light-guides in optical sensor systems
1630 optical filters: I/128
1631 optically thin plasma: II/153
1632 optical pumping: II/141
1633 optical quantities: II/157
1634 optical-signal modulation: II/193
1635 optical systems: II/190
1636 optimum spark conditions: II/134
1637 order of spectrum: I/100
1638 organic effect: I/136, II/170
1639 orifice: II/124
1640 oscillating reaction*: a reaction in which the concentration of one component increases and decreases in a regular (oscillatory) manner
1641 oscillating spark: II/131
1642 oscillator: II/35
1643 oscillator strength: II/121
1644 oscillography: an indexing error for oscillopolarography
1645 oscillopolarography: I/57, II/75
1646 Ostwald ripening: I/21, II/86
1647 outer zone: I/122, II/166
1648 outflow velocity: II/61
1649 overall distribution constants: I/59, II/88
1650 overall formation constant: I/55, II/11
1651 overcritically damped discharge: II/131, 133
1652 oxidant: I/122, II/165
1653 oxidation–reduction indicator: I/40
1654 oxidation–reduction titration: I/42, 47
1655 packed column: I/78, II/97
1656 packing: I/78, II/97
1657 pair attenuation coefficient: II/234
1658 pair production: II/235
1659 paper chromatography: I/75, II/94
1660 parallel ignition: II/133
1661 parallel–parallel mirror configuration: II/141
1662 paralysis circuit: II/219
1663 paramagnetic compounds: II/190
1664 parent ion: I/32, II/205
1665 partial decay constant: II/234
1666 partial evaporation: II/125
1667 particle annihilation: II/226
1668 particle bombardment: II/153
1669 particle detector: II/215
1670 particle flux*: particle flux density
1671 particle flux density: II/225
1672 particle-induced X-ray emission spectrometry*: determination of species by means of particle-induced emission of X-rays
1673 particle-induced X-ray emission spectroscopy (PIXES): II/250
1674 particle radiation: II/237
1675 particle size*: size of particles, usually as indicated by sieve classification
1676 particle size analysis*: determination of particle size, usually by sieve classification
1677 particle size distribution*: the relative amounts of a sample occurring within each size class
1678 partition chromatography: I/75, II/94
1679 partition coefficient: I/68, 70< 84, II/90
1680 partition constant: I/61, II/90
1681 partition function: I/144 (called the partition ratio on II/105)
1682 peak: I/66, 77, II/43, 61, 96, 101
1683 peak analysis: II/234
1684 peak area: I/66, 81, II/45, 101, 102
1685 peak base: I/66, 81, II/101, 102
1686 peak capacitor voltage: II/130
1687 peak current: II/56, 129
1688 peak diffraction coefficient: II/178
1689 peak elution volume: I/83, II/104
1690 peak fitting: II/234
1691 peak height: I/33, 66, 81, II/43, 101, 102
1692 peak intensity: II/121
1693 peak maximum: II/102
1694 peak resolution: I/70, 85, II/107
1695 peak spark current: II/131
1696 peak width: I/66, 82, II/101, 102
1697 peak width at half height: I/66, 82, II/101, 102
1698 Penning gas mixture: II/148
1699 percentage error: I/10
1700 percentage recovery: I/10
1701 periodic current: II/56
1702 periodic voltage: II/64
1703 permeation chromatography: I/75, II/94
1704 permselectivity: I/92
1705 pH: I/44, II/15
1706 pH electrode: II/26
1707 pH indicator*: a mixture of two or more indicators, either in solution or impregnated on paper, and suitable for indicating the approximate pH of a test solution
1708 pH interpretation: I/44, 48, II/25
1709 pH measurement: I/45, 49, II/24
1710 pH operational scale: I/49, II/29
1711 pH standards: I/45

1712 pH titration*: a titration with end-point location by means of pH measurements
1713 phase analysis*: analysis by selective extraction of components individually or in groups
1714 phase fluorimetry: II/197
1715 phase ratio: I/81, II/100
1716 phase titration: I/42
1717 phase volume ratio*: ratio of the volumes of the two phases in a solvent extraction
1718 phosphorescence: II/184
1719 phosphorescence analysis: II/199
1720 phosphorescence excitation–emission spectrum: II/197
1721 phosphorescence spectrometer: II/190
1722 phosphorimtery: II/197
1723 phosphoroscopes: II/193
1724 photocurrent: I/129
1725 photodetector: I/128, II/193
1726 photoelectric attenuation coefficient: II/234
1727 photoelectric effect: II/234
1728 photoelectric peak: II/234
1729 photoelectron: II/250
1730 photoelectron spectra: II/246, 247
1731 photoelectron spectroscopy (PES): II/245, 246
1732 photoelectron yield: II/252
1733 photoemission spectroscopy: II/247
1734 photographic emulsion calibration: II/162
1735 photographic intensity measurements: II/160
1736 photographic parameter: I/109
1737 photographic photometry: II/160
1738 photographic transmittance: II/160
1739 photoionization: I/31, II/203
1740 photoionization spectroscopy: II/203
1741 photometric end-point: I/38
1742 photometric titration*: titration in which the end-point is located by monitoring the absorbance of one of the reactants or products, or of an indicator
1743 photometry: I/99
1744 photomultiplier tube: II/193, 234
1745 photon*: a quantum of electromagnetic radiation
1746 photon activation: II/234
1747 photon emission spectrum: II/250
1748 photon emission spectroscopy*: analytical use of photon emission
1749 photon emission yield: II/174
1750 photon flux*: photon flux density
1751 photon flux density: II/225
1752 photon radiation: II/217
1753 photopeak: II/235
1754 physical aging: I/21, II/86
1755 physical equilibrium: I/143
1756 physical interference: I/135
1757 physical quantity: I/94
1758 physiosorption: II/252
1759 picotrace: I/16
1760 pile-up: II/235
1761 pinch effect: II/124
1762 P.I.N. semiconductor detector: II/235
1763 π-radian magnetic field analyser: I/29, II/202
1764 π/2-radian magnetic field analyser: I/29, II/202
1765 π/3-radian magnetic field analyser: I/29, II/202
1766 pixel*: the smallest area displayed on a television screen, that can be individually controlled
1767 PIXES: II/250
1768 Planck's law: I/127, 119
1769 plane cathode glow-discharge source: II/152
1770 plane polarization: II/122
1771 plane polarized: II/122
1772 plasma II/119
1773 plasma gas: II/136
1774 plasma impedance: II/135
1775 plasma initiation: II/136
1776 plasma jet: I/125, II/124
1777 plasma load: II/135
1778 plasma parameters: II/138
1779 plasma temperature: II/119
1780 plasma torch: I/125, II/136
1781 plasmon states: II/249
1782 plateau: II/39,172
1783 platrode: II/127
1784 pneumatic nebulizer: I/121, II/165
1785 Pockels cell: II/143
1786 point-to-plane configuration: II/135
1787 point-to-point configuration: II/135
1788 polarity: II/125
1789 polarization: II/122, 198
1790 polarized: II/122
1791 polarizer: II/193
1792 polarography: I/157, II/76
1793 polarometric titration: I/154
1794 polyatomic: II/145
1795 polychromator: I/99, 128, II/156
1796 polyfunctional ion-exchanger: I/89
1797 polynuclear complexes: I/56, 57, II/12, 14
1798 population inversion: II/141
1799 porodes: II/127
1800 porous cup electrodes: II/127
1801 positive column: II/147
1802 positive column current density: II/150
1803 positive ions: II/118
1804 positron: II/235
1805 post-filter effects: II/199
1806 post-precipitation: I/21, II/86
1807 potential*: the work done against electrical forces to bring unit charge from a reference point to the point concerned; often used to mean potential difference
1808 potential-step chronocoulometry: I/155
1809 potentiometric end-point: I/38
1810 potentiometric selectivity coefficient: I/170
1811 potentiometric titration: I/153, II/71
1812 potentiometric weight titration: II/65
1813 potentiometry: I/153, II/71
1814 powder sample: II/182
1815 powder techniques: II/135
1816 power-compensation differential scanning calorimetry: II/45
1817 power efficiency of fluorescence: I/142
1818 power of detection: II/134
1819 practical measurement of pH: I/49
1820 practical resolution: II/158
1821 practical specific capacity: I/90
1822 pre-arc period: I/107

1823 precipitate (noun): I/19, II/84
1824 precipitate (verb): I/19, II/84
1825 precipitation from homogeneous solution: I/19, II/84
1826 precipitation indicator: I/40
1827 precipitation titration: I/42, II/48
1828 precision: I/9, 115, 134, II/6
1829 precision null-point potentiometry: I/153
1830 precision of a balance: I/13, II/36
1831 precision of a weighing: I/13, II/37
1832 precision of indication: I/12, II/35
1833 precolumn reactor*: a reactor column placed before the main column of a chromatographic system
1834 precursor: I/32, II/62, 235
1835 precursor ion: I/32, II/205
1836 precursor isotope: II/222
1837 prefilter effects: II/199
1838 pre-ionization: II/206
1839 premix burner: I/122, II/165
1840 prespark period: I/107, II/134
1841 pressure: II/120
1842 pressure effects: II/120
1843 pressure-gradient correction factor: I/67, 81, II/101
1844 pressure–voltage–current relationship: I/146
1845 primary combustion: I/122, II/165
1846 primary discharge: II/152
1847 primary ion: II/249, 251
1848 primary ion-selective electrode: I/171
1849 primary pH standards: II/19, 16
1850 primary radiation: II/174
1851 primary reference standards: II/19
1852 primary source: II/117
1853 primary standard: I/41, II/47
1854 primary standard solution: I/41, 48
1855 primary substance: I/41 (indexing error for primary standard)
1856 principal gas: II/134
1857 prism base: I/100, II/157
1858 progenitor ion: II/205
1859 program (noun)*: a set of instructions expressed in a computer language
1860 program (verb)*: write a program
1861 programme (noun)*: a series of instructions
1862 programme (verb): I/23
1863 programmed current chronopotentiometry: I/156, II/75
1864 projection screen: II/133
1865 prolate trochoidal mass spectrometer: I/30, II/203
1866 prompt coincidence: II/215
1867 prompt neutron: II/232
1868 proportional counter tube: II/235
1869 proton: II/13
1870 protonation constant: I/57, II/13
1871 protonation equilibria: I/57, II/13
1872 proton number: II/212
1873 P-type delayed fluorescence: II/185
1874 pulse amplifier: II/211
1875 pulse amplitude analyser: II/235
1876 pulse amplitude distribution: II/179
1877 pulse amplitude selector: II/235
1878 pulsed discharge lamp*: a discharge lamp run at low current, with regular short pulses of high current
1879 pulsed laser sources: II/142
1880 pulse duration: II/62, 142
1881 pulse energy: II/142
1882 pulse height analyser: II/235
1883 pulse polarography: I/159
1884 pulse power: II/142
1885 pumped system: II/146
1886 pumping cavity: II/142
1887 pumping energy: II/142
1888 pumping lamp: II/142
1889 pumping period: II/142
1890 pump power: II/142
1891 pyroelectric detectors: II/193
1892 Q-switch: II/142
1893 Q-switched operation: II/142
1894 quadratic electro-optic effect: II/143
1895 quadrupole mass analyser: II/202
1896 qualitative analysis*: identification of the components of a sample
1897 quality factor: II/236
1898 quantitative analysis*: determination of the percentage composition of a sample, or of one or more of its components
1899 quantitative differential thermal analysis: II/44
1900 quantity: I/102
1901 quantum counters: II/193
1902 quantum efficiency of fluorescence: I/142, II/198
1903 quantum efficiency of luminescence: II/198
1904 quantum yield: II/197
1905 quantum yield of fluorescence: II/188, 197
1906 quantum yield of luminescence: II/197
1907 quarter-transition-time potential: II/51, 62
1908 quarter wavelength lines: II/136
1909 quasi-monochromatic photometry: I/110
1910 quench gas: II/179
1911 quenching: I/142, II/120, 188, 198, 236
1912 quenchofluorimetry: II/189
1913 rad: II/236
1914 radial electrostatic field analyser: I/29, II/201
1915 radiance: I/97, 98, II/155
1916 radiance temperature: II/119
1917 radiant emissivity: I/98, II/155
1918 radiant energy: I/97, II/154, 155, 194
1919 radiant energy density: I/98, II/155, 194
1920 radiant exposure: I/98, II/155, 160
1921 radiant flux: I/97, 98
1922 radiant intensity: I/98, II/141, 155
1923 radiant particles: II/128
1924 radiant plume: II/135
1925 radiant power per solid angle: II/141
1926 radiant quantity: I/97
1927 radiation: II/118, 147
1928 radiation chemistry: II/236,241
1929 radiation counter: II/236
1930 radiation detector: II/236
1931 radiation energy: II/221
1932 radiationless transition: II/187
1933 radiation pulses: II/142
1934 radiation quantities: II/154
1935 radiation sources: II/111, 237

1936 radiation spectrometer: II/235, 243
1937 radiation spectrum: II/237
1938 radiative capture: II/237
1939 radiative de-excitation: II/121
1940 radiative excitation: II/121
1941 radiative processes: II/120
1942 radiative recombinations: II/119
1943 radiative transitions: II/189
1944 radioactive*: exhibiting, or referring to, radioactivity
1945 radioactive age: II/237
1946 radioactive chain: II/237
1947 radioactive cooling: II/83
1948 radioactive dating: II/218
1949 radioactive decay: II/237
1950 radioactive effluent: II/221
1951 radioactive equilibrium: II/237
1952 radioactive fall-out: II/237
1953 radioactive half-life: II/228
1954 radioactive indicator: I/38
1955 radioactive isotope: II/214, 238
1956 radioactive label: II/237
1957 radioactive materials: II/238
1958 radioactive mean life: II/231
1959 radioactive nuclei: II/242
1960 radioactive nuclide: II/231
1961 radioactive series: II/237
1962 radioactive substances: II/218
1963 radioactive tracer: II/237
1964 radioactive waste material: II/238
1965 radioactivity: II/238
1966 radioanalytical chemistry: II/211
1967 radioanalytical purification: II/238
1968 radiochemical purity: II/238
1969 radiochemical separation: II/238
1970 radiochemical yield: II/238
1971 radiochemistry: II/238,241
1972 radiochromatograph: II/238
1973 radiocolloid: II/238
1974 radiofrequency plasmas: II/135
1975 radiofrequency polarography: I/162, II/81
1976 radioisotope: II/238
1977 radiolysis: II/239
1978 radiometric end-point: I/38
1979 radiometric titration: II/47, 244
1980 radionuclide*: a radioactive nuclide (the term is indexed in II as occurring in connection with several other terms, but no defiinition is given)
1981 radionuclidic purity: II/239
1982 radio-reagent*: a radioactively labelled reagent
1983 raffinate*: the phase remaining after an extraction has been performed and the extract separated (see note on I/2)
1984 Raman radiation: II/150
1985 Raman scattering: II/122
1986 Raman spectroscopy: II/117
1987 random coincidence: II/239
1988 range: I/9, II/5
1989 ratemeter: II/239
1990 rate of liquid aspiration: I/123, II/166, 167
1991 rate of liquid consumption: I/123, II/166, 167
1992 rate of nucleation: I/19, II/84
1993 Rayleigh scattering: II/122 (incorrectly indexed and given in text as Raleigh scattering)
1994 R_B value: I/84, II/106
1995 reaction interval: II/41
1996 reaction layer: II/63
1997 readability: I/13, II/36
1998 reading: I/130
1999 real absorption: I/138
2000 real time*: pertains to a system yielding a response virtually simultaneously with the event concerned
2001 rearranged molecular ion: I/32, II/205
2002 rearrangement ion: I/32, II/205
2003 receiving time: II/217 (incorrectly given in the text and indexed: should have been resolving time)
2004 reciprocal linear dispersion: I/101, II/158
2005 recoil: II/239
2006 record: II/43
2007 recovery*: the amount of an analyte found, expressed as a fraction of the amount known to be present
2008 recovery factor: I/61, II/90
2009 recovery test: I/134
2010 rectified alternating current arc: II/124
2011 rectified source: II/127
2012 redox indicator: I/40
2013 redox ion-exchangers: I/92
2014 redox polymers: I/92
2015 redox titrations: I/42
2016 red shift*: *Pure Appl. Chem.*, 1983, **55**, 1292
2017 reference beam: I/128
2018 reference electrode: I/151, 169, II/27, 60
2019 reference element: I/102, 111, II/139, 163, 171
2020 reference-element technique: I/137, II/171
2021 reference holder: II/42
2022 reference intensity: I/111, II/163
2023 reference line: II/139
2024 reference material: II/42
2025 reference solutions: I/131, II/138
2026 reference value pH standard: I/45, 53, II/17, 26
2027 reflected power: II/136
2028 reflection electron energy-loss spectroscopy (REELS): II/249
2029 reflection factor: I/101, II/158
2030 reflection high-energy electron diffraction (RHEED): II/248
2031 reflux nebulizers: I/122, II/165
2032 refraction effects: II/199
2033 refractive index: I/100, II/157
2034 regeneration current: II/53
2035 reignition voltage: II/132
2036 relative activity: II/9
2037 relative atomic mass: II/212
2038 relative biological effectiveness: II/239
2039 relative counting: II/239
2040 relative density*: specific gravity
2041 relative error*: the difference between the observed and true values, expressed as a fraction of the true value
2042 relative retention: I/68, II/105
2043 relative standard deviation: I/9, II/5, 183

2044 releasers: I/137, II/172
2045 releasing agents*: reagents which can release a metal from a complex
2046 releasing mechanism*: the way in which a releaser or a releasing agent achieves its effect
2047 rem: II/239
2048 remotely coupled: II/136
2049 reprecipitation: I/21, II/86
2050 resident time: II/137
2051 residual current: II/56
2052 residual gases: II/153
2053 residual ionization: II/147
2054 residual liquid junction: II/28
2055 residual liquid-junction potential: II/21, 28
2056 resin matrix: I/89
2057 resistor: II/131
2058 resolution: I/70, II/107, 108, 158, 178, 180
2059 resolution 10% valley (definition): I/30, II/203
2060 resolving power: I/30, 101, II/158, 203, 240
2061 resolving time: II/240 (should replace receiving time on II/217)
2062 resolving time correction: II/240
2063 resonance broadening: II/122
2064 resonance energy: II/240
2065 resonance line: I/138
2066 resonance neutron: II/232
2067 resonance spectrometer: I/128
2068 resonant cavity: II/140
2069 response constant: II/63
2070 response slope*: the slope of a linear calibration plot
2071 response time: I/130
2072 responsivity: I/129
2073 rest point: I/13, II/36
2074 restricted glow-discharge: II/150
2075 result: I/8, II/4
2076 retardation factor (R_f): I/83, II/105
2077 retention: II/240
2078 retention index: I/86, II/108
2079 retention parameters: I/66
2080 retention temperature: I/83, II/105
2081 retention time: I/66
2082 retention volume: I/66, II/103
2083 reversed direct-injection burner: I/122, II/166
2084 reversed-phase chromatography: I/76, II/93
2085 rigid matrix electrode: I/171
2086 ripening: I/21, II/86
2087 rise time*: the time needed for a signal to rise from a low to a high percentage of its maximum value (usually from 5 to 95 or from 10 to 90%)
2088 rise velocity: I/126
2089 roentgen: II/240
2090 rotating disc shutters: II/143 (but not indexed)
2091 rotating mirror shutters: II/143
2092 rotating platform: II/127
2093 rotational energy transition: II/128
2094 rotational temperature: II/119
2095 Rutherford back-scattering (RBS): II/250
2096 Saha–Eggert law: II/119
2097 salting-in*: increasing the solubility of a solute by addition of a salt
2098 salting-out: I/63, II/91
2099 salting-out chromatography: I/76
2100 sample: I/101, 102, 130, II/39, 181
2101 sample beam: I/128
2102 sample electrode: II/125, 126
2103 sample holder: II/39
2104 sample injector: I/64, II/98
2105 sample interval: II/63
2106 sample of limited size: II/182
2107 sample solution: II/138
2108 sample weight classification: I/15
2109 sampling boat: I/124
2110 sampling cup: I/124
2111 sampling loop: I/124
2112 Sand equation: I/167
2113 saturable dye switch: II/143
2114 saturated solution: I/18, II/83
2115 saturation: I/18, II/83, 172, 240
2116 saturation chamber*: a tank for saturating a chromatographic paper with solvent or eluent vapour
2117 saturation plateau: I/137, II/172
2118 scale expansion: I/130
2119 scaler: II/240
2120 scaling circuit: II/240
2121 scanning electron microscopy (SEM): II/247
2122 scanning laser analysis: II/145
2123 scanning electron transmission microscopy (STEM): II/248
2124 scatter: I/132 (of results), II/122 (of radiation)
2125 scattering: I/138, II/241
2126 scattering of radiation: I/138, II/122, 217
2127 scavenger: I/19, II/241
2128 scavenging: II/241
2129 scintillating material: II/241
2130 scintillation: II/241
2131 scintillation counter: II/241
2132 scintillation detector: II/179, 180, 241
2133 scintillation spectrometer: II/241
2134 scintillator: II/241
2135 scintillator detector: II/220
2136 screened indicator: I/40 (an alternative definition might be that it is a mixture of an indicator and an indifferent dyestuff chosen so that all three complementary colours are present at the transition point, in proportions resulting in a grey colour)
2137 screened off: II/151
2138 scrubbing: I/62
2139 scrubbing solution: I/62
2140 sealed hollow cathode lamp: I/127, II/152
2141 secondary combustion: I/122, II/166
2142 secondary discharge: II/152
2143 secondary electron multiplier: I/31
2144 secondary electrons: II/247
2145 secondary fluorescence: II/175
2146 secondary ion: II/249
2147 secondary ion mass spectrometry (SIMS): II/251
2148 secondary ion yield: II/252
2149 secondary radiation: II/174, 241
2150 secondary standard: I/41, II/48
2151 secondary standard solution: I/41, II/48
2152 second derivative potentiometric titration: I/153

2153 secular equilibrium: II/242
2154 seeded arcs: II/128
2155 selected area electron diffraction (SAED): II/248
2156 selective elution: I/76, II/92
2157 selective volatilization: II/118
2158 selectivity coefficient: I/90, 170
2159 self-absorption: I/107, 139, II/122, 242
2160 self-absorption broadening: II/122
2161 self-absorption effects: II/199
2162 self-absorption factor: II/242
2163 self-electrode: II/125
2164 self-ignited spark discharge: II/132
2165 self-reversal: I/107, 139, II/122
2166 self-shielding: II/242
2167 semiconducting surface: II/150
2168 semiconductor: II/242
2169 semiconductor detector: II/242
2170 semi-integral polarography: I/151, II/66
2171 semimicro: I/15
2172 semimicro sample: I/16
2173 semi-Q-switched: II/142
2174 semitransparent mirror: II/141
2175 sensitive volume of a detector: II/242
2176 sensitivity: I/114, II/5
2177 sensitivity for a stated load: I/13, II/36
2178 sensitivity of a direct probe: I/34
2179 sensitivity of an inlet system: I/34
2180 sensitivity of an ion source: I/34
2181 sensitized luminescence: II/188
2182 separated flame: I/123, II/166
2183 separate solution method: I/173
2184 separation factor: I/20, 85, 91, II/86, 107
2185 separation process: II/97
2186 separation temperature: I/80, II/100
2187 sequestration*: prevention of interference by a metal, by formation of a sufficiently stable complex of the interferent
2188 series: I/9, II/4
2189 series ignition: II/133
2190 shielded electrical discharge: II/128
2191 shielded flame: I/123, II/166
2192 shields: II/151
2193 short circuit discharge: II/153
2194 Shpol'skii spectra: II/190
2195 sievert: II/242
2196 sifter electrode: II/126
2197 sign conventions: I/166
2198 silicon photodiodes: II/193
2199 simulation technique: I/137, II/172
2200 simultaneous analytical systems: II/139
2201 simultaneous differential thermal analysis: II/39
2202 simultaneous techniques: II/38 (called multiple techniques on I/27)
2203 single-beam spectrometer: I/128, II/191
2204 single-channel pulse height analyser: II/242
2205 single-element lamp: I/127
2206 single escape peak: II/223
2207 single-focusing mass spectrometer: I/28, II/201
2208 single pulse: II/142
2209 single-sweep oscillographic polarography: I/158
2210 singlet–singlet absorption: II/187
2211 singly ionized element: II/138
2212 sliding spark: II/134
2213 slit: I/100, II/156
2214 slit-height: I/100, II/156
2215 slit-width: I/100, II/156
2216 slot-burners: I/122, II/166
2217 slow neutrons: II/232
2218 solar-blind photomultiplier*: a photomultiplier that has maximum response at about 220 nm, but practically zero response at wavelengths longer than about 360 nm
2219 solid angle: I/130
2220 solid sample: II/125, 182
2221 solid solution: I/20, II/85
2222 solid solution sample: II/182
2223 solid state crystal lattice: II/149
2224 solid state detector: II/180
2225 solid state lasers: II/142
2226 solid support: I/65, II/98
2227 solubility product: I/18, II/83
2228 solute–solvent interaction*: specific interaction between solute and solvent
2229 solute-volatilization interferences: I/136, II/170
2230 solution: I/18, II/83
2231 solution equibria: I/54, II/9
2232 solvent: I/18, II/83, 89, 91
2233 solvent blank: I/131
2234 solvent effects: I/136, II/200
2235 solvent extraction: I/60, II/87
2236 solvent front: I/80, II/99
2237 solvent migration distance: I/80, II/99
2238 sorption: I/91
2239 sorption isotherm: I/91
2240 source: I/126, II/150 (spectroscopic), II/227 (radioactive)
2241 source efficiency: II/242, 243
2242 space charge: II/149
2243 space-resolving techniques: II/144
2244 spark chamber: II/133
2245 spark cross-excitation: II/144
2246 spark discharge: II/129
2247 spark frequency: II/131
2248 spark gap: II/129
2249 spark generator: II/130
2250 spark ignition: II/129
2251 sparking equilibrium: II/134
2252 sparklike discharges: II/133
2253 spark repetition rates: II/132
2254 spark source ionization: II/204
2255 sparks per half-cycle: II/132
2256 spark stands: II/133
2257 spatial distribution interferences: I/136, II/141, II/171
2258 spatial resolution*: degree of ability to distinguish between two objects very close together
2259 spatial stabilization: II/124
2260 specific activity: II/243
2261 specific burn-up: II/243
2262 specific interferences: I/135, II/170
2263 specific ionization: II/243
2264 specific retention volume: I/68, 83, II/104
2265 specimen: II/42

2266 specimen-holder assembly: II/43
2267 spectral background: II/134
2268 spectral band-width: I/100
2269 spectral characteristics: II/134, 138
2270 spectral continuum: I/137, II/173
2271 spectral intensity: I/102, II/174
2272 spectral interferences: I/135, ii/169
2273 spectral line shape: II/121
2274 spectral line shift: II/121
2275 spectral period: I/97
2276 spectral radiance: I/97, II/154
2277 spectral radiance of the black body: II/154
2278 spectral radiant energy density: II/121
2279 spectral radiation quantities: I/97, II/154
2280 spectral response curve: I/129
2281 spectrochemical accuracy: I/116
2282 spectrochemical analysis: I/93, 114, 118, II/111
2283 spectrochemical applications: II/128
2284 spectrochemical buffer: I/107, 137, II/159, 172
2285 spectrochemical carrier: I/108, II/159
2286 spectrochemical diluent: II/159
2287 spectrochemical properties: II/128
2288 spectrogram: II/243
2289 spectrograph: I/99, II/156
2290 spectrographic instruments: I/99, II/156
2291 spectrography: II/111
2292 spectrometer: I/99, 120, 128, II/156
2293 spectrometer resolution: II/178
2294 spectrometry: I/120
2295 spectrophotometric titration: II/47
2296 spectroscope*: an instrument for displaying a spectrum
2297 spectroscopy: I/120
2298 spectrum*: a display of intensity of radiation as a function of the energy (or a related quantity such as wavelength) of the radiation
2299 spectrum of radiation: II/237
2300 spike power: II/142
2301 spikes: II/142 (can also mean known additions of analyte or other component of the system)
2302 spin conservation rule: II/185
2303 spontaneous electrogravimetry: II/155
2304 spontaneous emission transition probability: II/121
2305 spontaneous fission: II/243
2306 spot: I/77, II/96
2307 spray chamber: I/122, II/165
2308 spray discharge: II/150
2309 sprayer: I/121, II/165
2310 sputtering: II/117, 253
2311 sputtering rate: II/149
2312 sputtering yield: II/149, 252
2313 square-wave current: II/58
2314 square-wave polarography: I/161, II/80
2315 stability constant: I/56, II/11
2316 stability constants of metal-ion complexes: I/56, II/9, 11, 12
2317 stabilization rings: II/124
2318 stabilized arcs: II/123, 124
2319 stabilized current supplies: II/127
2320 staircase polarography: I/160
2321 standard addition method: I/170
2322 standard deviation: I/9, 10, 114, II/5
2323 standard deviation for counting: II/183
2324 standardized solution: II/47
2325 standardized titrant solution: II/47
2326 standard observer: I/99, II/156
2327 standard reference material*: synonymous with certified reference material
2328 standard reference solution: II/17, 24, 26, 29
2329 standard solution: I/40, II/47
2330 standard substances: I/41, II/47
2331 standard subtraction method: I/171
2332 Stark broadening: II/122
2333 start*: the start line on a chromatogram
2334 start line*: the starting line on a chromatogram
2335 starting line: I/77, II/96
2336 starting point: I/77, II/96
2337 static field mass spectrometer: II/202
2338 stationary-electrode voltammetry: II/76
2339 stationary phase: I/65, 79, II/97,98
2340 stationary-phase fraction: I/81, II/100
2341 stationary-phase volume: I/80, II/100
2342 stationary radiation: II/126
2343 statistical weight: I/144
2344 steady state: II/139
2345 step: I/66, 78, II/96, 101
2346 step height: I/66, 78, II/96, 101
2347 step-sector: I/111, II/162
2348 stepwise elution: I/76, II/92
2349 stepwise formation constant: I/56, II/11
2350 stepwise line fluorescence: I/140
2351 Stern–Volmer law: II/198
2352 stoichiometric end-point: I/38, II/47
2353 stoichiometry*: numerical relationship between reactants and products of chemical reactions or between components of compounds
2354 Stokes fluorescence: I/140, II/189
2355 stopping power: II/210, 243
2356 stray light: I/103
2357 stray radiation: II/159
2358 striations: II/147
2359 stripping: I/62, II/91
2360 stripping analysis: I/150, II/65
2361 stripping solution: I/62, II/91
2362 strobe interval: II/63
2363 sublimation: II/118
2364 submicro analysis: I/17
2365 substoichiometric isotope dilution analysis: II/230
2366 subtrace analysis: I/17
2367 suction nebulizer: I/122, II/165
2368 summit: II/53, 56
2369 summit current: II/56, 58
2370 summit potential: II/56, 62
2371 superposition: II/145
2372 supersaturated solution: I/19, II/84
2373 supersaturation: I/19, II/84
2374 supplementary discharge: II/152
2375 supplementary electrodes: II/152
2376 support: II/145
2377 supporting electrolyte: II/60
2378 support plate: I/79, II/98
2379 suppressors: I/137, II/172
2380 suprathermal luminescence: I/138

2381 surface: I/18, II/251
2382 surface analysis: II/245
2383 surface barrier semiconductor detector: II/243
2384 surface concentration: II/251
2385 surface contamination: II/252
2386 surface coverage: II/252
2387 surface reaction: II/63
2388 surface spark: II/134
2389 synchronously excited fluorescence spectrum: II/197
2390 synchronously excited phosphorescence spectrum: II/197
2391 synchronously rotating gap: II/132
2392 Szilard–Chalmers effect: II/243
2393 T-tubes: I/124
2394 tailing: I/77, II/96
2395 tandem gap: II/132
2396 tandem mass spectrometry*: use of two or more mass spectrometers in series, sometimes with a reaction chamber in the field-free region between them
2397 tank circuit: II/135
2398 tapered cavity: II/140
2399 target: II/141
2400 Tast interval: II/63
2401 Tast polarography: I/158, II/77
2402 temperature effects: I/69, II/200
2403 temperature-programmed chromatography: I/76, II/92
2404 temporal resolution*: ability to separate events occurring close together in time
2405 temporal spacing: II/142
2406 temporal stabilization: II/124
2407 tensammetry: I/152
2408 Tesla coil: II/127, 133
2409 test electrode: II/60, 66
2410 theoretical end-point: I/38, II/47
2411 theoretical-plate number: I/69, 85, II/47
2412 theoretical retention volume: I/67, 71 (I/71 says that the term occurs in Section 13.6.06, but this is incorrect – it should be Section 13.6.05)
2413 theoretical specific capacity: I/90
2414 thermal aging: I/21, II/86
2415 thermal analysis: I/24, 25, 26, II/38
2416 thermal column: II/244
2417 thermal conductivity: II/125
2418 thermal conductivity detector*: a gas chromatographic detector based on the difference in thermal conductivity between the carrier gas alone and the carrier gas plus solute
2419 thermal decomposition: I/24
2420 thermal equilibrium: I/143, II/120
2421 thermal evaporation: II/134
2422 thermal fission: II/244
2423 thermal ionization: I/31, II/203
2424 thermally ignited arc: II/123
2425 thermal neutrons: II/233
2426 thermal processes: II/150
2427 thermal radiation: I/138, II/120
2428 thermal source: II/141
2429 thermoacoustimetry: II/40, 46
2430 thermoanalytical techniques: II/39, 40
2431 thermobalance: II/39
2432 thermochemical reactions: II/125
2433 thermodilatometry: II/40, 45
2434 thermodynamic constants: II/9
2435 thermodynamic equilibrium: I/143, II/119
2436 thermodynamic temperature: I/143, II/119
2437 thermoelectrometry: II/40, 46
2438 thermogram: I/24, II/38
2439 thermography: I/24
2440 thermogravimetric aanalysis: I/25
2441 thermogravimetric curve: I/24, 26, II/39
2442 thermogravimetry: I/25, 26, II/39, 40
2443 thermoluminescence: II/46
2444 thermomagnetometry: II/40, 46
2445 thermomechanical analysis: II/40, 45
2446 thermomechanometry: II/40, 45
2447 thermometric end-point: I/38
2448 thermometric titration: II/47
2449 thermomicroscopy: II/46
2450 thermoparticulate analysis: II/40, 42
2451 thermophotometry: II/46
2452 thermopile: II/193
2453 thermoptometry: II/40, 46
2454 thermorefractometry: II/46
2455 thermosonimetry: II/40, 46
2456 thermospectrometry: II/46
2457 thermovaporimetric analysis: I/25
2458 thickness of the reaction layer: II/63
2459 thin film: II/251
2460 thin-film electrode*: an electrode (usually carbon) coated with a thin film of mercury
2461 thin-layer chromatography: I/76, II/94
2462 three-electrode configuration: II/65
2463 three-electrode plasma: II/125
2464 three-slot burner: I/122, II/166
2465 threshold energy: II/244
2466 thyristor control: II/127
2467 tilt of the photographic plate: I/100, II/157
2468 time-constant: I/130, II/138
2469 time-of-flight mass spectrometer: I/29, /202
2470 time-of-wait curve: II/134
2471 time-resolved spectroscopy: I/107, II/134
2472 time-resolving techniques*: techniques distinguishing between analytes by means of differences in the times at which their signals appear
2473 titrant: I/41, II/47
2474 titration: I/41, II/47
2475 titration curve: II/49
2476 titration error: I/42, II/48
2477 titre: I/42, II/48
2478 titrimetric analysis: I/42, II/47
2479 titrimetric conversion factor: I/42
2480 titrimetry: II/47
2481 torsional braid analysis: II/45
2482 total absorption peak: II/244
2483 total ion current: I/33, II/206
2484 total quantum efficiency of fluorescence: I/142
2485 total retention volume: I/82, II/103
2486 Townsend first coefficient: II/148
2487 trace: I/16
2488 trace analysis: I/16
2489 trace constituent: I/16
2490 tracer: II/244
2491 transferred plasmas: II/125

2492 transformation: I/109, II/161
2493 transformation constant: I/109, II/161
2494 transition interval: I/43, II/48
2495 transition probability: II/216
2496 transition probability for absorption: I/141, II/121
2497 transition probability for spontaneous emission: I/141, II/121
2498 transition probability for stimulated emission: II/121
2499 transition time: II/64
2500 transit time: I/126, II/137
2501 transmission electron energy loss spectroscopy (TEELS): II/249
2502 transmission electron microscopy (TEM): II/247
2503 transmission factor: I/104, 141, II/158
2504 transmission high energy electron diffraction (THEED): II/248
2505 transmission ratio: II/162
2506 transmittance: I/108, II/160
2507 transportation:II/118
2508 transport interferences: I/136, II/170
2509 transverse electromagnetic mode: II/140
2510 trapping: II/149
2511 travel time: I/126
2512 triangular wave polarography: I/158
2513 triangular wave voltammetry: I/158, II/77
2514 triggerable stationary gap: II/132
2515 triplet–triplet absorption: II/187
2516 true coincidence: II/216
2517 tube assembly: II/135
2518 tuned amplifier measuring system: II/253
2519 tuned-line oscillator: II/136
2520 tungsten-filament lamp: I/126
2521 tungsten–halogen lamp: II/142
2522 tuning stub: II/140
2523 turbidimetric end-point: I/38
2524 turbidimetric titration: II/47
2525 turbidimetry*: measurement of the intensity of light transmitted through a suspension, preferably with collimation of the emergent beam to avoid scattering interference
2526 turbulent flame: I/122
2527 turbulent flow*: flow at a rate fast enough to ensure uniform mixing of the stream
2528 twin nebulizer: I/122, II/165
2529 two-colour indicator: I/39, II/48
2530 two-dimensional chromatography: I/76, II/93
2531 2200-metre-per-second flux density: II/225
2532 Ulbricht's sphere: I/113
2533 ultimate focal spot diameter: II/143
2534 ultramicroanalysis: I/16
2535 ultrasonic nebulizer: I/122, II/165
2536 ultratrace analysis: I/16
2537 ultraviolet photoelectron spectroscopy (UPS): II/250
2538 uncontrolled alternating current arc: II/123
2539 uncontrolled high-voltage spark generator: II/132
2540 undercritically damped oscillating discharge: II/133
2541 unidirectional discharge: II/132
2542 unidirectional spark: II/124
2543 unpolarized: II/122
2544 unresolvable band spectra: II/119
2545 unresolved band spectra: II/118
2546 Untergrund: I/103, II/159
2547 uptake rate: II/138
2548 vacuum-cup electrode: II/126
2549 vacuum phototube: I/128
2550 vacuum sparks: II/134
2551 valence electrons: II/250
2552 value of a division: I/12, II/35
2553 vaporization: II/143
2554 vapour cloud: II/132
2555 vapour jet: II/132
2556 vapour-phase interferences: I/136, II/171
2557 vapour plume*: vapour jet (or cloud) in form of a plume
2558 variance: I/9, 115, 133, II/5
2559 variate: I/9
2560 vibrating reed electrometer: I31
2561 vibrational energy transition: II/118
2562 vibrational temperature: II/119
2563 visual end-point: I/38
2564 visual indicator: I/38
2565 Voigt function: II/121
2566 Voigt profile: II/121
2567 volatilization: I/121, II/117, 165
2568 volatilization rate: II/118
2569 volatilizers: I/108, 137, II/159, 172
2570 voltage: II/64
2571 voltage–current characteristics: II/123
2572 voltage–time relationship of an arc: II/123
2573 voltammetric analysis*: voltammetry
2574 voltammetric constant: II/64
2575 voltammetry*: analysis based on measurement of the current flowing through the indicator electrode of an electrochemical cell, as a function of the potential applied to the cell
2576 voltammogram*: the current–voltage curve obtained in voltammetry
2577 voltamperogram*: an etymologically more correct term for voltammogram
2578 volume capacity: I/90
2579 volume distribution coefficient: I/91
2580 volume swelling ratio of ion-exchangers: I/90
2581 volumetric flow-rate I/82, II/102
2582 wall-stabilized arc: II/124
2583 wave height: II/64
2584 wavelength configuration: II/140
2585 wavelength dispersion: II/176
2586 wavelength modulation: II/193
2587 wavenumber: I/97
2588 weak potential gradient: II/148
2589 weighing procedure: I/13 (indexed as weighting procedures)
2590 weight-change determination: I/26
2591 weight titration: I/42
2592 Westcott cross-section: II/244
2593 width: II/178
2594 width of light-beam: I/100
2595 working electrode: II/60, 66
2596 working gas: II/133 (indexed incorrectly as working gap)
2597 xenon lamp: I/127
2598 X-radiation: II/222, 239
2599 X-ray analysis: II/247

2600 X-ray data interpretation: II/182
2601 X-ray diffraction: II/176
2602 X-ray emission spectrometry: II/177
2603 X-ray emission spectroscopy: II/173
2604 X-ray escape peak: II/223
2605 X-ray fluorescence: II/174
2606 X-ray generation: II/173
2607 X-ray generation by electrons: II/173
2608 X-ray generation by photons: II/174
2609 X-ray generation by positive ions: II/174
2610 X-ray intensity: II/182
2611 X-ray measurements: II/175
2612 X-ray photoelectron spectroscopy (XPS): II/250
2613 X-ray photons: II/223
2614 X-ray spectrum: II/173, 223
2615 Zeeman background correction*: use of the Zeeman effect for background correction; in the source-shift method, the π component of the split source line is absorbed by both the analyte and the background, but the σ_+ and σ_- components are absorbed by only the background species.
2616 Zeeman effect: II/122
2617 zero point of a glass electrode: II/28
2618 zero point of the scale: I/13, II/36
2619 zone: I/77, II/96

ADDENDA

A01 acousto-optical shutters: II/143 (incorrectly indexed as rotating acousto-optical shutters)
A02 atomic absorption spectroscopy: II/111, 117
A03 atomic emission spectroscopy: I/93, II/111, 154
A04 atomic fluorescence spectroscopy: II/111, 117
A05 blackening: I/110
A06 flame spectroscopy: II/111, 163
A07 internal reference electrode: I/170
A08 molecular luminescence spectroscopy: II/111
A09 optical emission spectroscopy: II/111
A10 X-ray fluorescence spectroscopy: II/111

Cross-reference table

E	F	G	S	R	C	J
0001	0654	0015	0703	0092	0536	0081
0002	0455	0019	2177	0004	1266	1186
0003	0893	0021	0918	0002	1270	1188
0004	0891	0020	0917	0003	1265	1187
0005	2364	0018	2364	0001	1271	1185
0006	0005	0649	0001	1475	2121	0468
0007	0819	0024	0873	0471	2124	0475
0008	0972	0025	0995	1480	2120	0478
0009	0007	0023	0003	0005	2123	0476
0010	0008	0026	0004	1476	2119	0469
0011	0376	0030	0374	0882	2127	0471
0012	1951	0029	2204	1923	2131	0477
0013	1473	2533	1632	0464	2130	0493
0014	1000	0027	1024	2552	2126	0470
0015	1130	0028	1250	2281	2132	0480
0016	0011	0775	0007	1477	0768	0413
0017	0010	2366	0006	1478	1423	2532
0018	1684	0032	1633	1687	2122	0481
0019	1896	0031	1885	1688	2125	0472
0020	1917	2383	2037	0172	2133	0479
0021	0016	0650	0012	1474	2110	0467
					2127	
					2129	
0022	1881	1934	1226	1604	2545	1149
0023	1291	1994	1440	0769	1941	0819
0024	1336	1996	1480	0767	1940	0820
0025	2441	1995	2488	0768	1939	0821
0026	2442	0035	2487	0129	1942	0837
0027	0020	0034	0019	0130	1942	0837
0028	0492	1998	0467	0825	1937	0836
0029	1201	1997	1384	0770	1938	0835
0030	2324	0061	2309	1672	1846	0216
0031	1497	0051	1663	1011	1289	0020
0032	0023	0052	0023	0023	1029	2250
0033	0092	0055	0107	0022	1030	2256
0034	2120	0056	2271	1549	1031	2255
0035	1563	0050	1698	0027	1032	1778
0036	2146	0049	2319	0030	2106	0349
0037	0027	0057	0027	0031	1026	0351
						2300
0038	0377	0058	0376	0871	1028	0352
0039	0465	0059	0440	0026	1027	2301
0040	0032	2600	0035	0468	1980	1542
0041	0033	0036	0034	0014	1154	0344
0042	2596	1886	2595	1849	1986	1418
0043	0037	0037	0038	0015	2108	0489
0044	0036	0038	0037	0016	2106	0487
0045	0038	0039	0039	0020	2104	0485
0046	0334	0040	0579	0017	2107	0486
0047	0573	0043	0544	0019	2105	0490
0048	1295	0042	1443	0018	2109	0488

E	F	G	S	R	C	J
0049	0039	0044	0041	0131	1650	0133
0050	1250	0045	1396	0313	1651	0134
0051	0045	0046	0046	0011	0777	0496
0052	0047	0047	0048	0012	1261	0499
0053	0048	0048	0047	0013	1260	0498
0054	2549	0070	1058	2064	1355	0132
						0994
0055	1735	1331	1910	0216	1315	0543
0056	2443	0063	2489	0033	1172	0030
0057	0050	0062	0053	0034	1172	0030
0058	0051	0064	0054	1491	1171	A01
0059	0563	0067	2265	0038	0647	0031
0060	0375	0065	0373	0035	2153	0032
0061	0720	0069	0732	0037	1890	0035
0062	1673	0066	1856	0036	0152	0034
0063	1685	0068	1864	0039	1411	0033
0064	0067	2521	0073	0046	1193	0735
0065	0066	2528	0182	0486	1190	0729
0066	0168	2525	0621	1406	1192	0734
0067	0352	2526	1960	1403	1191	0736
0068	1817	2519	2386	1402	1186	0730
0069	2404	2521	0075	0045	1189	0732
0070	0355	2522	0625	2407	1188	0731
0071	1832	2523	1977	1540	1187	0733
0072	0019	0071	0020	0040	0004	0029
0073	0746	0073	0762	0041	0392	1601
0074	2444	0074	2490	0042	0388	1604
0075	0058	0072	0067	0043	0389	1602
0076	2172	0075	0839	0047	1438	2541
0077	2025	2450	2160	2253	0565	1270
0078	0540	0077	0525	0076	1510	0127
0079	0985	0091	1006	1298	0681	2161
0080	0986	0092	0132	0063	0675	2170
0081	2348	0191	2351	1053	0676	2171
0082	0190	0093	0223	0127	0678	2169
0083	1395	0094	1536	0699	0677	2168
0084	2347	0192	2350	0049	0666	2163
0085	0213	0090	0239	0069	0674	0282
						0285
0086	1073	0099	0273	0064	0668	0669
0087	1203	0096	1385	0065	0669	0666
0088	0603	0085	0639	0066	0672	0667
0089	2351	0086	2352	1048	0673	0668
0090	0936	0097	0940	0075	0662	2165
0091	1555	0098	1736	0068	0663	2166
0092	0618	0095	0656	0070	0671	0667
0093	1202	0084	1383	0074	0665	0666
0094	1362	0102	2429	0073	0667	2158
0095	1962	0100	2154	0067	0664	2164
0096	2105	0083	2256	1862	0670	2160
0097	0851	0087	1775	0071	0680	2162
0098	1075	0088	1877	2058	0661	2167
0099	2332	0089	2313	0072	0679	2159
0100	0137	0460	0158	2236	0521	2226
0101	0792	2544	0844	2232	1199	0318
0102	1606	2545	1786	2239	1198	0312
0103	0867	0109	0887	0081	2323	2449
0104	0831	0107	1482	0080	2320	0106
0105	0838	0108	0286	0078	2321	0107
0106	0439	0110	0424	0079	2322	0108
0107	0142	0111	0167	0077	0616	0102
0108	0146	2446	0166	0082	2214	1044
0109	0144	0112	0164	0084	2215	1043
0110	1997	0113	2112	0083	2254	1045

E	F	G	S	R	C	J
0111	0147	1937	0168	0793	0998	2554
0112	1062	0114	1094	2124	2261	2457
0113	1363	0116	2430	0091	2263	2458
0114	1477	0115	1636	0090	2262	2459
0115	2044	0353	0612	0058	2176	1436
0116	2045	1730	0637	0087	2266	1524
0117	2582	1012	2585	0086	2265	0028
0118	2539	0117	2535	0089	2264	2460
0119	1991	0149	2158	0093	0543	0036
0120	2228	0150	1162	0094	0544	0037
0121	0148	0152	0169	0096	0546	1878
0122	1175	0151	1296	0095	0554	1859
0123	0759	0153	0792	0426	1321	0236
0124	2174	0154	2558	0097	0704	1422
0125	0463	2000	0438	0724	0102	2392
0126	1702	2001	1901	0726	0103	2391
0127	1001	0194	1025	0218	0101	0997
0128	1859	0195	2006	0219	0230	0999
0129	2365	0196	2365	0220	0104	0998
0130	1856	0105	2004	1631	2040	0109
0131	0151	0294	0174	0480	0360	0001
0132	0169	0297	0211	0102	0364	0009
0133	0423	1726	0413	2080	0365	0005
0134	1364	0300	2431	0499	0979	0002
0135	1418	0298	0197	0496	0975	0006
0136	0649	0296	0693	0500	1364	0003
0137	1971	0299	1614	0498	0980	0004
0138	1958	1226	2144	0713	1752	1072
0139	0185	0138	0207	0100	2112	0464
0140	0184	0139	0205	0101	2113	0463
0141	2554	0140	2540	1926	2114	0465
0142	0183	0158	0206	0099	2115	0022
0143	1063	0159	1096	2126	0001	0021
0144	1860	0160	2007	1578	0136	2040
0145	1763	1555	1930	0627	0194	1298
0146	0009	0161	0005	0112	2373	0633
0147	2251	0162	1123	0119	2374	0634
0148	0232	0246	0262	0154	2366	0639
0149	0996	0163	1016	0118	2360	0644
0150	2253	0164	1125	0121	2361	0645
0151	1176	0165	1297	0117	2377	0635
0152	2258	0166	1129	0120	2378	0636
0153	1970	0172	1607	0114	2375	0642
0154	1492	0173	1659	0115	2379	0637
0155	2529	0174	2478	0113	2380	0638
0156	1652	1593	1826	0125	2376	0646
0157	2195	0176	1164	0126	2362	0640
0158	1976	0175	1618	0123	2364	0641
0159	2271	0177	1176	0116	2365	0647
0160	1777	0167	1897	0124	2372	0656
0161	0191	0169	0214	0109	2367	2188
0162	2062	0171	0923	2573	2370	0629
0163	1219	0168	1339	0110	2368	0631
0164	0192	2579	0216	0104	2369	0630
0165	0198	0017	0208	1315	1896	0658
0166	0381	2011	0382	0880	1897	0659
0167	0864	0199	0888	1287	0494	0202
0168	0969	0200	0992	1285	0489	0204
0169	2274	0202	1190	1286	0491	0206
0170	2070	0201	2231	0300	0490	0205
0171	1010	0203	1033	2551	0492	0207
0172	2272	0204	1177	1284	0493	0208
0173	2059	0198	2222	1283	0490	0203
0174	0205	2036	0230	0006	2558	0935

E	F	G	S	R	C	J
0175	0206	0216	0233	0008	2555	0937
0176	0207	0215	0231	0007	2555	0936
0177	0208	0218	0232	0007	2556	0936
0178	0209	0217	0234	0008	2556	0937
0179	0211	0219	0236	0010	0610	0209
0180	0210	0220	0235	0009	2572	0938
0181	0641	0899	0685	0261	0775	2339
0182	0931	0898	0941	0262	0774	2338
0183	1848	0900	2432	0260	0773	2337
0184	1591	1431	2064	2036	1610	2196
0185	2571	1439	0334	2048	1614	2202
0186	1893	1442	2033	2038	1616	2198
0187	2546	1433	2561	2037	1617	2195
0188	1640	1437	1835	2040	1615	2201
0189	0933	0221	0942	0021	2518	0865
0190	1122	1966	2188	1877	0538	0454
0191	1128	1965	2189	1876	0538	0453
0192	1123	1967	2187	1878	0539	0452
0193	0242	2415	1309	2311	0054	1816
0194	0547	2418	0533	1557	0056	1819
0195	0240	2419	1114	2315	0057	1818
0196	0239	2420	2114	2314	0055	1820
0197	2110	1975	2262	1247	0547	0725
0198	0976	1971	0999	1243	0555	0726
0199	2288	1974	1189	2020	0548	0727
						1890
0200	2485	1977	2532	1639	0542	0457
0201	2458	1976	2517	1246	0541	0456
0202	2499	0222	2442	1626	0049	2194
0203	0212	2504	0238	0183	1975	1591
0204	2214	0224	1165	1511	0255	1869
0205	1436	0223	0144	2455	1629	1870
0206	0220	0225	0245	0132	0008	1784
0207	0221	0226	0247	1316	1169	0188
0208	0965	1270	0988	1323	1066	0896
0209	1457	0228	1609	1319	1072	0440
						2211
0210	2352	0229	2353	1049	2363	0648
0211	1733	0230	1904	1322	1067	0435
0212	0224	0231	0250	1324	1170	0189
0213	0535	0232	0520	1601	0048	2205
0214	0700	0233	0741	0133	2211	1073
0215	0225	0235	0251	0138	0050	2220
0216	2598	0261	2604	1255	2530	2222
0217	0260	2489	0310	0510	2529	2223
0218	0564	0258	2266	0141	1449	2212
0219	0721	0260	0733	0140	1891	2214
0220	1686	0259	1865	0142	1413	2213
0221	0747	0272	0763	0143	1900	1476
0222	2445	0273	2491	0144	1898	1478
0223	0226	0271	0252	0145	1899	1477
0224	1803	2249	2300	1961	1598	0348
0225	0062	2496	0072	0612	1599	1785
0226	0837	0274	0289	0149	1907	1691
0227	0418	2605	0352	0146	0505	1693
0228	0632	0277	1045	1900	1886	0609
						1288
0229	1009	0278	1027	2555	1218	0610
0230	0443	2607	0427	0150	1911	2033
0231	2446	0279	2492	0147	1903	1490
0232	0227	0280	0253	0148	1902	1489
0233	0544	2012	0631	2432	0957	0743
0234	1999	2013	2118	0572	0959	0745

E	F	G	S	R	C	J
0235	0001	0286	0254	2388	1303	0370 2080
0236	0238	2417	1311	2313	1304	2081
0237	0548	0288	0532	1559	1307	2086
0238	1524	0287	1692	0575	1305	2082
0239	0796	0289	0853	2386	1309	2084
0240	2166	0285	0817	1836	1308	2087
0241	2477	2329	2505	2387	1306	2083
0242	0138	0283	0159	2235	1814	2091
0243	1474	0282	1631	0463	1813	2092
0244	2302	0284	2337	0184	1293	1797
0245	0228	0292	0258	0152	1873	2107
0246	1905	0510	2046	1051	0857	2028
0247	0710	0281	0739	1915	1354	0018
0248	0181	0293	1763	0854	0115	2224
0249	1780	2048	2075	2211	0625	2046
0250	2366	2049	2367	2136	0626	2047
0251	0231	0301	0261	0153	0762	2363
0252	0276	2448	0330	2265	2407	1250
0253	0653	2447	0705	2198	2406	1248
0254	0652	0856	0702	2264	2408	1249
0255	1903	1083	2044	1052	0812	1673
0256	0136	0303	0156	2234	0143	2071
0257	1028	0302	0877	2256	0142	2075
0258	0722	2577	0737	1467	0684	2118
0259	1220	2471	1347	1466	0683	2119
0260	2345	2472	2203	1348	0682	2117
0261	1320	2393	1896	1698	1051	1183
0262	2396	2395	2417	0251	1050	1181
0263	2408	2394	2572	1166	1049	1182
0264	0259	0488	0305	0507	1467	0418
0265	2593	0489	2603	1250	1468	2629
0266	2000	0306	2115	2210	1753	1166
0267	2152	2234	0832	1833	0394	0190
0268	2153	0561	0833	1834	1912	1338
0269	0911	2430	1048	2268	0085	2000
0270	0647	0324	0699	0771	0413 1888	1805
0271	2339	1787	2344	0158	1001	0386 0389
0272	2342	1790	2347	0161	0999	0393
0273	2357	1792	2356	1050	1002	0391
0274	1875	1789	0308	0162	1003	0395
0275	0180	0823	0202	0160	1000	0390
0276	1313	1791	1460	0163	1004	0392
0277	0466	0830	0441	1258	2439	1841
0278	0848	0831	1769	1259	2440	1840
0279	0247	0323	1681	0367	0058	2172
0280	0870	0205	0895	2561	1817	2435
0281	1782	0310	2074	0705	1700	0191
0282	2561	0312	2543	1929	1702	1767
0283	2413	0315	2569	1162	1701	1768
0284	1026	0003	2098	0274	1695	1769
0285	2346	1909	1346	1902	1696	1770
0286	1317	0325	1523	1232	1562	1790
0287	0608	1082	0641	0384	1209	0704
0288	0252	1084	0275	0727	1437	1760
0289	0182	0326	1764	1807	1276	0382
0290	1753	1094	1922	0504	0395	2491
0291	1758	1093	1927	0503	0397	2493
0292	2405	1164	2573	1163	0396	0779
0293	0257	0240	0302	1613	2399	1991
0294	0568	1092	0546	0505	0398	0359
0295	0389	1091	1253	0506	1756	0461

E	F	G	S	R	C	J
0296	0163	1095	0175	0733	1494	2426
0297	0952	1097	0971	0738	1495	2427
0298	2518	1098	2465	0735	1493	2425
0299	0266	0517	0312	0555	0726	2324
0300	2122	0519	2272	1550	0728	2326
0301	2008	0518	2126	0558	0727	2325
0302	0193	1146	0217	0108	1950	0228
0303	0195	1147	0219	0107	1951	0229
0304	1216	0370	1425	1425	1949	0230
0305	2544	2346	1991	1227	2395	1361
0306	0159	2348	0189	0489	2398	1362
0307	0950	2349	0968	1210	2397	2332
0308	2112	2350	1598	1899	2086	2408
0309	1249	2351	1395	0312	2394	0462
0310	0569	1101	0538	0739	0252	1184
0311	0275	1102	0324	0740	2268	1552
0312	1061	1103	1095	2125	2309	0110
0313	2614	0795	2614	1236	2318	0114
0314	2410	2074	2568	0746	2317	0111
0315	1481	1104	0268	0748	2312	0113
0316	0165	1105	0180	0484	2310	0118
0317	0537	2503	0522	1602	2311	0112
0318	1840	1106	1974	0749	2316	0117
0319	1940	1107	2068	0747	2313	0116
0320	2046	1108	2185	0744	2315	0347
0321	2540	1109	2536	0745	2314	0115
0322	0869	1112	0890	0742	2271	0105
0323	0834	1110	1483	0741	2269	2450
0324	0842	1111	0287	0743	2270	2451
0325	0289	2569	0342	2410	1230	1192
0326	0732	2374	0771	0423	1646	1372
0327	0865	2375	0891	2567	1647	1373
0328	1992	2376	2113	0571	1645	1374
0329	1499	2589	1667	0128	1168	1973
0330	2064	2068	2228	0307	1427	0322
0331	2114	1088	2269	0729	0157	0919
0332	1603	1089	1788	1819	1671	1382
0333	1280	1090	0889	0730	0119	1381
0334	0464	0327	0439	2362	1958	1662
0335	1494	0328	1661	2363	1959	1660
0336	1858	0330	2005	2365	1956	1661
0337	2022	0329	2138	2364	1957	1659
0338	0024	0053	0026	0024	0254	0361
0339	1690	0809	1869	0552	0253	0360
0340	0835	1233	1484	1238	0358	0357
0341	1831	1235	1975	1538	0359	1543
0342	2494	1236	2434	1408	0357	1538
0343	0365	1229	0361	2097	0223	0975
0344	0576	1230	0547	2202	0222	0359
						0981
0345	2090	1231	2239	0551	0224	0980
0346	0512	1232	0491	0821	0225	0979
0347	2448	0331	2524	2366	0006	0524
0348	0021	0343	0022	2376	0989	0271
						0276
0349	2550	0336	1060	2383	0994	0278
0350	0820	0345	0871	2378	0992	0279
0351	1036	0346	1066	2382	0995	0290
0352	1051	0347	1079	2381	0982	0284
0353	1171	0338	1628	2374	0990	0273
0354	1342	0340	1496	2373	0988	0274
0355	1386	0339	1546	2372	0983	0272
0356	0836	0342	1485	2379	2012	0281
0357	0701	0341	0749	2380	0996	0277

E	F	G	S	R	C	J
0358	2060	0337	2235	2377	0981	0280
0359	0304	0332	2103	2367	0983	0283
0360	0309	0333	2104	2370	0984	0286
0361	0095	0334	0092	2368	0985	0288
0362	1294	0335	1453	2369	0986	0287
0363	0310	0347	2105	2371	0997	0275
0364	1360	1962	1510	1627	0457	1446
					1680	
0365	0311	0348	0606	2390	1799	0579
0366	0314	0349	0605	2392	1801	0576
0367	0343	0351	0604	2393	2338	0578
0368	0315	0350	0576	2394	1800	0577
0369	0344	0352	0609	2397	1136	0568
0370	0487	0357	0465	2399	1141	0570
0371	0346	0355	0613	2400	1142	0569
0372	0488	0361	0466	2402	1137	0573
0373	1785	0360	2081	0818	1140	0572
0374	2447	0362	2496	2404	1138	0574
0375	0350	0358	0617	2405	1139	0571
0376	1805	2590	1946	0910	2381	0198
0377	1671	1898	1703	2443	1267	0559
0378	0297	1899	0284	2438	1269	0560
0379	2317	1897	2332	2439	1268	1162
0380	1615	0248	0040	2440	1246	0558
0381	2056	2508	1089	1763	0363	0008
0382	0374	1140	0372	0773	1550	0497
0383	1266	1141	1419	0774	2022	1639
0384	0402	2429	0402	0872	0086	2237
0385	2005	1142	2121	0777	2199	0761
0386	1993	1143	2108	0776	2200	0762
0387	2188	1144	1366	0778	2201	0760
0388	2323	1145	2310	0775	2198	0098
0389	0361	1150	0362	2102	0730	1641
0390	0454	1149	2178	1971	0732	1640
0391	2392	1148	2412	1972	0731	1642
0392	0372	0363	0406	0779	0837	0495
0393	0280	1086	0329	2385	1366	2588
0394	1624	1087	1806	2384	1370	2589
0395	0264	1984	2172	0865	1879	2334
0396	0265	1983	0407	0787	0132	0770
					1880	
0397	0414	1151	0410	0790	1623	0767
0398	0413	1152	0411	0789	2546	0768
0399	2095	1153	2249	1793	2547	0769
0400	0415	2175	0349	2081	1578	1030
0401	0914	2180	1054	2270	1585	1036
0402	0292	2177	0272	2411	1582	1037
0403	2123	2182	2273	1552	1581	1034
0404	0717	2279	0729	0469	1584	1032
0405	0680	2176	2282	1522	1579	1035
0406	1908	2178	2050	1696	1580	1031
0407	0699	2181	0748	1962	1583	1033
0408	0422	1989	0412	0791	2528	0377
0409	0341	1990	0593	0792	2533	0379
0410	1693	1991	2223	2360	2535	0380
0411	2371	1992	2366	2137	2534	0378
0412	2592	1993	2589	1252	2532	0381
0413	0934	1154	0943	0795	2589	2036
0414	0430	2432	0420	0372	1699	1766
0415	0437	1155	0423	0796	0074	0781
0416	2149	2444	0818	1835	0071	1907
0417	1102	2480	1225	1501	2045	0399
0418	0449	1161	0044	0797	1570	0786
0419	0140	1158	0162	0798	1567	0807

E	F	G	S	R	C	J
0420	0438	1156	1319	0799	1563	0808
0421	1046	1157	1073	1714	1566	0809
0422	2449	1160	2493	0800	1564	0810
0423	0448	1159	0425	0801	1565	0811
0424	0450	1162	0434	0802	1570	0787
0425	0451	0250	0435	0804	0209	1167
0426	0471	1163	0447	0805	1609	1314
					2418	
0427	0135	0367	0157	2237	1283	0782
0428	0866	0364	0892	2563	1285	0783
0429	0970	0365	0993	0807	1282	0785
0430	0775	0366	0845	0806	1284	0784
0431	0462	1186	0437	0864	1552	1776
0432	0518	1190	0496	0863	1554	1781
0433	1983	1192	2199	0861	1555	1780
0434	0495	1187	0476	0826	1556	1782
0435	1310	1189	1457	0862	1557	1235
0436	1980	1191	2197	0860	1553	1779
0437	1607	1193	1787	0866	2018	1638
0438	2308	0236	0443	1982	0778	0500
0439	0173	1164	0176	0813	2242	1379
0440	0477	1273	0453	1638	0281	1562
						1654
0441	2450	1167	2494	0814	0349	1561
0442	0474	1166	0450	0815	0350	1560
0443	1750	0237	1919	1270	2164	1664
0444	0486	1171	0464	0820	0193	1470
0445	0571	1173	0542	1576	0963	1483
0446	0161	1170	0191	0491	0965	1465
0447	2406	1172	2567	1573	0964	1475
0448	0522	0251	0500	1984	0210	1167
0449	0370	0252	0370	0772	2587	1169
0450	0529	0800	0513	1973	2588	1168
0451	0651	0238	0713	2092	2243	0966
0452	1747	0529	0504	2085	0362	0965
0453	0527	2464	0511	0535	1219	1669
0454	0526	2465	0510	0534	0207	0214
0455	1707	1174	1271	2585	0206	0213
0456	0085	0490	0101	1200	1434	2622
0457	1902	1177	2043	1199	1430	2625
0458	1994	1175	2109	1198	1431	2621
0459	0052	1176	0055	1197	1433	2619
0460	1661	0383	1844	1178	1428	2623
0461	1200	0384	1382	1292	1429	2624
0462	0530	1178	0514	0844	1432	2620
0463	0217	1179	0241	0849	0468	0416
0464	1416	2156	1092	0851	1334	1150
0465	0164	0841	0184	0847	1335	1152
0466	0186	1180	0210	0845	1337	1154
0467	0558	2227	0635	0915	1325	1488
0468	2468	2229	2519	1591	1323	1486
0469	1869	2228	2018	1594	1324	1485
0470	0821	2565	0690	0846	1333	0856
0471	1604	2582	1789	1820	1336	2536
0472	1251	0840	1404	0285	1331	1151
0473	2453	1731	2498	0914	1329	1481
0474	0559	1729	0634	0917	1330	1482
0475	0960	1732	0983	2495	1327	1479
0476	0981	1733	1002	2519	1326	1480
0477	1829	1734	1955	1534	1328	1525
0478	1087	2532	0691	0701	1322	1804
0479	0857	2155	2433	0848	1332	1153
0480	2454	1181	2500	0850	0467	0780
0481	0534	1182	0519	0810	0460	1305

E	F	G	S	R	C	J
0482	0348	1183	0610	0808	0461	1306
0483	0359	1184	0614	0809	0462	1307
0484	0971	1185	0994	0811	0084	1550
0485	1133	2406	1254	2278	2538	2228
0486	2587	0668	2580	0232	1019	0788
0487	1422	0808	1567	0949	1367	2582
0488	0622	0012	0655	0904	1368	2581
0489	0623	0011	0658	0905	1369	2583
0490	0373	1426	0528	1977	0835	0501
0491	0825	1197	2425	1944	1210	2343
0492	0826	1198	2426	1945	1207	2347
0493	2196	1203	1103	1951	1203	2334
0494	2198	1202	1105	1950	1204	2351
0495	1326	1199	1469	1942	1208	2342
0496	2206	1205	1101	1947	1205	2349
0497	2207	1204	1102	1499	1206	2350
0498	2076	1200	2220	1946	1202	2348
0499	2394	1201	2415	1943	1200	2335
0500	2597	1206	2596	1948	1201	2336
0501	0549	0022	0531	1558	2450	1431
0502	1996	1207	2110	0870	2349	0124
0503	2452	0369	2497	0913	1338	0550 1609
0504	0557	0368	0633	0916	1339	0549 1610
0505	0457	2557	2176	2111	1150	0270 0596
0506	0932	0797	0516	1682	0459	1302
0507	0531	0798	0517	1680	1286	1295
0508	2520	2556	2466	2112	1144	0598
0509	0894	2558	0929	2581	1149	0599
0510	1260	2553	1411	0354	1145	0603
0511	1699	2559	1884	1595	1148	0346
0512	1886	2552	2025	2217	1146	0600
0513	2560	2554	2550	1939	1147	0604
0514	2573	2555	2556	1361	1147	0604
0515	2359	0806	2361	1981	1561	0555
0516	0896	1196	0922	2580	1560	0611
0517	2528	1195	1753	1969	1559	0612
0518	0624	1209	0574	0891	1302	0563
0519	1921	1212	2058	0362	0976	0566
0520	0761	1210	0791	0427	0978	0564
0521	1214	1211	1315	2317	0977	0565
0522	0635	1218	0683	1801	1450	2547
0523	1064	0371	1097	2127	1300	0561
0524	0300	0807	0295	1412	2445	1413
0525	2119	1827	2270	1548	1220	1370
0526	2121	1058	2274	1551	0379	0052
0527	0272	1214	0318	2361	1240	0617
0528	1666	1215	1849	0909	1239	1095
0529	0766	0262	0798	0451	1241	0613
0530	2226	1213	1142	2014	1242	0614
0531	2086	1216	2182	0906	1237	0615
0532	0949	1217	0944	0908	1238	0616
0533	0812	1224	0865	0921	1358	2556
0534	2061	1222	2216	0922	1359	2555
0535	0489	1220	0483	0827	1360	1225
0536	0490	1221	0487	0919	1362	1238
0537	0491	1223	0479	0920	1361	1193
0538	2173	1225	2328	2093	1363	2557
0539	0942	0234	0945	2429	0047	0350
0540	0627	0372	0638	0923	1251	0494
0541	0565	2220	0537	2199	1460	1597
0542	1757	2226	1923	2204	2393	1606

E	F	G	S	R	C	J
0543	0351	0359	0623	2408	1988	1603
0544	1755	2225	1931	0139	2080	1605
0545	0353	1014	0624	2406	1546	1607
0546	1833	2233	1957	1537	0390	1599
0547	0604	2223	0643	0899	0391	1600
0548	0545	1227	0536	0868	1687	0521
0549	0219	0413	0244	1126	1685	0849
0550	0356	0373	0619	2416	2234	0792
0551	0357	0376	0620	2417	2237	0795
0552	1834	0374	1956	2414	2236	0793
0553	2584	0375	2578	2415	2235	0794
0554	0630	2616	0359	2418	1020	0796
0555	2320	0377	1134	1007	1018	0797
0556	0411	2618	0408	2420	2383	2003
						2006
0557	1211	2619	1314	2419	2382	0196
0558	0493	0380	0468	0831	2585	1366
0559	0575	0483	0570	2130	0005	0041
0560	2535	0382	0660	0390	1884	1462
0561	1228	2336	2056	0478	2551	2405
0562	1909	0406	2048	1693	1174	0911
						1329
0563	1784	0385	2094	1036	1932	0892
0564	2399	2342	2422	1037	1929	2026
0565	0551	2344	0535	1556	1930	2027
0566	2603	2340	2607	1039	1931	1025
0567	1029	0386	0878	2257	0293	1511
0568	0719	2570	0731	1805	1889	2240
0569	0290	2575	0270	2412	1893	2242
0570	0500	2571	0470	0834	1892	2244
0571	0611	2573	0645	0896	1894	2241
0572	2115	2576	1207	2101	1895	2243
0573	0677	0456	1413	2073	1374	0268
0574	0676	1045	1414	2075	0375	1594
0575	0675	2332	1416	2076	0323	1507
0576	0827	0625	0049	0437	1780	1327
0577	1180	2467	1302	0547	1005	1375
0578	1627	2470	1809	0548	1006	1380
0579	0561	0387	0630	2430	0294	1518
0580	0678	0388	0728	0405	1211	1331
0581	0828	0389	0050	0406	1780	1328
0582	1822	0390	1966	0407	1212	2037
0583	0697	0391	0675	0408	0880	1777
0584	0712	0392	0751	0409	1691	2043
0585	1916	2326	1887	1689	1843	2023
0586	2097	2325	2246	1795	1842	2022
0587	0715	0393	0680	1645	0282	1958
0588	0358	0395	0622	1653	0287	1953
0589	0784	0396	0808	1647	0289	1961
0590	1787	0397	2082	1649	0292	1954
0591	1836	0398	1964	1650	0286	1962
0592	2470	0394	2520	1654	0283	1955
0593	1838	0399	1980	1648	0288	1959
0594	2265	0400	1118	1651	0284	1960
0595	0111	0401	0117	1655	0290	1957
0596	2425	0402	2397	1652	0291	1956
0597	2585	0403	2581	1646	0285	1963
0598	0082	0405	0099	0411	0176	1516
0599	0714	0404	0681	0412	0176	1515
0600	2599	1511	2608	2246	0095	1971
0601	1222	0408	1341	0413	1693	1335
0602	0723	0407	0752	0414	1692	1334
0603	0745	1493	0761	0415	1158	0649
0604	0897	0410	0924	2576	1167	0599

E	F	G	S	R	C	J
0605	1461	1494	1601	1610	1166	0654
0606	0724	0411	0766	0418	1161	0650
0607	2400	2343	2423	1038	1163	0652
0608	0901	0412	0930	2575	1165	0509
0609	2098	0184	2247	1792	1162	0653
0610	0755	0633	2263	1789	2410	1537
0611	0756	0033	0755	1336	1597	2225
0612	0149	2439	0760	1485	2036	1040
0613	0763	2603	0795	0396	1918	1692
0614	0515	0414	0494	0460	1214	1968
0615	2456	0416	2506	0461	1215	2441
0616	0764	0415	0796	0462	1213	2440
0617	0765	0417	0797	1784	2052	0812 0887
0618	0059	0429	0068	1770	0179	0882
0619	0312	0419	0607	1781	0189	0877
0620	0324	0420	0591	1782	0188	0876
0621	0736	0421	0783	1785	0184	0878
0622	0785	0422	0809	1772	0186	0889
0623	1837	0431	1978	1775	0183	0890
0624	2471	0433	2523	1783	0180	0881
0625	1871	0432	2020	1776	0181	0880
0626	1839	0430	1981	0457	0185	0888
0627	1846	0423	0266	1773	0174	0813
0628	0254	0424	0276	1777	0187	0879
0629	0112	0425	0119	1786	0175	0886
0630	0113	0426	0121	1787	0309	1464
0631	2422	0427	2395	1780	0190	0885
0632	0118	0428	0125	1788	0178	0884
0633	2426	0418	2398	1779	2053	0883
0634	2586	0434	2585	1771	0182	0891
0635	0974	0006	0997	0450	2260	0245
0636	0741	0439	0781	1519	1348	0301
0637	0773	0435	0806	0459	1345	0299
0638	0391	0437	0389	0876	1349	0300
0639	0581	0440	0552	0458	1346	0302
0640	0498	0441	0469	0824	1347	0303
0641	0542	0442	0527	2421	1885	1461
0642	0783	0443	0807	0430	1577	2310
0643	0786	2341	0811	1762	2138	0431
0644	2415	2342	1056	2189	2139	0432
0645	0725	0444	0773	0436	0502	1297
0646	0572	0848	0543	1576	2454	1439
0647	0155	0849	0187	0488	2456	1440
0648	0658	1599	0710	1964	2458	1442
0649	0475	0850	0452	1574	2455	1441
0650	1820	0851	1963	1575	2457	1443
0651	2069	0445	2227	0302	2449	1427
0652	0243	0447	1682	0371	2453	1429
0653	1174	0448	1298	2304	2459	1428
0654	2175	0446	2322	0564	2451	1430
0655	0634	0626	0682	1798	0564	2311
0656	0187	0628	0212	0103	0567	2315
0657	0366	0632	0363	0853	0569	2312
0658	0579	0631	0549	1804	0568	2314
0659	1242	0629	1388	0329	0571	2313
0660	1821	0630	1965	1803	0570	2316
0661	0661	0627	0722	0166	0572	1068
0662	1906	0450	2047	0438	1180	2101
0663	2360	0449	2362	0439	1181	2100
0664	0789	0451	0813	0440	1157	2239
0665	0501	2572	0471	0822	2034	0261
0666	2563	2574	2546	1930	2035	0262
0667	0790	0452	0841	0441	0659	2141

E	F	G	S	R	C	J
0668	0791	0453	0855	0443	0660	2001
0669	0794	0454	0852	0444	2092	2045
0670	0335	2340	0580	0546	0436	1377
0671	0804	0455	0814	0447	1376	0265
0672	0502	0457	0473	0823	1372	0267
0673	1350	2170	1497	1545	1375	0266
0674	1864	0458	2010	1580	1373	0269
0675	0805	0182	0815	1828	1767	2455
0676	0809	0409	0753	0449	2434	1047
0677	2303	0459	2321	1811	0657	2178
0678	0811	2453	0864	1808	0649	2173
0679	0392	2459	0391	0885	0658	2176
0680	0503	2461	0474	0835	0651	2180
0681	1045	2463	1072	1716	0654	2183
0682	0496	2458	0477	0833	0655	2184
0683	1472	2456	1596	0543	0653	2179
0684	1981	2463	2198	1813	0650	2182
0685	1030	0461	0879	2258	0478	1517
0686	0913	0468	1049	0477	0479	1667
0687	0681	0467	2283	0475	0480	1666
0688	0704	0469	0744	0476	0481	1665
0689	0818	0470	0872	0470	1152	1244
0690	1052	0471	0874	2463	1153	1245
0691	0166	0462	0193	0392	0783	1707
0692	2248	2609	1143	0398	1909	2035
0693	2321	2610	2304	0397	1910	2034
0694	1734	0463	1905	0393	1914	1337
0695	2236	0464	1151	1002	1913	1708
0696	0582	0465	0554	2201	1901	0359
0697	0345	0354	0611	2397	1904	1339
0698	0347	0356	0615	2401	1905	1340
0699	2249	2608	1144	0398	1908	1711
0700	2250	2611	1145	2013	1915	1709
0701	1823	0466	1967	0400	1906	1341
0702	0716	0477	0679	0479	1601	1670
0703	0947	0487	0966	1671	0804 2591	0027
0704	1255	2372	1408	0342	0332	0156
0705	1608	2583	1790	0732	0338	0155
0706	0560	2371	0636	0918	0334	1499
0707	0924	1821	0950	0731	0333	1498
0708	2387	2373	2407	1419	0834	1497
0709	0040	2370	0042	2094	0804	0400
0710	1464	0482	1604	0867	0453	1514
0711	0628	1905	0358	1267	0782	1513
0712	1450	0678	1581	0942	1703	0863
0713	0256	0492	0279	0431	0449	1651
0714	2234	0495	1153	0434	0450	1254
0715	2328	0496	2311	0432	0446	1304
0716	0119	0493	0126	0435	0451	1653
0717	2430	0494	2402	0433	0452	1652
0718	0863	0500	0886	0193	2218	0674
0719	1008	0502	1043	2569	0828	0913
0720	1053	0507	0875	2583	2342	0914
0721	1025	0505	1205	2568	2345 0916	0895
0722	1639	0504	1831	2437	2343	2438
0723	2127	0506	2278	2571	2346	0915
0724	0890	2546	0916	2572	2217	0728
0725	0072	0785	0083	0053	0627	1832
0726	0748	0787	0764	0417	0628	1831
0727	1235	0508	1355	1968	0061	0764
0728	0386	0532	0395	0888	0003	0946
0729	1901	0533	2042	0178	0002	0017

E	F	G	S	R	C	J
0730	1002	0599	1029	2556	1571	2308
0731	0416	0541	0350	2255	1947	1357
0732	0778	0540	0854	2254	1948	1356
0733	0167	0549	0194	2485	0360	1551
0734	0478	0548	0454	2486	0351	1519
						1654
0735	0643	0544	0696	2489	0354	1563
					0566	
0736	1107	0543	1231	2493	0369	1559
0737	1677	0550	1859	2488	0347	1558
0738	1676	2232	1858	1372	0411	1566
0739	1098	0547	0721	2487	0636	1556
0740	0587	0551	0561	2491	0387	1597
0741	2304	0542	2338	2492	0368	1554
0742	1425	0556	1570	0135	2077	1557
0743	1112	0555	1232	0137	2078	2410
0744	1761	1916	1926	0136	2079	2411
0745	1760	0553	1925	2505	0371	2409
0746	1339	0554	1489	0386	0370	1564
0747	0919	0557	0978	2524	0373	1565
0748	0958	0558	0981	2496	0393	0185
0749	0959	0559	0982	2494	0414	1536
0750	0964	0560	0987	2506	0372	1534
						1535
0751	2004	0562	2120	2507	0348	1583
0752	1207	0563	1374	2504	0353	0448
0753	0267	0567	0314	2517	0419	1584
0754	0726	0568	0768	0424	0420	1585
0755	0767	0564	0799	0455	0432	1577
0756	1007	0569	1036	2557	0422	1569
0757	2268	0571	1184	2022	0424	1570
0758	1907	0570	2051	1695	0423	1571
0759	1391	0575	1543	0674	0426	1574
0760	0999	1136	1023	0764	0416	1580
0761	1552	0572	1735	2508	0430	1573
0762	1594	0579	1778	2512	0415	1578
0763	1892	0566	2032	2511	0431	1568
0764	1538	0577	1718	2513	0428	0043
						1582
0765	1539	0576	1720	2514	0427	1581
0766	2220	0573	1172	2516	0425	1576
0767	2275	0574	1191	2509	0987	0158
						0289
0768	2367	0578	2377	2510	0429	1572
0769	1655	0580	1840	2515	0356	1555
0770	0978	0582	0979	2522	0409	1552
0771	0979	0581	0981	2523	0410	1553
0772	0980	0583	1001	2518	0355	1539
0773	0982	0584	1003	2520	0403	1480
					1326	
0774	0983	0585	0939	2521	0352	0186
0775	0984	0586	1004	2525	2355	0660
0776	0084	0587	0100	2529	2356	0664
0777	1687	0588	1866	2528	1063	1294
0778	1807	0589	1947	2530	2037	1311
0779	0991	0590	1013	2531	2141	2467
0780	0990	0591	1015	2532	2150	2485
0781	0992	0592	1014	0276	2144	2488
0782	0216	0593	0242	1510	2145	2489
0783	0337	0594	0582	2533	2149	2487
0784	0607	0595	0647	0894	2147	2486
0785	1730	0596	1906	1427	2146	2490
0786	2590	0597	2590	1257	2148	2492
0787	0116	0598	0118	2534	1821	0183

E	F	G	S	R	C	J
0788	1587	0600	1761	2536	0524	1821
0789	1918	0601	1886	2537	0527	1826
0790	2211	0602	1168	2535	0525	1824
0791	0609	2014	0642	2539	1774	1713
0792	1195	1670	1377	2540	1773	1712
0793	0998	0609	1022	2541	2116	0491
0794	1736	0610	1911	2542	2117	0492
0795	1795	0611	2079	0817	2510	0973
0796	0793	0622	0846	0446	1535	0178
0797	0099	0615	0086	1870	1536	0179
0798	0689	0616	0668	1458	1542	0176
0799	1630	0617	1813	2543	1532	0174
0800	2094	0612	2244	1796	1534	0177
0801	0852	0619	1086	2459	1533	0181
0802	2142	0618	2475	2545	1537	0861
0803	2493	0621	2438	1410	1539	0173
0804	2501	0620	2447	1396	1538	0175
0805	2074	0614	2232	0309	0516	2559
0806	0103	0082	0088	0052	1457	1251
0807	0063	2449	1479	2263	2405	1247
0808	1020	0135	1047	1239	0786	1773
0809	1132	0136	1257	2279	0787	1774
0810	1788	0623	2083	0819	0931	0195
0811	2459	0624	2508	2558	1436	0194
0812	1270	0538	0062	0294	1776	1715
0813	1442	0537	0152	2451	1777	1716
0814	0921	0634	0960	2501	1497	0705
0815	1622	0635	1803	2559	0196	0182
0816	1623	0636	1804	2560	0200	1724
0817	0494	0847	0745	0832	1607	2204
0818	1781	0156	2076	2213	0280	1658
0819	1050	0155	1077	2462	0279	1657
0820	1056	0679	1081	1481	2098	0746
0821	1920	0214	2059	0363	2304	1330
0822	1181	2468	1303	0515	2220	0042
0823	1101	2433	1222	0277	2432	1038
0824	1038	2436	1064	1712	2433	1042
0825	0083	0786	0089	0048	1377	1832
0826	0749	0787	0765	0416	1378 2101	1831
0827	1482	0641	1637	2475	1120	0141
0828	1105	0120	1228	0213	1078	2568
0829	2213	0122	1170	1632	1081	2569
0830	1003	0126	1028	2546	1088	2570
0831	1348	2171	1499	1544	1082	2572
0832	1631	0130	1814	2262	1084	2587
0833	1513	0128	1677	1068	1085	2573
0834	1570	0125	1746	1114	1087	2580
0835	1588	0129	1762	1143	1079	2578
0836	1861	0131	2011	1579	1080	2579
0837	2049	0135	2191	1234	1090	2585
0838	2179	0132	1359	0719	1094	2575
0839	2212	0133	1169	2027	1083	2577
0840	2368	0134	2368	2131	1092	2571
0841	2199	0124	1106	1988	1093	2574
0842	2200	1656	1107	1989	1086	2586
0843	1084	0106	1214	0214	1091	2576
0844	1004	0127	1037	0212	1089	2570
0845	0470	0642	0446	2469	0575	0180
0846	1113	0643	1235	2466	0576	1833
0847	1737	0645	1912	2468	0578	1835
0848	2032	0644	2167	2467	0577	1834
0849	2315	0646	2330	1705	1867	0912
0850	0673	0647	0734	1806	2467	0905

E	F	G	S	R	C	J
0851	1115	0648	1237	2476	2413	1951
0852	0880	0211	0912	0197	2044	0257
0853	0637	0748	0689	0709	2043	0259
0854	2296	0212	1872	0198	2038	0260
0855	1129	0653	1246	2481	1966	1401
0856	1127	0652	1247	2479	1964	1398
0857	1119	0651	1248	2480	1970	1399
0858	1120	0658	1245	2478	1968	1396
0859	1121	0654	1241	2483	1965	1394
0860	0387	0656	0396	0889	0658 1969	1395
0861	0497	0657	0480	0842	1967	1400
0862	1296	0655	1444	2482	1971	1397
0863	1794	0660	2073	1697	2041	0246
0864	0807	0659	1517	1237	2042	0247
0865	1778	0662	1899	0180	2308	2519
0866	0396	0669	1258	2286	0535	2004
0867	0588	0676	0562	2289	0531	2009
0868	0580	0672	0551	2288	0533	2010
0869	0584	0674	0559	2287	0534	2007
0870	0504	0671	0481	0841	0530	2008
0871	0412	0670	0409	0788	0532	2006
0872	1065	0673	1098	2126	0529	2005
0873	0233	0247	0263	0155	1342	0707
0874	1626	2009	1808	0164	1343	1280
0875	2322	1893	2307	1919	0545	2011
0876	1501	0319	1669	0237	1295	0220
0877	1649	0320	1825	0236	1296	0219
0878	1388	0680	1540	1494	0195	2012
0879	0318	0661	0594	0737	1738	2015
0880	1243	0755	1393	0311	1981	0976
0881	2151	0966	0826	1830	2541	0974
0882	1153	0691	0870	2601	1194	2017
0883	1155	0692	1275	2298	1472	2016
0884	1156	0005	1274	1362	1470	2628
0885	1160	2062	1277	1840	1445	0324
0886	1648	2064	1823	1839	1446	0325
0887	1161	2071	1278	0402	1442	0319
0888	1159	2063	1282	0404	1291	0321
0889	1227	2066	1348	1314	1447	0329
0890	1621	2069	1803	1186	1448	0327
0891	1915	2070	2057	1642	1444	0326
0892	2063	2061	2226	0301	1443	0323
0893	1233	0686	1353	2296	0847	0756
0894	1397	0796	1537	2297	0846	0755
0895	1165	0693	1622	1448	1044	2093
0896	1205	0696	1312	2312	1045	2096
0897	1343	2166	1492	1542	1047	2095
0898	2257	0694	1130	1447	1046	2097
0899	2372	0695	2371	2139	1048	2094
0900	2109	2599	2261	1674	0549	0460
0901	1184	0291	1301	1018	1805	2076
0902	1426	0290	1576	1656	1804	2077
0903	1234	0852	1354	2300	1602	2058
0904	1172	0206	1284	2299	2223	0498
0905	0100	0699	0114	1685	1464	2102
0906	0317	0704	0586	2396	0217	2537
0907	2558	0698	0331	1934	1465	2535
0908	1173	0702	1295	2303	2327	0580
0909	0098	0703	0085	2301	2332	0589
0910	0898	0709	0925	2582	2336	0582
0911	2072	0705	1038	1149	2333	0592
0912	2077	0707	0627	0757	2335	0590
0913	2231	0708	1146	2302	2331	0588

E	F	G	S	R	C	J
0914	1950	0832	0629	1266	2337	1252
0915	2065	0704	2229	0304	2329	0583
0916	1158	0710	2085	2305	2328	0585
0917	2460	0711	2510	2306	2330	0584
0918	0215	0839	1605	1337	0226	0341
0919	1187	0712	1285	1596	1991	0706 2072
0920	0687	2190	0666	1455	1995	1229
0921	0713	0717	0676	1486	1994	1228
0922	1323	0992	1467	0643	1992	1227
0923	0533	0714	1757	1686	1993	2074
0924	1700	0716	1895	0217	1997	2073
0925	1188	2197	1286	1598	0273	2432
0926	0808	0318	0863	2310	1197	1026
0927	2338	0313	1654	2096	1195	1028
0928	0760	0314	0793	0428	1196	1029
0929	1455	1982	1590	2309	1263	0970
0930	1566	1169	1200	2308	1262	0968
0931	2575	2006	2539	0272	1781	0366
0932	0728	0736	0775	0420	1224	1796
0933	1209	0737	1320	2320	1301	0864
0934	0505	0276	0482	0828	2222	1163
0935	2241	0738	1157	1005	0789	2019
0936	0810	0740	0754	2352	0648	2185
0937	1375	0739	1526	2351	1944	2059
0938	2507	0741	2452	1897	2574	0969
0939	0154	0743	0181	1895	2562	0984
0940	0975	0744	0998	1898	2573	0978
0941	2506	0742	2453	1896	2576	0961
0942	1571	0745	1381	1207	2575	0983
0943	1229	0750	1349	2424	1604	0985 1089
0944	1595	0751	1777	2428	0051	0986
0945	1231	1453	1351	2426	1605	2235
0946	0326	0752	0597	2355	2326	1234
0947	1926	0754	0894	1767	A10	2110
0948	2316	0753	2331	2354	1667	1983
0949	0428	0316	0418	2209	1697	1771
0950	0629	1117	0357	2413	1698	0314
0951	0987	0317	1005	0374	1878	1772
0952	1168	0311	1624	1449	0788	1336
0953	1731	2478	1907	2544	1694	1197
0954	0895	2479	0931	1500	2580	1198
0955	1447	0878	0155	1499	0023	1861
0956	1237	2008	1386	1440	1259 1772	2484
0957	0274	0771	0323	0331	0237	0409
0958	1949	0772	0628	0332	1439	0415
0959	2007	0777	2125	0330	0736	0410
0960	0269	0778	0313	0556	0742	0414
0961	1989	0774	2157	0333	1822	0410
0962	2229	0776	1161	0335	1636	0412
0963	2218	0773	1116	0334	1830	0411
0964	1361	1329	2428	0553	1177	0384
0965	2089	1330	2241	0317	1178	0385
0966	0064	0794	0069	0316	1652	0339
0967	0319	0782	0587	0319	1660	0337
0968	0665	0784	0723	0323	1653	0444
0969	1215	0789	1318	0536	1656	0442
0970	2605	0793	2598	0318	1657	0340
0971	1451	0790	1583	0322	1654	0342
0972	0327	0780	0598	0326	1659	0419
0973	1518	0783	1702	0320	2026	1453
0974	0738	0781	0785	0321	1658	0443

E	F	G	S	R	C	J
0975	1664	2443	1220	0324	1572	0338
0976	0328	0779	0599	0328	1649	0429
0977	0178	0791	0199	0327	2070	0336
0978	2373	0792	2375	2132	1655	0441
0979	1713	2337	1326	1909	2225	0607
0980	2521	0803	2467	2113	0790	0231
0981	2615	0801	2192	1235	0791	0232
0982	2143	0802	2476	1562	0792	0233
0983	1435	0804	1579	0952	0793	0244
0984	1154	0810	1273	0338	1548	0621
0985	0320	0811	0585	0337	1549	0619
0986	1947	0820	1651	0349	1133	0595
0987	0199	0821	0209	0351	1134	0424
0988	1137	0819	1265	0350	1135	0423
0989	1977	0828	1620	0366	0827	0620
0990	1263	0842	2214	0325	2111	0618
0991	1283	1936	1642	0607	1264	0953
0992	0941	0845	0962	2072	0122	0372
0993	1057	0846	1083	1482	0126	0373
0994	0170	1219	0195	0361	1686	0575
0995	0650	0854	0701	2196	1011	0567
0996	0994	0858	1010	0381	1962	0594
0997	0554	0860	0298	0382	1960	0556
0998	2058	0859	2212	0380	1961	0593
0999	0194	0861	0218	0105	1852	0740
1000	1217	0864	1426	0387	1851	0742
1001	1093	0862	0346	0700	1854	0557
1002	1218	0863	1424	0388	1853	0741
1003	1602	2017	1785	0379	2557	0991
1004	2325	1796	2308	0378	2511	0990
1005	1264	0865	1417	0385	0829	0562
1006	1081	0869	1211	1320	1068	0446
1007	0610	0060	0644	0895	1033	2302
1008	2533	0770	2485	0532	2460	0850
1009	1438	0880	0153	1523	2128	0474
1010	1439	0881	0154	2458	0528	1825
1011	0682	0886	2289	1516	0025	1854
1012	1855	0884	2014	1585	0018	1877
1013	0679	0882	2286	1507	0021	1852
1014	1023	0885	0300	1506	0030	1851
1015	1022	0883	1204	1505	0031	1852
1016	1863	0877	2013	1584	0013	1876
1017	1194	0887	1376	2356	0930	1838
1018	1298	0888	1445	0619	0929	1837
1019	2037	0889	2318	0185	2219	0701
1020	0253	2514	0278	1774	1729	1906
1021	0751	0181	0787	1297	1731	0364
1022	0620	0180	0640	0903	1730	0363
1023	2559	0179	2542	1931	1155	0365
1024	0868	2016	0897	2566	2515	0964
1025	1275	0503	0065	2465	2344	2439
1026	1274	2282	0064	2464	0296	2543
1027	1267	0240	0057	0295	0860	0401
1028	1718	0896	1333	0355	0483	0082
1029	0926	0897	0952	2498	2306	2032
1030	0156	0915	0177	0483	1672	1303
1031	1818	1574	1961	1404	0826	0711
1032	1819	1575	1962	1405	2204	0084
1033	1786	0909	2080	0816	0819	0698
1034	2451	0908	2495	0290	0817	0700
1035	0476	0907	0451	0288	0818	0699
1036	0469	0906	0445	0292	0820	0696
1037	2497	0905	2444	0291	0821	0695
1038	2440	0911	2499	0289	0816	0697

E	F	G	S	R	C	J
1039	0278	0912	0327	0286	0822	0684
1040	2039	0913	2174	2290	0813	0739
1041	0160	0904	0178	0482	0823	0670
1042	1431	0903	1575	0912	0825	0671
1043	2047	0916	0056	0283	0814	0723
1044	1876	2494	0711	0287	0811	0686
1045	1096	0756	0718	0712	0810	0685
1046	2281	0902	1185	2021	0815	0129 0724
1047	1097	0914	0719	0284	0824	0672
1048	1066	0901	1099	2129	2137	A11
1049	2516	1972	1988	0215	0560	2331
1050	2604	2341	2597	2244	2487	2323
1051	0636	0918	0686	1799	1318	1391
1052	1420	0919	1565	0950	1317	1392
1053	2187	0920	1367	0723	1319	1390
1054	0935	0234	0963	1525	1316	1393
1055	0393	0924	0393	0874	1275	0527
1056	0299	0925	0294	1278	1274	0526
1057	0927	0923	0953	0364	1272	0530
1058	1974	0926	1615	0365	2389	1636
1059	0953	0927	0974	0373	1919	1097
1060	0305	2547	2099	2375	2522	2340
1061	1503	2548	1670	1013	1073	2436
1062	0189	0895	0222	0376	1743	2353
1063	0285	0894	0285	0375	1737	2354
1064	0279	0893	0326	0377	1723	0673
1065	0612	0928	0648	0900	1637	1783
1066	2588	0929	2586	0356	1466	1309
1067	0923	2518	0949	0211	1684	1096
1068	0705	0930	0745	0357	2409	1775
1069	0706	0931	0746	0358	1175	1355
1070	0707	0932	0747	0359	2212	1224
1071	1276	0933	1423	0360	2486	1937
1072	0152	0844	0183	0487	0344	1540
1073	0054	2598	1018	1488	0341	1533
1074	0363	2596	0365	2099	0346	1549
1075	1281	2595	1429	1489	0345	1547
1076	1089	2594	0345	0638	0342	1544
1077	0364	2597	0368	2100	0343	1541
1078	0536	0934	0521	1603	2300	0104
1079	0803	0002	0861	1352	2033	0345
1080	1068	0935	1464	0603	1499	0103
1081	0541	0936	0526	0605	2584	0121
1082	1723	0193	1335	1354	A08	1714
1083	0543	2530	0632	0726	0221	1799
1084	1998	2531	2127	0569	0220	1800
1085	1936	0515	2000	0170	1779	1718
1086	2009	0516	2128	0171	1778	1717
1087	1289	0528	1438	0192	0034	0776 2247
1088	1282	0965	1431	0606	0623	1934
1089	1826	1234	1959	1536	2403	1545
1090	1290	0948	1439	0618	2162 2464	0900
1091	2363	0950	0257	2389	2466	0901
1092	0546	0951	0530	1555	2465	0902
1093	0943	0949	0964	0620	2463	0898
1094	0967	1271	0989	0622	0487 2448	2404
1095	2461	0952	2514	0869	1176	0398
1096	1959	0957	2145	1154	0801	2446
1097	2033	0958	2168	0629	2347	2445
1098	0229	0954	0259	0624	0803	2443

E	F	G	S	R	C	J
1099	1766	0955	1934	0623	0802	2442
1100	1314	0956	1461	0625	2348	0119
						2444
1101	0417	2410	0351	1213	0619	1943
1102	0779	2409	0856	1212	0620	1942
1103	0188	0959	0213	0630	0488	2024
1104	0589	0103	0564	1172	1642	1051
1105	2382	0104	2378	1171	1643	1050
1106	1315	0119	1463	0637	2324	0237
1107	1241	0963	1390	0631	2524	1060
1108	2369	1763	2369	2134	2527	1059
1109	2519	0964	2469	0634	2525	1061
1110	1316	1765	1521	0632	2526	1061
1111	2326	0526	2303	1917	1883	1656
1112	2010	0971	2116	0201	1525	1179
1113	2618	0976	2618	0207	1522	1686
1114	1540	0947	1721	1079	2357	1293
1115	1546	0946	0801	1087	2163	1292
1116	0566	1480	0565	1022	1923	1011
1117	0633	1478	0332	1017	1924	1012
1118	1527	1479	1696	1020	1922	1010
1119	0086	0984	0108	0658	2301	0427
1120	1309	0986	1588	0657	2302	1282
1121	0087	0985	0109	0659	2504	0426
1122	1321	0987	1465	0098	2542	0606
1123	0012	0990	0008	0639	1060	1176
1124	0313	0998	0608	0640	1059	1177
1125	0737	0989	0784	0641	1061	1178
1126	0399	0991	0398	0877	1057	1180
1127	1069	0980	1873	0642	1056	1684
1128	2292	1676	1091	0644	1058	1175
1129	1325	0993	1468	0646	1673	0502
1130	0553	0999	2067	0645	1674	0504
1131	1136	1000	2202	1351	1675	0505
1132	0800	0996	0859	2267	1676	0503
1133	1146	0998	1420	0977	1677	0506
1134	1421	0997	1566	0951	1678	0507
1135	1329	2201	1470	0647	0759	2262
1136	1328	0994	1471	0651	1632	1137
1137	1334	0995	1478	0648	0466	0436
1138	2385	2529	2408	0245	2203	1179
1139	0420	1001	0354	1024	2371	0632
1140	2503	1002	2449	1030	0522	1829
1141	2617	1004	2617	1026	1531	1388
1142	0887	1003	0913	1031	0836	0661
1143	1338	0866	1488	1029	1221	0264
1144	1341	2165	1490	1073	0805	0387
1145	0619	2164	0657	0902	0807	0388
1146	1340	2162	1507	1075	0808	0394
1147	1978	2161	1621	1074	0806	2245
1148	2305	2163	2339	1072	0809	2246
1149	1620	1427	1802	1663	2496	1410
1150	1762	2613	1392	1665	2494	1389
1151	2523	2614	2471	1661	2495	1383
1152	2509	1005	2450	1027	0694	2145
1153	0006	0973	0002	0200	1526	1175
1154	0539	0974	0524	0208	1529	1682
						1685
1155	0390	0978	0387	0873	1530	1683
1156	0961	1006	0985	0210	1524	1681
1157	0963	0970	0986	0206	1523	1680
1158	2160	0967	0828	0202	1521	1177
1159	2161	0968	0829	0203	0125	0376
1160	0065	0975	0074	0044	1528	1178

E	F	G	S	R	C	J
1161	0939	0969	0959	0205	0124	0375
1162	1975	0970	1611	0209	1520	1176
1163	1070	0980	1874	0204	1518	1174
1164	0176	0016	0201	1628	1179	1360
1165	1223	0921	1342	1662	2154	0544
1166	2568	0837	2553	1659	2155	0545
1167	2569	0838	2554	1660	0235	1509
1168	2600	0922	2605	1664	2156	0546
1169	2504	2615	0626	1028	2152	0682
1170	2050	1007	2193	1025	0256	1388
1171	2510	1008	2448	0660	0523	1811
1172	0903	0977	0934	2374	1164	1074
1173	0902	0983	0932	1965	1519	1074
1174	0904	0982	0935	1966	0060	1076
1175	0905	0979	0933	1967	0059	1075
1176	0621	1009	0659	1249	1545	0459
1177	2472	1010	2521	1592	0540	0458
1178	2480	1016	2529	0661	0335	2473
1179	1371	1015	2611	0662	0337	2474
1180	2481	1018	2530	0663	0336	2471
1181	1372	1017	2612	0664	0322	2472
						2475
1182	1373	1019	1524	0665	1379	0044
1183	0234	1027	0265	0679	1384	0070
1184	0329	1028	0600	0684	1399	0059
1185	0590	1041	0566	0690	1395	0077
1186	2239	2617	1155	1001	1386	0068
1187	0832	1022	1486	0689	1388	0062
1188	0336	1023	0588	0693	1393	0063
1189	1404	1026	1555	0590	1389	0065
1190	1517	1025	1701	0692	1392	0066
1191	0840	1024	0288	0694	1390	0064
1192	0378	1021	0377	0879	1387	0053
1193	1969	1029	1616	0680	1402	0075
1194	1564	1052	1699	0683	1394	0074
1195	1476	1042	1634	0465	1383	0067
1196	2221	1035	1174	0687	1381	0071
1197	1979	1034	1619	0686	1382	0072
1198	1208	1038	1375	0682	1396	0057
1199	2340	1788	2345	0159	1397	0058
1200	1385	1056	1539	0671	0374	0046
1201	2343	1049	2348	0669	1385	0050
1202	1387	1037	1544	0668	1184	1128
1203	0294	1046	0281	0667	0381	1595
1204	0388	1057	0397	0878	0384	1593
1205	0606	1043	0646	0901	0385	0051
1206	1006	1044	1031	2549	0380	0047
1207	1349	2172	1500	0670	0377	0049
1208	0049	1047	0051	0666	0386	0056
1209	1862	1048	2012	1582	0382	0055
1210	1894	1051	2035	0173	0378	0048
1211	2370	1050	2370	2135	0383	1592
1212	1688	1055	1867	0675	2471	1612
1213	2011	1054	2129	0676	0376	1596
1214	2409	1053	2574	0677	2469	0054
1215	1556	1030	1738	0681	1403	0061
1216	2028	1020	2163	0678	1380	0045
1217	1265	1059	1418	0691	2470	0073
1218	1541	1039	1719	0685	1400	0016
						0079
1219	2247	1040	1124	2016	1398	0015
						0069
1220	0954	1032	0976	0697	1404	0076
1221	2181	1031	1362	0688	1405	0060

E	F	G	S	R	C	J
1222	0951	1036	0973	0698	1401	0078
1223	1398	0257	1550	0978	0772	2264
1224	1399	0255	1549	0567	0771	1024
1225	1287	1060	1435	1205	0137	2029
1226	0753	1061	0789	0584	2004	1613
1227	1401	1062	1552	0587	2305	0528
1228	1085	1063	1215	0585	2002	0295
1229	2505	1064	2451	0586	2020	0297
1230	1796	1065	2093	0588	0298	1654
1231	1403	1066	1554	0589	0310	1628
1232	0752	1067	0127	0592	0311	1629
1233	1406	1068	1557	0593	2021	1647
1234	1409	1069	1558	0594	2007	1614
						1617
1235	0788	1077	0810	0596	2014	1619
1236	0096	1078	0113	0050	2015	1620
1237	0833	1072	1487	0601	2011	1621
1238	1410	1070	1561	0122	2359	0643
1239	0002	1073	0014	1815	2010	1622
1240	0088	1071	0104	0597	2009	1624
1241	2517	1079	1990	0600	2016	1626
1242	1376	1074	1527	0599	2013	1618
1243	2139	1076	2296	1764	2008	1625
1244	2491	1075	2463	0595	2017	1623
1245	1835	1085	1968	1533	1277	0383
1246	0286	1134	0339	2602	1299	0227
1247	0070	1135	0076	0765	0447	1874
1248	0570	1138	0540	0766	0445	1875
1249	2349	1208	2349	1054	1287	0754
1250	1487	1347	1655	0560	1866	1969
1251	1488	1346	1658	0991	1862	1970
1252	1166	1238	1625	0944	0171	1277
1253	0591	1239	1290	0945	0170	1275
1254	1239	1240	2482	0943	0172	1276
1255	1417	1241	1562	0946	0295	2513
1256	1449	1244	1580	0924	1095	2591
1257	0022	1258	0017	0930	1119	2599
1258	2301	1248	2336	0028	1098	2592
1259	0101	1249	0077	0935	1096	2612
1260	1108	0170	0215	0925	1118	2596
1261	0197	1250	0221	0933	1114	2597
1262	1149	1266	2159	0939	1109	2608
1263	1390	1252	1545	0673	1110	2615
1264	1055	0638	1080	0929	1097	1606
1265	0090	1253	0106	0938	1104	2595
1266	1559	1256	1727	0927	1111	2512
						2610
1267	1558	1257	1726	0926	1112	2511
						2609
1268	0074	1246	0078	0940	1113	2613
1269	1567	1265	1201	0932	1100	2593
1270	2177	1253	1999	0310	1107	2600
1271	1014	1254	1035	0308	1108	2601
1272	1775	1259	1216	0931	1101	2611
1273	1645	1251	1818	1379	1106	2605
1274	1284	1260	1430	0936	1103	2607
1275	1542	1247	1722	1080	1102	0184
						2614
1276	2103	1263	2253	0941	1115	2594
1277	2438	1261	0840	0934	1105	2602
1278	2183	1267	1361	0937	1117	2598
1279	1802	1245	1436	0928	1099	2603
1280	2144	1264	2477	1563	1116	2604
1281	1344	2167	1491	1543	0968	2495

E	F	G	S	R	C	J
1282	1467	1363	1591	0538	2478	0927
1283	1569	2005	1062	1570	0173	1273
1284	1475	0501	1630	0466	1791	2089
1285	1629	1188	1812	1010	0331	1508
1286	1456	1281	1599	0953	1569	1786
1287	0441	1282	0430	0803	1679	1787
1288	1633	1194	1827	2434	1568	1788
1289	1579	1278	1750	1112	0908	1916
1290	2601	0239	2606	1268	2340	1048
1291	0595	0041	0545	1614	1130	0624
1292	0593	1100	0539	1617	1124	0625
1293	0592	0867	0568	1620	1125	0626
1294	0596	0436	0553	1616	1128	0623
1295	0594	1137	0541	1618	1126	0627
1296	0597	1404	0556	1619	1129	0622
1297	1462	1495	1602	1610	1166	0654
1298	1463	0254	1603	1611	0164	1491
1299	1465	1816	1600	1612	0441	1491
1300	1968	1301	1606	0967	2166	0238
1301	2233	1284	1152	1004	2180	1206
1302	0384	1297	0385	0966	2186	1216
1303	0671	1285	0661	0958	2174	1432
1304	0797	1286	0857	0954	2177	1240
1305	0018	1283	0018	0965	0433	1213
1306	0873	1291	0900	0966	2179	1212
1307	2496	1287	2440	0964	2168	1199
1308	1143	1292	1264	2276	2169	1200
1309	2570	1289	0333	0957	2181	1243
1310	0403	1294	0375	0962	2178	1207
1311	1808	1288	1949	0955	0517	2561
1312	1812	1295	1953	0963	2170	1437
1313	2543	1296	0246	0956	2175	1434
1314	0061	1293	0071	0959	2182	1999
1315	0404	1298	0390	0961	2184	1208
1316	2581	2482	2579	0231	2185	1435
1317	0615	1290	0652	2032	2173	1433
1318	1674	1305	1857	1371	1635	1237
1319	1197	1304	1380	2357	1631	1239
1320	1212	1303	1317	2318	1634 2172	1205
1321	0702	2089	0742	1963	1633	1221
1322	1445	1302	0147	2452	2167	1236
1323	0330	0732	0589	0520	2292	0144
1324	1124	0718	1243	0521	2287	0152
1325	0331	0725	0601	0518	2284	0145
1326	1413	0729	2479	0522	2282	0157
1327	1865	0438	2015	1581	2283	0139
1328	1452	0733	1584	0525	2281	0154
1329	0332	0720	0602	0526	2297	0138
1330	0816	0724	0869	0529	2294	0136
1331	1049	0722	1076	0528	2296	0137
1332	1125	0721	1244	0527	2293	0135
1333	1415	0723	2481	0530	2295	0157
1334	1703	0726	1268	0517	2291	0142
1335	0849	0727	1773	0516	2289	0148
1336	1319	0962	1520	0635	2290	0149
1337	0458	2254	0507	0524	2285	0150
1338	0729	0730	0770	0523	2286	0151
1339	0333	0719	0603	0531	2279	0140
1340	1706	0728	1267	0519	2280	0146
1341	2602	0735	2593	1251	2288	0153
1342	0302	0241	0320	1157	2583	0335
1343	2091	1237	2240	1156	2392	2025
1344	0089	1310	0105	0976	A06	0519

E	F	G	S	R	C	J
1345	0906	1314	0936	0974	1257	0515
1346	1224	1312	1343	0111	1256	0514
1347	1225	1309	1344	1829	1254	0517
1348	1226	1313	1345	0715	1252	0516
1349	2374	1311	2379	0973	1253	0513
1350	1043	1315	1070	0975	1255	0518
1351	0672	1306	0662	0554	0465	1299
1352	0394	1308	0394	0971	0464	2181
1353	0616	1307	0635	0970	0463	1300
1354	1031	1316	0880	2259	1475	2630
1355	0799	1242	0860	2266	0192	1412
1356	0912	1317	1050	2271	1474	2517
1357	0157	1550	0179	0485	0312	1484
1358	0419	1548	0353	1220	0315	1463
1359	0769	0263	0804	0456	0314	1473
1360	1145	0827	0237	1218	0316	1474
1361	1433	1545	1577	0497	0319	1454
1362	1434	1546	1578	1802	0318	1457
1363	0644	0545	0697	2490	0317	1456
1364	1754	1547	1924	0628	0320	1458
1365	1095	1549	0720	1219	0313	1455
1366	1480	1332	1635	0983	0506	2558
1367	1679	1334	1861	1373	0507	2562
1368	1117	1333	1240	2231	0509	2560
1369	2227	1335	1149	0982	0510	2564
1370	2259	1336	1131	0980	0511	2565
1371	2280	1337	1183	0981	0512	2563
1372	0907	2083	0937	1892	0518	1823
1373	0908	1275	2234	2579	0519	1822
1374	1193	1280	1292	1893	0909	0706
1375	0556	1274	0301	1894	0508	1827
1376	1945	1277	1647	0786	0514	1828
1377	0368	1615	0364	2103	1340	0763
						2514
1378	2125	1344	2276	0988	0969	2379
1379	0179	1340	0200	0493	0246	0940
1380	0757	1341	0756	0985	0245	0947
1381	0298	1338	0293	0986	0243	0939
1382	0639	1339	0704	2197	0244	0941
1383	0876	1342	0904	0987	0247	0948
1384	1485	1343	1645	0984	0248	2378
1385	2407	1521	2570	1165	0412	0993
1386	0525	0891	0503	1325	2523	0995
1387	1489	1345	1656	0990	0088	2370
1388	1490	1348	1046	0992	2258	2380
1389	0042	1350	0043	0993	2259	2381
1390	0382	1355	0386	0997	2477	0926
1391	1988	1354	2208	1349	2473	0925
1392	0814	1361	0867	0995	2474	0932
1393	1982	1362	2200	1980	2475	0933
1394	1634	1364	1833	0996	2476	0928
1395	2232	1356	1150	1000	2481	0930
1396	2260	1358	1132	1008	2480	0931
1397	2203	1359	1112	0998	2479	0929
1398	0514	2159	0493	0837	0236	2044
1399	0468	0118	0444	0812	1596	1157
1400	1505	1365	1673	1014	1069	2384
1401	0882	1367	0907	1016	1071	2385
1402	1035	1366	0032	1970	1258	2386
1403	1577	1368	1745	1106	1070	2387
1404	1261	1369	1409	0352	1477	2383
1405	1468	1370	1592	0539	1478	2377
1406	2035	1371	2460	1392	1479	2382
1407	1590	1432	1689	2039	1611	2200

E	F	G	S	R	C	J
1408	0406	1441	0379	2042	1612	2197
1409	2572	1440	2555	2047	1619	2199
1410	0407	1443	0378	2041	1613	2216
1411	2547	1434	2559	2043	1617	2218
1412	0708	1438	1518	2044	1618	2219
1413	0709	1444	1512	2046	1620	2217
1414	1109	0812	1229	0579	0165	1284
1415	1333	0813	1476	0581	0163	1283
1416	1110	0814	1230	0580	0162	1286
1417	2081	0815	2221	0582	0161	1287
1418	1521	1391	1690	0574	0166	1285
1419	1746	1373	0225	1070	1053	2052
1420	1658	1372	1841	1071	1054	0422
1421	1512	1376	1676	1067	1055	0428
1422	1510	1375	1679	1065	1052	0420
1423	1511	1374	1680	1066	1052	0421
1424	1562	1377	1697	2035	1216	1789
1425	0158	1430	0188	0492	2491	1386
1426	1937	1436	2001	2045	2493	1387
1427	1088	1428	0694	0702	2497	1384
1428	1254	1429	1406	0341	2498	1385
1429	0248	1383	1684	0368	2046	1423
1430	1516	1384	1700	1035	1508	2376
1431	0928	1820	0947	1880	0833	1094
1432	1519	1386	1706	1032	1527 2490	2421
1433	1520	1387	1707	1033	0028	2419 2422
1434	0073	1388	0090	1034	0020	2418 2420
1435	0196	1394	0220	0106	1226	0536
1436	1301	1395	1449	1043	1231	0533
1437	1302	0706	1450	1042	1227	0532
1438	0162	1392	0192	1040	1228	0534
1439	1419	1393	1563	1041	1229	0535
1440	1528	1396	1708	1046	2249	1007
1441	1529	1397	1709	1045	2248	1004
1442	0668	0001	0726	1047	2250	1009
1443	1732	1624	1914	1428	2251	1005
1444	1738	1398	1915	1429	2253	1008
1445	1082	1399	1212	1044	2252	1006
1446	1537	1406	1717	1078	2058	1997
1447	1543	1415	1723	1081	2060	1996
1448	0214	1416	0240	1094	2059	1993
1449	1544	1407	1724	1084	2061	1994
1450	1549	1408	1731	1085	2062	2371
1451	1545	1409	1725	1086	2056	2393
1452	1550	1410	1728	1093	0167	2396 2397
1453	1551	1412	1732	1089	2051	2375
1454	1547	1411	1729	1088	2057	2398
1455	2126	1413	2277	1090	2054	2394
1456	1560	1414	1739	1092	2055	2395
1457	1548	1417	1730	1082	2048	2372
1458	1764	1418	1929	0626	2050	2374
1459	1767	1419	1935	1083	2049	2373
1460	0776	1402	0849	1633	1498	2390
1461	1561	1403	1740	1077	1664	0080
1462	0583	1405	0555	1076	1665	0130
1463	1565	1420	1741	1096	0933	2401
1464	1054	1421	1078	1095	0932	2501
1465	1446	1422	0148	1097	2594	0800
1466	0523	1506	0501	1194	1818	1049
1467	0483	2115	0458	0857	1247	0293

E	F	G	S	R	C	J
1468	0251	1505	1811	2225	2093	2399
1469	0445	0816	0431	1958	1022	0772
1470	0626	1425	0575	1960	1025	0777
1471	1303	1423	1451	1959	1024	0773
1472	0440	0817	0428	1768	1022	0774
1473	0444	0818	0429	1769	1021	0771
1474	1532	1401	1711	1056	1023	0775
1475	0829	0552	0957	2500	1463	0362
1476	1704	1445	1269	1487	1462	0101
1477	1574	0307	1742	0544	2489	0662
1478	1573	1447	1743	0545	1007	0663
1479	1575	1446	1748	1113	0611	2428
					1511	
1480	2147	1450	2320	0029	0794	1984
1481	1033	1449	0882	1107	0795	0967
1482	1576	1448	1744	1105	0796	1419
1483	1232	1454	1352	2427	1983	2235
1484	1580	1451	1751	1109	2183	1438
1485	1827	1455	1969	1108	1987	2236
1486	0029	1456	0029	1130	1502	0355
1487	0380	1457	0380	1132	1505	0353
1488	1582	1458	1754	1131	2514	0989
1489	0017	1462	0013	1135	1507	2433
1490	0028	1459	0030	1133	1503	0356
1491	0379	1461	0381	1137	1504	0354
1492	1017	1460	1039	1134	1501	2144
1493	1583	1463	1755	1136	1506	2434
1494	2252	1465	1137	1116	0698	2147
1495	2273	1466	1196	1117	0699	2146
1496	0141	1467	0163	1127	0697	2153
1497	0218	1468	0243	1125	0692	2152
1498	0281	1472	0325	1129	0696	2148
1499	2254	1469	1138	1121	0688	2151
1500	2277	1470	1197	1122	0689	2150
1501	1377	1471	1528	1128	0695	2143
1502	2263	1473	1139	1118	0687	2157
1503	1536	1474	1714	1124	0686	2149
1504	2012	1477	2130	1123	0690	2156
1505	2262	1475	1136	1119	0691	2155
1506	2289	1476	1195	1120	0693	2154
1507	1584	1464	1756	1115	0685	2142
1508	2495	0944	2439	1409	0448	0126
1509	1585	1481	1758	1138	0267	1343
1510	2013	1482	2131	1141	0271	1353
1511	1586	1483	1760	1142	0272	2431
1512	0839	1484	0290	1140	0262	0085
1513	1589	1486	1759	1139	0259	1358
1514	0258	1485	0303	0509	0260	1367
1515	0442	0524	0426	1144	0264	1346
1516	0447	0525	0433	1145	0265	1345
1517	0874	1487	0902	2564	1516	2417
1518	0133	1380	0142	1101	0472	2389
1519	0134	1381	0135	1102	0473	2388
1520	1592	1488	1776	1150	0471	1324
1521	1530	1379	2306	1100	0477	1320
1522	1535	2466	1715	1064	0052	1271
1523	2189	2413	1368	1103	0485	1342
1524	0250	1832	1687	1104	0470	1323
1525	1828	1489	1958	1535	0482	1319
1526	1345	0799	1494	0186	2202	1253
1527	1599	1498	1781	1160	1517	1687
						1688
1528	0004	1499	0016	1814	1978	1589
1529	0915	1504	1052	0514	2564	0906

E	F	G	S	R	C	J
1530	1279	1500	1428	0224	1955	2369
1531	0003	1501	0015	0513	1979	1590
1532	2014	1502	2132	0511	2565	0907
1533	1960	1503	2146	0512	1977	0908
1534	1600	2586	1783	1816	2094	2402
1535	0899	2587	0928	2578	2097	2403
1536	1601	2580	1784	1818	2095	1762
1537	2334	2584	2315	1920	2096	1763
1538	2536	1497	2533	0565	2419	1690
1539	1483	1508	1638	1355	2312	2038
1540	1379	1509	1530	1357	2319	2013
1541	2204	1360	1113	0999	0781	2014
1542	2619	1507	2619	1356	0779	2090
1543	1613	1510	1797	1175	0780	0120
1544	2545	1513	2562	1191	1558	1720
1545	0754	1515	0790	1215	2550	1940
1546	2462	1516	2515	1216	2549	1939
1547	1614	1514	1796	1217	2548	1941
1548	1032	1517	0881	2260	1540	1765
1549	2084	1518	2254	1206	1541	1764
1550	0599	0675	0563	1211	1245	1013
1551	2395	1520	2416	2441	1243	1014
1552	2606	1519	2600	2442	1244	1015
1553	1616	1522	1798	1179	2492	1402
1554	1617	1523	1799	1180	2499	1403
1555	0025	1524	0025	0025	2502	1407
1556	0069	1525	0110	1183	2503	1406
1557	0268	1527	0315	0557	2500	1408
1558	0693	1526	0672	1181	2505	1409
1559	1189	1528	1293	1185	2506	1404
1560	0690	1529	0671	1454	2507	1417
1561	1596	1530	1779	1766	0053	1405
1562	1635	1533	1828	2435	2508	1415
1563	2001	1531	2133	1184	2501	1416
1564	2357	1532	2380	1182	2509	1414
1565	1262	1551	1410	0353	1543	1689
1566	1245	0499	1398	0151	2140	0425
1567	1308	1268	1587	1360	1320	0371
1568	0598	1552	1291	1221	0832	2415
1569	0600	1536	0569	1173	0612	1908
1570	2457	2330	2507	1174	0618	1936
1571	2006	1540	2121	1187	0622	1933
1572	2192	1541	1372	1189	0624	1932
1573	1872	1543	1987	1202	0135	1945
1574	0946	1542	0965	1190	0615	1929
1575	1769	2230	1937	0139	2085	1598
1576	0091	2588	0111	1176	2084	1950
1577	1995	1539	2110	1177	0617	2626
1578	0035	1535	0036	1214	0613	1952
1579	0301	0960	0296	1192	0139	2030
1580	1502	0961	1671	1193	0138	2031
1581	1355	2414	1506	1209	0621	2054
1582	1346	1544	1495	1208	0614	1938
1583	1660	1553	1843	1224	2442	1161
1584	0656	1556	0706	1225	2441	1160
1585	1642	1557	1816	1222	0923	0447
1586	2162	1558	0830	1223	0924	0445
1587	2383	1589	2381	1226	0098	1782
1588	0093	1115	0112	2594	0938	2256
1589	0306	1118	2101	2588	0937	0298
1590	0674	1133	0735	2595	0948	0628
1591	1162	1125	1280	0403	0945	0320
1592	2026	1126	2161	1844	0946	0328
1593	0429	1116	0419	2591	0947	0313

E	F	G	S	R	C	J
1594	1238	1119	1387	2598	0942	0331
1595	2027	1120	2162	1845	0943	0332
1596	1400	1121	1551	2592	0951	0311
1597	1402	1122	1553	2593	2003 2019	0296
1598	1689	1130	1868	2589	0944	0333
1599	2029	1123	2164	2586	0935	0315
1600	2023	1124	2171	2597	0936	0657
1601	2329	1129	2312	2587	0952	2272
1602	2489	1127	2156	2599	0939	0316
1603	0735	1128	0778	0425	0940	0317
1604	2500	1132	2443	2590	0954	0330
1605	2511	1131	2455	2596	0953	0310
1606	1646	1113	1819	2600	0211	0308
1607	1651	1560	1820	1229	0955	0304
1608	1636	1561	1834	2436	0956	0305
1609	1647	1562	1822	1228	0949	0306
1610	1495	1563	1662	0994	0950	0307
1611	1654	0242	1836	1153	0859	0403
1612	1268	0244	0058	0295	0860	0402
1613	2613	0245	2615	1152	0304	2067
1614	1659	1578	1842	1291	0033	0484
1615	1304	0520	1452	1280	0274	1354
1616	1759	0511	1928	1281	0266	1364
1617	2180	0536	1360	1279	2298	0096
1618	1987	0509	2196	1978	1371	0222
1619	0277	1576	0328	1524	1278	0263
1620	0667	2578	0172	0258	0641	2116
1621	0323	1096	0596	0736	1281	0235
1622	0427	1099	0417	0734	1496	0234
1623	0288	1581	0341	1707	0158	0922 1258
1624	0854	1579	1087	1706	0159	0921 1257
1625	1077	1582	1879	1708	0160	0923 1259
1626	0473	1590	0449	1302	0863	1913
1627	0694	1585	0673	1301	0879	0681
1628	2255	1586	1128	1303	0526	0678
1629	1151	1587	1272	1305	0912	0217
1630	1157	1588	1276	1308	1471	0680
1631	1765	1584	1932	1304	0906	0677
1632	1847	1592	0267	1300	0862	1926
1633	1948	1589	1652	1306	0878	0679
1634	1581	1452	1752	1110	0910	0676 1912
1635	2335	1591	2316	1307	0911	0675
1636	0472	1583	0448	1299	2593	0803
1637	1662	2090	1846	1566	0896	1138
1638	0875	2174	0903	0195	2339	2437
1639	1663	1577	1847	1167	2216	0221
1640	2030	1600	2165	0780	2426	1091
1641	1092	1601	0708	1331	0010	1090
1642	1665	1596	1848	0339	2428	1830
1643	1206	1598	1373	1912	2431	1088
1644	1668	1602	1851	1333	1863	0210
1645	1669	1604	1852	1334	1864	0211
1646	1507	1595	1644	1975	0007	0212
1647	2616	0210	2616	0199	2039	0258
1648	2557	0213	2548	1928	1461	2533
1649	0519	0322	0497	1265	2578	1241
1650	0516	0321	0484	1264	2582	1226
1651	0640	2389	0684	1891	0198	0343
1652	1670	1605	1853	1288	2275	1052

E	F	G	S	R	C	J
1653	1297	1606	1446	1289	2274	0822
1654	2475	1607	2502	1290	2272	0823
1655	0425	0822	0414	0550	1982	0977
1656	2057	1611	2211	1151	2543	0982
1657	0383	1610	0383	0881	0418	1451
1658	1913	1609	2054	1241	0417	1579
1659	0342	1612	0595	0157	2468	2215
1660	0057	1614	1020	1368	0113	2210
1661	0482	1613	0456	0856	0114	2203
1662	0369	2107	0366	2098	1476	1839
1663	0453	1617	0436	1370	1925	1023
1664	1380	1490	1533	1879	1512	0218
1665	0499	1620	0472	2422	0133	2056
1666	1103	1619	1223	2423	0134	2055
1667	0145	2275	0165	0085	1424	2526
1668	0235	2267	0264	0156	1417	2528
1669	0733	2268	0776	0422	1419	2525
1670	1190	2269	1287	1599	1421	2530
1671	0691	2270	0669	1457	1422	2531
1672	2256	1621	1127	1872	1425	1929
1673	2278	1622	1179	1874	1418	2534
1674	2002	2274	2117	0570	1416	2527
1675	2337	2271	2343	1765	1414	2522
1676	0080	2272	0103	0056	1412	2523
1677	0815	2273	0868	1810	1415	2524
1678	0338	2454	0590	1812	0656	2175
1679	0397	2460	0399	0885	0658	2176
1680	0507	2462	0488	0836	0651	2180
1681	1196	2455	1378	2358	0652	2174
1682	1728	1623	1902	1426	0703	1892
1683	0076	1625	0091	1479	0716	1893
1684	2314	1632	0204	1463	0712	1904
1685	0222	1628	0248	1317	0710	1903
1686	2411	1636	2576	1432	0715	1896
1687	0577	1638	0558	2203	0714	1899
1688	0405	1629	0388	0876	0720	0798
1689	2591	1631	2591	1254	0719	1905
1690	0077	1626	0052	1901	0718	1901
1691	1272	1634	0060	0296	0709	1898
1692	1327	1635	1472	0649	0713	1894
1693	1509	1637	1675	0989	0707	1895
1694	2096	1627	2245	1794	0708	1902
1695	0578	1633	0550	1433	0717	1897
1696	1443	1630	0149	2453	0711	1900
1697	1444	0879	0150	2454	0019	1862
1698	1515	1639	1716	0315	1576	2238
1699	1059	1786	1084	1343	2099	0747
1700	2042	2540	2181	1692	1014	0243
1701	0601	2524	0571	1423	2425 2516	0963
1702	2414	2520	2575	1422	2517	0962
1703	0339	1640	0584	1668	1844	1087
1704	1698	1641	1894	1905	2228	1233
1705	1701	1648	1900	1701	1683	1880
1706	0945	1642	0956	1703	1591	1886
1707	1299	1643	1447	0621	1594	1881
1708	1358	1644	1509	0655	1592	1887
1709	1522	1645	1691	0576	1589	1884
1710	0853	1580	1088	1706	1588	1882 1883
1711	1076	1646	1878	2063	1586	1889
1712	2483	1647	2503	1702	1590	1885
1713	0075	1649	0084	2274	2197	1272
1714	1186	1650	1307	2273	2206	0083

E	F	G	S	R	C	J
1715	1984	1652	2201	1350	2187	1269
1716	2455	1651	2501	2275	2188	1268
1717	1986	1653	2206	1261	2205	1274
1718	1768	1654	1321	2323	1451	2548
1719	0102	1655	0087	2321	1453	2551
1720	2201	0123	1108	2538	1456	2553
1721	2242	1657	1147	2322	1452	2550
1722	1709	1658	1322	2324	1454	2552
1723	1710	1659	1323	2325	1455	2549
1724	1711	1688	1324	2342	0867	1914
1725	1712	1660	1325	2328	0866	1911
1726	0385	1663	0384	0887	0868	0712
1727	0877	1661	0901	2345	0869	0713
1728	1739	1662	1913	2344	0872	0720
1729	1715	1664	1328	2346	0870	0714
1730	2217	1666	1173	2349	0873	0716
1731	2282	1667	1192	2348	0874	0718
						2221
1732	2066	1665	2233	0306	0871	0715
1733	2285	1669	1181	2350	0865	1915
1734	1074	1080	0274	0383	0799	0957
1735	1525	1671	1694	0578	0798	0958
1736	1678	1673	1860	2327	1841	0959
1737	1719	1672	1332	2334	2414	0955
1738	2514	2359	2458	2326	0800	0956
1739	1716	1674	1329	2330	0913	1909
1740	2286	1675	1182	2329	0914	1910
1741	1789	1678	2086	2331	0876	0722
1742	2464	1679	2511	2332	0875	0721
1743	1717	1677	1330	2333	0877	1291
1744	2524	1668	2470	2343	0864	0717
1745	1722	1680	1334	2335	0915	0687
1746	0026	0054	0024	2336	0920	0738
1747	2197	1683	1104	2340	0918	0691
1748	2276	1682	1180	2337	0917	0693
1749	2067	1681	2224	0305	0916	0692
						0694
1750	1191	1684	1289	2339	0921	0689
1751	0692	1685	0670	1460	0922	0690
1752	2003	1686	2124	2338	0919	0737
1753	1725	1687	1337	2341	0872	0712
1754	2551	1689	1059	2295	2088	2049
1755	1039	1692	1065	2294	2089	2051
1756	1347	1691	1493	2292	2087	2050
1757	1946	1690	1646	2291	2090	2053
1758	1727	1693	1283	2293	2091	2048
1759	1742	1694	1916	1434	1595	1931
						1935
1760	0046	1695	0066	1925	0458	1791
1761	0871	1696	0896	1436	1881	2002
1762	0742	1997	0782	1435	0016	1879
1763	0126	0078	0137	0059	0972	1209
1764	0127	0079	0138	0060	0973	1211
1765	0128	0080	0139	0061	0974	1210
1766	1748	0275	1007	2526	2213	1928
1767	2279	1697	1918	1882	1418	1927
1768	1469	1698	1593	0540	1628	2085
1769	2182	0875	1358	0721	1606	2207
1770	1809	1299	1948	1528	1621	2208
1771	1749	1300	1951	1453	1622	2209
1772	1752	1699	1921	1441	0300	2060
1773	1246	1701	1391	1446	0299	2063
1774	1277	1702	1427	0604	0308	2061
1775	1210	1706	1462	0636	0306	2065

E	F	G	S	R	C	J
1776	1411	1704	0355	1443	0305	2064
1777	1278	1703	0321	1442	0303	2069
1778	1683	1618	1863	1374	0301	2068
1779	2381	1705	2376	2138	0307	2062
1780	2487	1700	0170	1444	0302	2066
1781	1086	1707	1213	1445	2472	2070
1782	1773	1708	1704	1450	1624	2078
1783	1772	1959	0967	1452	0012	2079
1784	1610	1709	1793	1465	1648	0523
1785	0287	1710	0340	2603	1627	2352
1786	0484	1711	0459	0858	0340	1588
1787	0485	1712	0460	0859	0339	1587
1788	1813	1716	1944	1531	1131	0520
1789	1805	1713	1945	1527	1121	2230
1790	1810	1715	1950	1529	1600	2233
1791	1811	1714	1952	1526	1641	2232
1792	1814	1717	1954	1532	1123	2322
1793	2465	1718	2516	1530	1122	1219
1794	1843	1378	1984	1495	0486	1318
1795	1844	1719	1985	1497	0484	2356
1796	0841	1720	0291	1496	0474	1313
1797	0446	1721	0432	1498	0475	1312
1798	1370	0249	1519	0615	1420	1864
1799	0956	1722	0975	1561	0476	2362
1800	0957	1723	0972	1561	1936	2321
1801	0424	1725	0415	1509	2443	2461
1802	0686	2224	0663	1459	2444	2462
1803	1396	1724	1538	1508	2267	1148
1804	1852	1727	1994	1492	2446	2476
1805	0888	1492	0914	1569	2488	2341
1806	1853	1491	1995	1567	0971	0683
1807	1854	1728	2003	1577	0399	1522
						2355
1808	0349	1735	0616	2402	0405	1531
1809	1790	1737	2087	1587	0400	1529
1810	0401	1741	0401	0883	0404	1530
1811	2467	1738	2518	1588	0401	1528
1812	2473	1740	2513	1590	0406	1526
1813	1868	1736	2017	1593	0402	1527
1814	0845	1795	1772	1565	0701	2187
1815	2356	1797	2357	1062	0700	2186
1816	0255	1269	0277	1778	0444	1611
1817	2073	0613	0927	2570	2333	0587
1818	1933	1496	1941	1146	1159	0655
1819	1523	1744	1693	0577	1593	1888
1820	2100	1743	2251	1607	1850	0920
1821	0261	1745	0306	1606	1849	0924
1822	1694	2492	1891	1609	2350	2483
1823	1879	0664	2022	1312	0205	1450
1824	1880	0663	2023	1311	0203	1448
1825	1878	0665	2021	1313	1273	0529
1826	1300	0666	1448	1310	0208	1447
1827	2463	0667	2504	1309	0204	1449
1828	1882	1746	2024	1630	1235	1165
1829	1870	1750	2019	1629	1236	1170
1830	1884	1748	2026	2215	1976	1793
1831	1885	1749	2028	2214	0214	1792
1832	1883	1747	2027	2216	2161	0894
1833	2024	2500	2170	1622	2531	2098
1834	1888	2501	2029	1624	1668	1204
						1217
1835	1382	2497	1531	0696	1513	1203
1836	1407	2499	1559	0602	1669	1218
1837	0889	2493	0915	1623	1670	2099

E	F	G	S	R	C	J
1838	1877	1742	2030	1608	2351	1201
1839	0245	0309	1685	0370	2352	2496
1840	1696	2495	1892	1621	2353	2497
1841	1891	0478	2031	0389	2238	0023
1842	0883	0480	0908	0194	2241	0026
1843	1135	0481	1256	1560	2240	0025
1844	2053	0479	2209	1979	2239	0024
1845	0432	1755	0421	1382	0232	0090
1846	0659	1752	0712	1388	2299	1000
1847	1383	1757	1532	1386	2358	0087
1848	0948	1033	0969	1387	0231	1296
1849	1078	1753	1880	1384	1064	0086
1850	2016	1758	2134	1383	2521	0094
1851	1079	1751	1881	1385	1062	0088
1852	2184	1754	1363	0720	2520	0089
1853	1071	1756	1875	1389	1074	0091
1854	2157	2424	0822	1390	1076	0093
1855	0452	1759	2340	1381	1077	0092
1856	1247	0892	1397	1321	1065	0992
1857	0223	0227	0249	1318	1459	2088
1858	1381	2498	1534	0695	2590	0218
1859	1923	1775	2060	1640	0215	2103
1860	1925	1777	2062	1641	0079	2106
1861	1922	1776	2061	1640	0218	2104
1862	1924	1778	2063	1641	1544	2105
1863	0354	2231	0618	2409	0216	1608
1864	0860	1779	1854	1643	2032	1627
1865	2238	1357	1158	1658	0191	1672
1866	0410	1781	0404	1019	1920	1077
1867	1619	1780	1801	1021	1921	1289
1868	2522	1782	2468	1669	0072	1998
1869	1930	1783	2066	1683	2482	2463
1870	0508	1785	0486	0830	2483	2108
1871	1047	1784	1074	1715	2484	2109
1872	1638	1594	1832	1684	2485	2464
1873	1182	2469	1304	1704	2221	1891
1874	0060	0945	0070	0611	1480	1808
						1845
1875	0129	0937	0136	0057	1483	1809
1876	0813	0939	0866	1809	1482	1843
1877	2128	0938	2281	1908	1484	1807
1878	1424	0824	1569	0609	1481	1846
1879	2191	0825	1370	0610	1487	1850
1880	0822	0940	0876	0467	1489	1844
						1848
1881	1005	0942	1030	2548	1490	1842
1882	0130	0943	0133	0058	1485	1810
1883	1825	1794	1979	0608	1488	1849
1884	1931	1793	1998	1147	1486	1847
1885	2330	0207	2302	1918	0063	1069
1886	0283	1799	0337	1860	0066	2365
1887	1011	1798	1026	2550	0065	2364
1888	1432	1801	1564	0947	0064	2368
1889	1695	1802	1889	1420	0067	2366
1890	1934	1800	1996	1148	0062	2367
1891	0744	1803	0786	1437	1710	1027
1892	0435	1804	0463	1398	1279	0482
1893	1199	1805	1845	1293	1280	0483
1894	0872	1806	0899	0754	0495	0904
1895	0132	1807	0141	0755	1935	0903
1896	0106	1808	0079	0751	0443	1471
1897	1138	1809	1252	1493	1603	1220
1898	0107	1817	0080	0781	0440	1493
1899	0114	1818	0120	0782	0439	1492

E	F	G	S	R	C	J
1900	1944	1385	0297	0783	1435	2538
1901	0461	1815	0509	0763	1441	2539
1902	2078	1813	0919	0759	2335	0590
1903	2082	1814	0920	0758	0515	2566
1904	2075	1810	2217	0760	1440	2540
1905	2079	1811	2218	0762	2334	0591
1906	2080	1812	2219	0761	0513	2567
1907	1857	2473	2009	1586	1934	0951
1908	1466	2474	1617	2433	1933	0950
1909	1720	1819	1331	0756	1232	0430
1910	1240	1318	1389	0314	1132	0910 1022
1911	1118	1319	1239	0336	2102	0537 0909 1021
1912	1185	1825	1306	2307	2103	1020
1913	1952	1828	2106	1717	1350	2506
1914	0125	0081	0140	0062	0967	1214
1915	1954	0256	2140	2029	0735	2257
1916	2376	2210	2372	2133	0767	2258
1917	0997	2192	1017	2031	0738	1921
1918	1018	2193	1040	2547	0753	2248
1919	0685	2195	0664	1461	0756	2249
1920	1116	2187	1238	2477	0755	2307
1921	1192	2196	1288	1597	0765	2261
1922	1332	2209	1477	0979	0760	1918
1923	1691	2185	1870	0566	0749	1923
1924	1776	2186	1218	1464	0766	1922
1925	1935	2202	2002	1677	0743	2521
1926	1675	2198	1648	0784	0750	1924
1927	1991	2188	2107	1727	0734	2260 2286
1928	0308	2183	2100	1719	0745	2288
1929	0460	2212	0508	2115	0747	2289
1930	0734	2189	0777	1724	0764	2291
1931	1016	2194	1032	2553	0754	2287
1932	2508	2203	2454	0134	2081	2412
1933	1286	2200	1434	0613	0752	2295
1934	1680	2199	1649	0573	0751	2297
1935	2190	2205	1371	0722	0769	2259 2290
1936	2243	2206	1160	1726	0739	2294
1937	2208	2207	1115	1992	0757	2293
1938	0270	2191	0316	1725	0741	2306
1939	0718	1911	0730	1721	0761	1919
1940	1111	0121	1234	1720	0746	1925
1941	1911	2204	2052	1723	0744	1917
1942	2036	2184	2173	1718	0740	2263
1943	2512	2211	2456	1722	0770	1920
1944	1956	1829	2147	1738	0595	2265
1945	0041	1845	0884	0221	0601	2276
1946	0291	1837	0271	1729	0606	2284
1947	2048	0010	1044	1730	0748	2584
1948	0631	1831	1270	0391	0593	2277
1949	0669	1844	0736	1744	0588	2283
1950	0910	1839	0938	1740	0602	2285
1951	1040	1846	1067	1731	0586	2305
1952	2108	1841	1629	1985	0594	2270
1953	0683	1832	2290	1421	0579	2279
1954	1305	1842	1454	1742	0609	2271
1955	1408	1847	1560	1741	0607	2274
1956	1486	1834	1657	1739	0591	2280
1957	1504	1835	1672	1734	0592	2281
1958	2548	1435	2560	0254	0603	2282

E	F	G	S	R	C	J
1959	1643	1833	1821	1737	0597	2269
1960	1650	1848	1824	1743	0598	2266
1961	2141	1838	2299	1736	0589	2268
1962	2307	1836	2342	1733	0608	2281
1963	2492	1843	2464	1728	0605	2275
1964	0666	1840	1665	1735	0596	2278
1965	1957	1849	2143	1732	0590	2300
1966	0307	0101	2102	1747	0582	2303
1967	1943	1850	2097	1746	0583	2304
1968	1942	1853	2096	1755	0573	2253
1969	2138	1854	2297	1756	0574	2254
1970	2083	1852	2236	1757	0585	2252
1971	1961	1851	2153	1758	0584	2251
1972	1963	1855	2149	1759	0587	2504
1973	0421	1858	2148	1749	0600	2505
1974	1768	0910	1936	1760	1819	0702
1975	1830	1856	1973	1761	1820	0703
1976	1964	1857	2150	1748	0607	2274
1977	1965	1859	2151	1750	0733	2296
1978	1791	1861	2088	1751	0580	2299
1979	2474	1860	2525	1752	0581	2298
1980	1965	1862	2152	1753	0598	2266
1981	1940	1863	2095	1754	0599	2267
1982	1954	1830	2155	1745	0604	2273
1983	1966	1864	2190	1841	1972	2507
1984	2017	1866	2135	0568	1351	2508
1985	0780	1867	0850	0794	1353	2509
1986	2291	1865	1198	2019	1352	2510
1987	0409	2592	0403	1957	1943	0548
1988	0817	2076	1511	0429	0561	1944
1989	1324	0941	0505	2114	0442	0605
1990	2555	0731	2541	1927	2134	0143
1991	2562	0734	2545	1937	0934	0147
1992	2565	1114	2549	1941	0941	0309
1993	0781	1875	0851	1881	1356	2590
1994	2532	1876	2483	0561	2461	0013
1995	1366	1877	1514	1842	0559	1871
1996	0555	1878	0299	1843	0557	1872
1997	1466	0013	1589	1098	0454	2500
1998	1454	0014	1586	1359	0455	2499
1999	0015	2507	0009	0401	2424	0917
2000	2401	0498	2424	1846	1848	0918
2001	1378	2403	1529	1394	0227	1532
2002	1374	2405	1525	1393	0228	1523
2003	2389	0182	2409	0246	—	—
2004	0798	1932	0858	1244	1292	0450
2005	2040	1972	2260	1358	0537	1863
2006	1019	0197	2194	1848	1151	0525
2007	2041	0843	2180	2074	1015	0242
2008	0398	2542	1259	2277	1017	0240
2009	2416	2541	1057	1637	1016	0241
2010	0172	0295	0186	0481	2437	1171
2011	2185	1823	1364	0278	2438	1172
2012	1306	1883	1455	1850	2274	0822
2013	0843	1882	0292	1851	2273	2616
2014	1845	1884	1986	1852	0421	2617
2015	2482	1885	2528	1853	2272	0823
2016	0711	1964	0740	0890	0970	0019
2017	1147	2445	1422	1699	1075	0832
2018	0938	0264	0958	2502	0146	0434 0831
2019	0988	0265	1008	2527	0150	0828
2020	2350	0266	2355	1059	0151	0829
2021	1850	1887	1993	1295	0148	0438

E	F	G	S	R	C	J
2022	1331	0267	1474	2054	0469	0827
2023	1972	0268	1610	0968	0149	0830
2024	1498	1888	1666	1012	0145	0437
2025	2170	0269	0837	1837	0147	0833
						1978
2026	2534	0270	1871	2034	1587	0433
2027	1938	1889	1041	1353	0550	1856
2028	2287	1891	1186	2023	0551	0010
						1857
2029	1139	1892	1260	2280	0553	1858
2030	0772	1890	0802	0452	0552	0014
						0710
2031	1612	1970	1795	1681	1013	0417
2032	0884	1276	0909	1875	2417	0553
2033	1311	0304	1458	0616	2416	0554
2034	0585	1894	0560	1847	2401	0801
2035	2412	2543	2571	1164	2402	0804
2036	0030	1900	0031	1338	2191	1262
2037	1493	1902	1660	1339	2195	1265
2038	0892	1903	0885	1340	2194	0011
2039	0456	1901	2179	1347	2192	1264
2040	0695	1904	0674	1342	2193	1267
2041	1060	1910	1085	1344	2196	1263
2042	2107	1906	2258	1346	2189	1261
2043	0758	1907	0759	1345	2190	1266
2044	0043	0746	1597	0282	1877	2309
2045	0044	0748	0045	0281	1875	0397
						1333
2046	1514	0747	1678	1069	1876	0396
						1332
2047	2055	1912	2213	0165	1357	2618
2048	0562	0681	0021	0448	2386	0187
2049	2085	2402	2237	1411	2400	0802
2050	0824	0178	2410	0250	1988	1308
2051	0602	1925	0572	1329	0153	0844
2052	1248	1923	1399	1328	0154	0848
2053	1392	1924	1547	1326	1847	0847
2054	1248	1921	2480	1327	0155	0843
2055	1866	1922	2016	1330	0156	0842
2056	1506	0890	1674	1015	1887	1003
2057	2088	2539	2238	1855	0434	1467
2058	2092	0183	2242	1791	0629	2111
2059	2093	2562	2243	1363	0630	2115
2060	1874	0186	1943	1790	0631	2114
2061	2391	0188	2411	0253	0632	2112
2062	0550	0189	0534	1554	0633	2113
2063	0918	1920	1055	1857	0839	0511
2064	1012	1914	1034	2554	0840	0508
2065	1973	1917	1612	1856	0842	0509
2066	1618	1918	1800	1858	0843	0510
2067	2244	1919	1163	1859	0838	0512
2068	0284	1915	0338	1262	0841	0547
2069	0509	0143	0489	1572	2207	0201
2070	1692	0144	1883	2233	2209	0199
2071	2390	0145	2413	0247	2208	0200
2072	2132	0142	0304	2030	2210	0404
2073	1851	1976	2078	2212	1608	1158
2074	0655	0855	0707	1269	2165	1155
2075	2104	0637	2255	1861	1217	0608
2076	1140	1933	1261	2285	2586	1376
2077	2106	1926	2257	2245	0041	2327
2078	1312	1927	1459	0617	0046	2330
2079	1682	1928	1862	1376	0042	2333
2080	2377	1929	2373	2141	0045	2328

E	F	G	S	R	C	J
2081	2393	1931	2414	0255	0043	2329
2082	2595	1930	2594	1256	0044	2336
2083	0244	1013	1683	0369	1547	0455
2084	0321	2404	0583	1248	0556	0451
2085	0925	0690	0951	2499	0845	0752
2086	1597	1895	1643	1974	0212	0994
2087	2388	0147	2418	0249	1815	1325
2088	2564	0146	2551	1933	1816	1326
2089	2111	1939	2264	1863	1473	2627
2090	1656	1961	1838	0240	2232	0253
2091	1657	1963	1839	0241	2231	0255
2092	1771	1960	1939	0239	2539	0256
2093	1015	1957	2445	1416	2537	0252
2094	2378	1958	2382	0238	2233	0254
2095	0782	1979	2186	1245	1469	0012 2503
2096	1470	1980	1594	0541	1783	0814
2097	0200	0527	2259	0256	2255	0197
2098	2052	0208	0750	0279	2256	0192
2099	0340	0209	0581	0280	2257	0193
2100	0844	1760	1767	1634	1869	1055
2101	1148	1773	1421	2084	1871	1057
2102	0937	1764	0954	2497	1870	1058
2103	1849	1767	1992	0410	2277	1063
2104	1318	1766	1522	0633	1234	0841
2105	1365	1390	1513	0653	1233	1281
2106	0847	1762	1770	1635	2341	0294
2107	2154	1769	0820	1832	2278	1065
2108	0371	1770	0371	0181	1872	A09
2109	1598	1771	1782	0972	1690	1062
2110	0625	1774	0311	2431	1688	1056
2111	0236	1772	0269	1418	1689	1066
2112	1034	1985	0883	2261	1795	0840
2113	0434	1986	0462	1400	0036	2057
2114	2163	0835	0834	1170	0039	2320
2115	2113	1987	2268	1169	0035	2317
2116	0296	1988	0283	0728	0038	2318
2117	1672	0868	1705	1451	0037	2319
2118	1114	2056	1236	1838	1297	2424
2119	0684	2422	0738	1413	0438	0601 1101
2120	0367	2423	0367	0852	0437	0597
2121	1553	1869	1737	1922	1796	1191 1255
2122	0094	1999	0081	0055	1797	1260
2123	1554	1868	1734	1921	1798	1106 1256
2124	1928	2216	0842	1822	1793	0846
2125	0774	2217	0843	1823	1794	0845
2126	0777	2218	0848	1824	0763	2292
2127	1743	2122	0678	1364	1682	1099
2128	1744	2123	0677	1365	1681	1100
2129	1496	2258	1664	2104	1807	1084
2130	2116	2251	0344	2110	1806	1078
2131	0459	2256	0506	2109	1810	1079
2132	0730	2252	0769	2107	1811	1080
2133	2230	2255	1141	2108	1808	1081
2134	2116	2257	0343	2105	1809	1082
2135	0739	2253	0767	2106	1812	1083
2136	1292	1424	1441	2472	2115	0960
2137	0862	0007	0171	2471	1625	2416
2138	1027	2516	1585	1666	2142	1222
2139	2169	2517	0819	1667	2143	1223
2140	1423	0008	1568	0549	0702	2018

E	F	G	S	R	C	J
2141	0433	2025	0422	0264	0251	1701
2142	0660	2022	0714	0268	0249	1706
2143	1593	2026	1780	0271	0497	1700
2144	0977	2021	1000	0266	0496	1699
2145	1183	2023	1305	0263	0501	1697
2146	1384	2028	1535	0267	0499	1694
2147	2261	2030	1133	1009	0500	0952
						1696
2148	2071	2029	2230	0299	0498	1695
2149	2018	2031	2136	0265	0250	1705
2150	1072	2027	1876	0269	0784	1703
2151	2158	2024	0823	0270	0785	1704
2152	2469	1739	2522	1589	0503	1702
2153	1041	1981	1068	1713	1250	0128
						0131
2154	0175	0805	0203	0490	0520	1548
2155	0768	0565	0805	0273	2224	0806
						1156
2156	0995	2040	1012	1906	2230	1232
2157	2577	2041	2564	1907	2227	1231
2158	0400	2042	0400	0886	2229	1230
2159	0201	2033	0226	1885	2567	0868
2160	0916	2431	1053	2269	2568	0872
2161	0881	2036	0906	2584	2569	0870
2162	1131	2035	1251	2283	2570	0869
2163	0203	2037	0228	1889	2554	1646
2164	0642	2039	0688	1886	2566	0874
2165	0204	2038	0229	1884	2553	0875
2166	0202	2032	0227	1888	2563	0873
2167	2318	0872	2333	1520	0015	1867
2168	2129	0873	2284	1517	0014	1865
2169	0731	0874	0779	1518	0017	1866
2170	1842	2043	1983	1513	0022	1860
2171	2130	0875	2287	1514	0027	1189
2172	2131	0876	2288	1515	0029	1190
2173	0436	0871	2285	1399	0024	1853
2174	1568	0870	1202	1521	0026	1868
2175	2607	0603	2609	2445	1160	0657
2176	2133	0604	2291	2444	1458	0404
2177	2134	0608	2295	2448	2462	0099
2178	2136	0606	2294	2449	2452	0407
2179	2135	0607	2292	2446	1775	0406
2180	2137	0605	2293	2447	1406	0405
2181	1484	2044	1639	1910	1500	1246
2182	1169	0009	1626	0275	0643	2193
2183	1533	1400	1712	1055	0634	1365
2184	1141	2365	1262	2282	0646	2177
2185	1910	2366	2049	1694	0644	2192
2186	2379	2367	2374	2140	0645	2191
2187	1491	1349	2280	1904	2258	2190
					2325	
2188	2140	1896	2298	1883	2151	1947
2189	0056	2045	1021	1568	0238	1445
2190	0645	0546	0698	2473	1626	0122
2191	1167	0836	1623	2470	0040	0853
2192	0861	2010	1855	2474	2412	0852
2193	0648	1228	0700	1800	0456	1371
2194	2219	2046	1171	1994	1861	1001
2195	2145	2050	2301	1911	2100	0851
2196	0930	2047	0970	1673	1803	0949
2197	0538	2502	0523	1600	0729	2039
2198	1714	2051	1327	0892	0925	1054
2199	2353	1081	2354	1060	1509	1002
2200	2333	2052	2314	1271	2006	1645

E	F	G	S	R	C	J
2201	0115	2053	0122	1273	2005	1643
2202	2358	2054	2360	1272	2000	1644
2203	2245	0535	1148	1275	0263	1351
2204	0131	0523	0134	1274	0257	1071
2205	1427	0512	1573	1282	0278	1350
2206	1741	0640	1909	0501	0275	1070
2207	2237	0513	1159	1003	0268	1352
2208	1285	0539	1432	0502	0269	1344
2209	1839	1603	1982	1332	0270	1359
2210	0013	2055	0010	1914	0261	0095
2211	0989	0514	1009	1335	0258	0097
2212	1091	0853	0695	1924	1294	1143
2213	1150	2060	2215	2461	2157	1144
2214	1269	2067	0061	0298	2158	1145
2215	1441	2065	0151	2457	2159	1146
2216	0249	2007	1688	2460	0080	1147
2217	1625	1243	1806	1023	1492	0215 1472
2218	1721	2057	0573	1976	1754	1278
2219	0139	1874	0160	1679	1426	2520
2220	0850	0684	1774	2120	0856	0751
2221	2164	0682	0836	2123	0851	0765
2222	0846	0683	1771	1636	0852	0766
2223	2087	0687	2183	0907	0855	0748
2224	0743	0688	0774	2122	0854	0749
2225	1453	0689	1582	2121	0853	0753
2226	2311	0685	2324	2119	1288	0750
2227	1914	1320	2055	1644	1760	2456
2228	1337	1324	1481	0187	1771	2465
2229	1353	2168	1502	1547	1770	2466
2230	2148	1321	0816	1826	1768	2453
2231	1048	1322	1075	1711	1769	1972 2454
2232	2171	1323	0838	1831	1761	2477
2233	2362	1325	0256	2390	1763	2482
2234	0885	1326	0911	0196	1765	2479
2235	1126	1327	1242	2484	1762	2481
2236	1236	1328	1356	2353	1764	2480
2237	0806	0700	0862	1825	1766	2478
2238	2176	2058	2325	1983	2135	0971
2239	1405	2059	1556	0591	2136	0972
2240	2178	1822	1357	0718	2384	1215
2241	0900	1824	0926	2577	2385	0871
2242	0303	1871	0322	1678	1310	0539
2243	2354	1870	2358	1063	1311	0540
2244	0293	0764	0280	0703	1041	1118
2245	1106	1826	1233	0704	1038	1114
2246	0638	0759	0687	0708	1037	1122
2247	1230	0761	1350	2425	1039	1115
2248	0855	0767	1206	0707	0367	1112
2249	1253	0762	1407	0340	1036	1120
2250	0055	0769	1019	0537	1035	1120
2251	1037	0763	1063	0706	1040	1121
2252	0664	0758	0725	0711	1365	1116
2253	2574	0768	2557	1932	1034	1113
2254	1389	0765	1541	0672	1042	1111
2255	1099	0757	0724	0785	0032	1855
2256	0801	0766	1940	0710	1043	1117
2257	1351	2169	1498	1546	1313	0542
2258	2101	1872	2250	1676	1312	0541
2259	2294	1868	1208	1675	1314	0538
2260	0031	2108	0028	2240	0070	1966
2261	0431	2113	1415	2242	0073	1949
2262	1356	2112	1504	2033	1954	1663

E	F	G	S	R	C	J
2263	1393	2109	1542	2241	0069	1946
2264	2608	2114	2599	2243	0068	1967
2265	2193	1761	1768	1240	2276	1055
2266	0802	1768	1766	0843	1874	1064
2267	0241	2084	1313	2008	0882	1140
2268	1437	2078	0145	1997	0884	1141
2269	0273	2085	0319	2002	0899	2135
2270	1204	2086	0515	1201	0897	2620
2271	1335	2079	1475	1995	0898	1133
2272	1357	2080	1505	1999	0889	1132
2273	1213	2087	1316	2319	1634	1136
2274	0703	2088	0743	1903	1633	1135
2275	1697	2018	1893	2006	0905	1139
2276	1479	2110	2141	1996	0886	2139
2277	1953	2111	2142	1998	0958	0744
2278	0696	2081	0665	1462	0888	2138
2279	1681	2082	1650	1375	0887	2140
2280	0613	2077	0649	0897	0901	2127
2281	1887	1935	1227	1605	0895	2124
2282	0108	2091	0102	2003	0890	2136
2283	0150	2092	0173	1233	0893	2123
2284	2341	2094	2346	2004	0636	2120
2285	1889	2095	1989	2005	0894	2122
2286	0787	2096	0812	2007	0892	2121
2287	1929	2093	2065	2001	0891	2125
2288	2222	2097	1121	2009	0900	1131
2289	2223	2098	1120	2010	0903	1134 2132
2290	1322	2100	1466	2000	0904	2133
2291	2224	2099	1119	2011	0885	2134
2292	2225	2101	1140	2012	0638	2130
2293	2099	0185	2248	1797	0639	2129
2294	2246	2102	1122	2015	0883	2137
2295	2476	2103	2509	2026	0637	2131
2296	2266	2104	1199	2017	0640	2128
2297	2267	2105	1175	2018	0902	2126
2298	2194	2106	1100	1987	0881	1132
2299	2209	2208	1111	1993	0758	2293
2300	1932	2116	1997	0259	1173	1124
2301	1802	2117	1433	1438	2404	1123
2302	2051	2118	2195	1938	2571	1129
2303	0962	2119	0984	1887	2559	0943
2304	1900	2386	2039	0177	2561	0944
2305	1163	2120	1279	2028	2560	0942
2306	2336	0697	1635	1700	0011	1142
2307	0295	2585	0282	1821	1575	2189
2308	0646	2121	0692	1817	0413 1573	1130
2309	2537	2581	2069	1818	1574	0522
2310	1939	2128	1249	1816	1182	1127
2311	2566	2129	2547	1936	1185	1126
2312	2068	2127	2225	0303	1183	1125
2313	0567	1881	0557	0753	0562	0551
2314	1815	1880	1971	0752	0563	0552
2315	0510	2133	0478	0839	2066	0040
2316	0520	2134	0498	0840	1225	0531
2317	0143	2132	0161	2051	2068	0039
2318	0177	2130	0198	2049	2067	0038
2319	0053	2131	1369	2050	2069	1487
2320	1824	2368	1970	1539	1222	0251
2321	1531	2137	1710	1058	0100	1975
2322	0829	2135	0757	2056	0094	1977
2323	0830	2136	0758	2057	1143	0602
2324	2159	0521	0831	2052	0096	1981

E	F	G	S	R	C	J
2325	2168	0522	0827	2053	0091	1980
2326	1653	1554	1837	2060	0092	1974
2327	1500	2139	1668	2055	0090	1719
2328	2156	2138	0825	2062	0089	1978
2329	2155	2140	0821	2061	0097	1978
2330	2306	2141	2341	2059	0099	1976
2331	1534	2142	1713	1057	0093	1972
2332	0917	2143	1051	2272	1926	0996
2333	0698	2144	2072	2065	1639	0665
2334	1458	2145	1608	0969	1640	0665
2335	1459	2146	1613	2066	1644	0238
2336	1783	2147	2077	2067	0229	0239
2337	2235	2151	1154	1006	1248	1164
2338	2589	2483	2582	0233	1249	1159
2339	1705	2148	1266	1196	0848	0757
2340	1221	0148	1340	2069	0849	0758
2341	2594	2487	2592	1253	0850	0759
2342	2019	2149	2123	2070	0844	1469
2343	1779	2152	1898	2068	2023	1637
2344	1083	2150	1210	2071	2071	1468
2345	1080	2235	2267	2088	0144	1104
2346	1271	2236	0063	0297	1963	0250
2347	2118	2237	2279	2087	1223	0248
2348	0993	2238	1011	2083	0642	1105
2349	0506	0953	0485	0838	2519	1378
2350	1178	2239	1301	2086	0635	0249
2351	1471	2153	1595	0542	1927	1103
2352	1797	2158	2084	2077	0993	0292
2353	2297	2157	1219	2078	0991	0291
2354	1177	2160	1300	2079	1928	1107
2355	1873	0308	1942	1913	1990	1290
2356	1478	2214	1641	1367	1491	2413
2357	2015	2215	2119	1366	2391	2414
2358	2298	2213	1221	1512	1012	1102
2359	2043	1968	2184	2082	1757	1108
2360	0104	1012	0115	0614	1758	1109
2361	2150	1969	0824	1827	1759	1803
2362	1369	2219	1890	0654	2226	1110
2363	2299	2240	2326	2089	1845	1016
2364	2300	2241	2327	2090	2247	0816
2365	0097	2243	0082	0051	2245	2041
2366	0078	2242	0094	2091	2244	0815
2367	1605	0141	1791	0257	2118	0466
2368	1508	2002	0360	0179	2592	1420
2369	0586	2004	0548	2200	0705	1423
2370	1867	2003	2008	1583	0706	1421
2371	2309	2390	2334	1159	0435	0334
2372	2165	2391	0835	1415	0927	0369
2373	2319	2392	2317	1414	0926	0368
2374	0662	2602	0716	0474	0776	2339
2375	0955	2601	0977	0473	0130	2338
2376	2309	2347	2323	1490	2396	0893
2377	0966	1272	0990	2316	2448	0897
2378	1751	2352	1920	1294	2447	0899
2379	2312	2244	2335	1484	2307	2494
2380	0109	2245	1640	1158	0199	1424
2381	2313	1566	2329	1473	0105	1983
2382	0079	1567	0095	0054	0107	1990
2383	0740	1572	0780	0419	0111	1985
2384	0467	1570	0442	1469	0109	1987
2385	0528	1573	0512	1471	0112	1046
2386	2038	1568	2175	1472	0108	1989
2387	2031	1571	2166	1470	0106	1988
2388	1090	1569	0715	1468	0110	1986

E	F	G	S	R	C	J
2389	2202	2246	1109	1990	1999	1615
2390	2205	2247	1110	1991	1998	1616
2391	0856	2248	2434	1916	2001	1635
2392	0878	2250	0898	2565	1638	1053
2393	2525	2259	2472	2224	0858	1466
2394	1927	2260	0893	1242	1630	1496
2395	0858	2261	2435	1986	0240	1444
2396	2331	2262	1135	2116	0239	1363
2397	0362	1616	0369	1369	0233	1347
2398	0282	1168	0336	1439	2544	1495
2399	0360	2263	0255	1099	0009	1982
2400	0823	2264	1515	2117	0946	1321
2401	2344	2265	1976	2118	1945	1322
2402	0886	2276	0910	2142	2065	0223
2403	0322	2277	0577	2395	0219	0226
2404	2102	2567	2252	0244	1856	0858
2405	1067	1262	1093	0243	1859	0855
2406	2295	2568	1209	0242	1860	0857
2407	2402	2278	2385	2143	2411	1567
2408	0230	2279	0260	0750	1953	1510
2409	0944	2451	0948	0716	1868	0867
2410	1798	2283	2089	2144	1410	2544
2411	1637	2280	1830	2147	1407	2542
2412	2609	2284	2601	2146	1408	2546
2413	0262	2281	0307	2145	1409	2545
2414	2552	2286	1061	2159	1724	1736
2415	0110	2287	0116	2155	1713	1751
2416	0426	2291	0416	1155	1751	1740
2417	0479	2510	0455	2153	1705	1746
2418	0727	2511	0772	0421	1706	1747
2419	0670	2293	0727	2158	1720	1749
2420	1042	2294	1069	2149	1732	1753
2421	1104	2292	1224	2160	1745	1742
2422	1164	2072	1281	2157	1750	1739
2423	1394	2288	1548	2154	1711	1722
2424	0153	2285	0190	0495	1707	1743
2425	1628	2289	1810	2151	1749	1738
2426	1912	2290	2053	2152	1718	1741
2427	2020	2513	2137	2148	1730	1754
2428	2186	2512	1365	2150	1742	1729
2429	2417	2295	2387	0032	1736	1723
2430	2361	2296	2363	2161	1714	1752
2431	2418	2323	2388	2162	1739	1748
2432	2034	2297	2169	2187	1719	1725
2433	2419	2298	2389	2168	1731	1755
2434	0521	2299	0499	2170	1726	1756
2435	1044	2301	1071	2171	1712	1758
2436	2384	2300	2383	2169	1728	1757
2437	2420	2302	2390	2188	1708	1744
2438	2421	2303	2394	2166	0177	0790
2439	2423	2304	2393	2167	2064	0789
2440	0117	2306	0124	2164	1721	1734
2441	0617	2307	0654	2163	1722	1735
2442	2424	2305	2396	2165	1748	1733
2443	2427	2308	2399	2172	1747	1761
2444	2428	2309	2400	2173	1704	1732
2445	0120	2310	0128	2176	1727	1727
2446	2429	2311	2401	2177	1725	1726
2447	1792	2312	2090	2174	0169	0225
2448	2478	2313	2525	2175	0168	0224
2449	2431	2314	2403	2179	1741	1730
2450	0121	2316	0123	2156	1740	1759
2451	2432	2317	2392	2186	1715	0791
2452	2433	2319	1917	2185	1709	1745

E	F	G	S	R	C	J
2453	2434	2315	2404	2181	1716	1731
2454	2435	2318	2405	2182	1744	1728
2455	2436	2320	2406	2183	1735	1723
2456	2437	2321	2391	2184	1717	1750
2457	0122	2322	0129	2180	1746	1737
2458	1024	1879	1203	2205	0558	1873
2459	1152	0484	1882	2206	0128	1801
2460	0922	0485	0955	2207	0129	1802
2461	0325	0486	0592	2208	0127	1798
2462	0480	0474	0461	2221	1788	0838
2463	1756	0475	1933	2222	1787	0839
2464	0246	0476	1686	2220	1786	0834
2465	1013	2015	1042	1564	2354	0862
2466	0532	2324	0518	2190	1290	0805
2467	1288	1511	1437	2238	0797	0954
2468	0511	2566	0490	1571	1855	0934
2469	2240	0701	1156	0252	1341	1930
2470	0605	2515	0650	0893	0297	2498
2471	2270	2563	1188	2025	1857	0860
2472	2355	2564	2359	1061	1858	0859
2473	2167	1364	2531	2192	0327	1503
2474	2439	2328	2486	2193	0321	1500
2475	0614	2333	0651	0898	0329	1501
2476	1058	2331	1082	1483	0330	1502
2477	2484	2327	2427	2191	0328	2518
2478	0123	1351	0131	2194	0325	1505
2479	1134	2335	1255	2284	0326	1504
2480	2486	2334	2610	2195	0324	1506
2481	0105	2338	0096	2178	1551	1721
2482	1729	0829	1903	1430	2581	1202
2483	0574	2339	0567	1502	2579	1194
2484	0909	0832	0921	1503	2337	0586
2485	2610	0834	2602	1504	2477	1242
2486	1890	0639	2034	1391	1952	1310
2487	2488	2124	2461	1955	0960	0778
2488	0081	2125	0097	1953	0961	1995
2489	0524	2126	0502	1954	0962	1994
2490	2490	2345	2462	0598	1865	1671
2491	1770	2396	1938	1407	2540	1586
2492	2498	2408	2441	1625	0082	2227
2493	0513	2353	0492	0829	0083	2229
2494	1367	2407	1516	0652	2536	2234
2495	1895	2382	2036	0174	2387	1195
2496	1897	2384	2038	0172	2133	0473
2497	1898	2387	2040	0176	2561	0945
2498	1899	2385	2041	0175	1882	0866
2499	2398	2388	2419	1417	2388	1196
2500	2397	0491	2420	1657	0928	1452
2501	2269	2354	1187	2024	2028	1459
2502	1557	2355	1733	1091	2029	1631 1512 1630
2503	1142	2356	1263	0884	2031	1633
2504	0770	2360	0803	0453	2030	0708 1494
2505	1985	2357	2205	1395	2027	1632
2506	2513	2358	2457	1670	2025	1634
2507	2515	2361	2459	2218	2390	2447
2508	1352	2362	1501	2219	1666	2448
2509	1572	2364	1747	1854	0966	1460
2510	1745	0004	0224	2247	0131	1668
2511	2386	2363	2421	0248	2303	0100
2512	1816	0472	1972	1541	1790	0824
2513	2583	0473	2583	0234	1789	0825

E	F	G	S	R	C	J
2514	1368	2154	2436	1401	0234	1546
2515	0014	2369	0011	2223	1792	0826
2516	2611	0497	0405	0717	2420	1093
2517	1021	1938	1765	1890	0861	0408
2518	2327	1389	2305	0583	1985	1650
2519	1667	1597	1850	0344	2171	1649
2520	1428	2549	1571	0948	2076	1349
2521	1429	2550	1572	0235	2075	1348
2522	0237	0534	0033	1168	1984	1648
2523	1794	2378	2091	2226	0077	1316
2524	2479	2379	2527	2227	0075	1315
2525	2526	2377	2473	2228	0076	1317
2526	1170	2380	1627	2229	0242	2516
2527	0859	2381	1294	2230	0241	2515
2528	1609	2612	1792	0394	1916	2042
2529	1293	2606	1442	0399	0504	1710
2530	0316	2604	0578	0395	1917	1698
2531	0688	0713	0667	1456	1996	1979
2532	2293	2397	1090	2095	2074	0125
2533	0762	1139	0794	1615	1127	0799
2534	2527	2398	2474	2250	0202	1426
2535	1611	2399	1794	2249	0201	1411
2536	0071	2400	0098	2251	0197	1425
2537	2284	2401	1194	2252	2552	0854
2538	0171	2412	0185	1188	2083	2407
2539	1252	2411	1405	0343	2082	2406
2540	0657	2421	0709	0472	2246	2400
2541	0663	0530	0717	1277	0276	1369
2542	1094	0531	0347	1276	0277	1368
2543	1641	1534	1815	1195	2063	2231
2544	2215	1538	1166	1203	0140	1964
2545	2216	1537	1167	1204	0141	1965
2546	2530	2416	1310	2311	0054	1817
2547	2553	0190	2544	1935	1840	2335
2548	0920	2425	0946	2503	2421	1795
2549	1726	2426	1338	0168	2422	0719
2550	1100	0760	0348	0167	2423	1067
2551	0973	2427	0996	0169	1156	0358
2552	2531	1352	2484	0562	1298	1948
2553	2538	2434	2534	0714	1661	1038
2554	1644	0381	1817	1378	1663	1017
2555	1412	0378	0356	1377	2435	1018
2556	1354	2173	1503	1380	1662	1019
2557	1774	0379	1217	1424	2436	2206
2558	2542	2428	2538	0442	0081	2141
2559	2541	2593	2537	1956	2024	1098
2560	0968	1913	0991	0189	2430	1092
2561	2502	2019	2446	1397	2427	1085
2562	2380	2020	2384	0188	2429	1086
2563	1799	2475	2092	0190	1515	2430
2564	1307	0677	1456	0191	1514	2429
2565	1198	2476	1379	2359	2072	2020
2566	1919	2477	1888	1690	2073	2021
2567	2576	2435	2563	2248	1008	0449
2568	2567	2438	2552	1940	1009	1041
2569	0034	2437	2565	0222	1010	1039
2570	2403	2073	2566	1161	0407	1520
2571	0271	2222	0317	0226	0408	1521
2572	2054	2075	2210	0533	0361	0007
2573	0124	2484	0130	0229	0723	2361
2574	0517	2485	0495	0228	0722	2360
2575	2580	2481	2577	0230	0721	2359
2576	2578	2221	2587	0225	0724	2357
2577	2579	2486	2588	0227	0725	2358

E	F	G	S	R	C	J
2578	0263	2488	0309	0508	1755	2470
2579	0395	2491	0392	1263	1973	2468
2580	0408	1908	2207	1341	1391	2469
2581	2556	2490	0335	1260	1974	1301
2582	0174	2509	0196	0494	0078	0367
2583	1273	0076	0059	0293	0120	1806
2584	0481	2535	0457	0855	0118	1813
2585	0795	2534	0847	0445	0116	1814
2586	1578	2536	1749	1111	0117	1815
2587	1632	2537	1829	0223	0121	1812
2588	1144	0826	1412	1952	1782	2502
2589	1904	2503	2045	1691	0213	1992
2590	0750	0253	0788	1296	2512	0988
2591	2466	2505	2512	0182	2513	0987
2592	2124	2538	2275	1553	2047	0123
2593	1448	0305	0143	2450	1344	1836
2594	1440	1279	0146	2456	0907	0709
2595	0940	2452	0961	1710	0830	0818
2596	1244	0157	1394	1709	0831	0817
2597	1430	2551	1574	0911	2160	0439
2598	2021	1952	2138	1869	0737	0159
2599	0068	1940	0093	1867	1828	0171
2600	1359	1942	1508	0656	1835	0167
2601	0771	1941	0800	0454	1837	0161
2602	2264	1943	1126	1871	1825	0168
2603	2290	1944	1178	1873	1826	0169
2604	1740	1945	1908	1431	1836	0160
2605	1179	1946	1299	1864	1838	0581
2606	1256	1953	1400	0345	1827	0170
2607	1257	1954	1401	0348	1824	1575
2608	1259	1955	1403	0347	1832	0688
2609	1258	1956	1402	0346	1839	2452
2610	1330	1951	1473	0650	1834	0162
2611	1526	1947	1695	1865	1823	0166
2612	2283	1948	1193	2347	1829	0164
						0172
2613	1724	1949	1336	1866	1931	0163
2614	2210	1950	1117	1868	1833	0165
2615	0552	2561	0529	0559	1784	1174
2616	0879	2560	0905	2562	1785	1173
2617	1801	1565	2071	1230	0123	0374
2618	1800	1564	2070	1231	0087	1794
2619	2612	2591	2613	0563	1802	1279

ADDENDA

E	F	G	S	R	C	J
A01	A03	A07	A10	A06	A03	A02
A02	A04	A01	A03	A01	A12	A04
A03	A05	A02	A04	A02	A11	A06
A04	A08	A03	A06	A03	A13	A05
A05	A02	A10	A02	A10	A04	A08
A06	A07	A04	A08	A08	A05	A12
A07	A01	A05	A01	A04	A07	A10
A08	A10	A06	A09	A05	A02	A13
A09	A06	A08	A05	A07	A01	A07
A10	A09	A09	A07	A09	A09	A03

French

0001	0235	"à blanc" : relatif au signal observable en l'absence de l'objet de la mesure
0002	1239	abondance isotopique
0003	1531	abondance isotopique naturelle
0004	1528	abondance naturelle
0005	0006	absorbance
0006	1153	absorbance interne
0007	0009	absorbeur
0008	0010	absorption
0009	0146	absorption atomique
0010	0017	absorption de particules
0011	0016	absorption de rayonnement gamma
0012	1123	absorption intégrale
0013	2210	absorption singulet-singulet
0014	2515	absorption triplet-triplet
0015	1999	absorption vraie
0016	0021	absorptivité
0017	1489	absorptivité molaire
0018	1305	accélérateur linéaire d'électrons
0019	0072	acide aminopolycarboxylique
0020	0027	acidimétrie
0021	0348	actinomètre chimique
0022	1257	action laser
0023	0032	activation
0024	0338	activation de particule chargée
0025	1555	activation par neutron
0026	1746	activation par photons
0027	0037	activité
0028	1490	activité molaire
0029	1486	activité molale
0030	2036	activité relative
0031	2260	activité spécifique
0032	0040	additifs
0033	0041	additivité
0034	2569	adjuvants de volatilisation
0035	1578	admittance non faradique
0036	0044	adsorbant
0037	0043	adsorbat
0038	0045	adsorption
0039	0049	aérosol
0040	0709	aérosol sec
0041	1945	age radioactif
0042	1389	agent de masquage
0043	2044	agents libérateurs
0044	2045	agents libérateurs
0045	0051	agglomération
0046	1760	agglomération
0047	0052	agrégat
0048	0053	agrégation
0049	1208	aiguille d'ionisation
0050	0057	alcalimétrie
0051	0058	alcaliser
0052	0459	alimentation continue
0053	2319	alimentations à courant stabilisé
0054	1073	allumage
0055	2250	allumage d'étincelles
0056	2189	allumage en série
0057	1660	allumage parallèle
0058	0075	ampérométrie
0059	0618	ampérométrie différentielle
0060	1874	amplificateur d'impulsions
0061	1314	amplificateur linéaire d'impulsion
0062	0225	amplificateur linéaire d'impulsion à effet de seuil
0063	0807	amplification
0064	0966	amplification à gaz
0065	1160	amplitude d'oscillation interne
0066	0065	amplitude de courant alternatif
0067	0064	amplitude de tension alternative
0068	2599	analyse à rayons X
0069	1556	analyse avec activation par neutron
0070	1247	analyse cinétique
0071	2536	analyse d'ultratrace
0072	0725	analyse de gaz effluent(s)
0073	1434	analyse de méso-trace
0074	1268	analyse de microéchantillon par laser
0075	1713	analyse de phase
0076	1683	analyse de pic
0077	1690	analyse de pic par ajustement
0078	2366	analyse de sous-traces
0079	2382	analyse de surface
0080	1676	analyse de taille des particules
0081	2488	analyse de trace
0082	0598	analyse dérivatographique
0083	0825	analyse du gaz dégagé
0084	0776	analyse élémentaire
0085	0456	analyse en flux continu
0086	1119	analyse instrumentale par activation

0087	1121	analyse instrumentale par activation neutronique
0088	1240	analyse isotopique
0089	1344	analyse locale
0090	1265	analyse locale par laser
0091	1576	analyse non destructive par activation
0092	0033	analyse par activation
0093	1588	analyse par activation nucléaire
0094	2122	analyse par balayage laser
0095	0361	analyse par chimiluminescence
0096	1236	analyse par dilution d'isotopes
0097	2365	analyse par dilution isotopique sous-stoechiométrique
0098	0909	analyse par fluorescence
0099	0797	analyse par fluorescence X à dispersion d'énergie
0100	0905	analyse par injection dans le flux
0101	1259	analyse par laser
0102	1719	analyse par phosphorescence
0103	0806	analyse par phosphorescence amplifiée
0104	2360	analyse par redissolution
0105	2481	analyse par tresse de torsion
0106	1896	analyse qualitative
0107	1898	analyse quantitative
0108	2282	analyse spectrochimique
0109	2380	analyse suprathermale
0110	2415	analyse thermique
0111	0595	analyse thermique dérivée
0112	0629	analyse thermique différentielle
0113	0630	analyse thermique différentielle en environnement isotherme
0114	1899	analyse thermique différentielle quantitative
0115	2201	analyse thermique différentielle simultanée
0116	0787	analyse thermique par émanation (radioactive)
0117	2440	analyse thermogravimétrique
0118	0632	analyse thermogravimétrique différentielle
0119	0716	analyse thermogravimétrique dynamique
0120	2445	analyse thermomécanique
0121	2450	analyse thermoparticulaire
0122	2457	analyse thermovaporimétrique
0123	2478	analyse titrimétrique
0124	2573	analyse voltampérométrique
0125	1914	analyseur à champ électrostatique radial
0126	1763	analyseur à champ magnétique sur π radians
0127	1764	analyseur à champ magnétique sur π/2 radians
0128	1765	analyseur à champ magnétique sur π/3 radians
0129	1875	analyseur d'amplitude d'impulsions
0130	1882	analyseur de hauteur d'impulsion
0131	2204	analyseur de hauteurs d'impulsion à canal unique
0132	1895	analyseur de masse à champ quadrupolaire
0133	1518	analyseur multicanaux
0134	1519	analyseur multicanaux de hauteurs d'impulsions
0135	0427	angle Compton
0136	0256	angle de Bragg
0137	0100	angle de divergence
0138	0242	angle échelette
0139	2219	angle solide
0140	0419	anion complexe
0141	1496	anion moléculaire
0142	0107	anisotropie
0143	2317	anneaux de stabilisation
0144	0109	annihilation
0145	1667	annihilation de particules
0146	0108	annihiler
0147	0111	anode annulaire
0148	0121	antiparticule
0149	0612	antivolatilisant
0150	2283	applications spectrochimiques
0151	0131	arc
0152	1072	arc à allumage par courant alternatif
0153	2424	arc à allumage thermique
0154	0939	arc à combustion libre
0155	0647	arc à courant continu
0156	1030	arc à courant élevé
0157	1357	arc à courant faible
0158	1425	arc à courant moyen
0159	0306	arc à distillation du vecteur
0160	1041	arc à haute pression
0161	0446	arc à température constante
0162	1438	arc à vapeur de métal
0163	0296	arc capillaire
0164	0465	arc contrôlé en courant alternatif
0165	0316	arc de couche cathodique
0166	0691	arc double
0167	0733	arc électrique
0168	0066	arc électrique alternatif
0169	0132	arc en atmosphère contrôlée
0170	0994	arc en globule
0171	2538	arc non contrôlé, à courant alternatif
0172	2010	arc rectifié à courant alternatif
0173	0439	arc renforcé
0174	2582	arc stabilisé par paroi
0175	2154	arcs ensemencés
0176	1164	arcs interrompus
0177	2318	arcs stabilisés
0178	0977	arcs stabilisés par écoulement de gaz
0179	1379	arcs stabilisés par magnétisme
0180	0275	arcs tamponnés
0181	0248	arrangement de type Boersma
0182	0289	arrangement de type Calvet
0183	0142	aspirateur

0184	0140	aspiration
0185	0139	aspirer
0186	0466	atmosphère contrôlée
0187	0656	atmosphère de décharge
0188	1103	atmosphère inerte de décharge
0189	1062	atome chaud
0190	0082	atomes d'élément analysé
0191	0161	atomisation
0192	0164	atomiseur
0193	0302	atomiseur à creuset de carbone
0194	0999	atomiseur à creuset de graphite
0195	0303	atomiseur à filament de carbone
0196	1435	atomiseur à filament métallique
0197	1261	atomiseurs à laser
0198	0165	atténuation
0199	0987	atténuation géométrique
0200	2097	augmentation de solubilité par addition d'un sel, inverse de relargage
0201	2159	auto-absorption
0202	2166	auto-blindage
0203	2163	auto-électrode
0204	2165	auto-inversion (d'absorption)
0205	0174	autoionisation
0206	0175	automate
0207	0176	automation
0208	0177	automatisation
0209	0178	automatiser
0210	0180	autooxydation
0211	0179	autoradiographie
0212	0203	balance
0213	0085	balance analytique (balance de précision)
0214	1448	balance microchimique
0215	0918	balayé (par un courant)
0216	0782	bande d'élution
0217	0463	bande de contrôle
0218	1497	bandes moléculaires
0219	0549	bandes moléculaires du cyanogène
0220	0206	barn
0221	0207	base
0222	1685	base de pic
0223	1857	base du prisme
0224	0212	basicité
0225	0215	becquerel
0226	0223	biampérométrie
0227	0232	bipotentiométrie
0228	0245	bloc porte-échantillons
0229	1098	bobine d'induction
0230	2408	bobine Tesla
0231	0251	bolomètre
0232	0148	bombardement atomique
0233	0873	bombardement par atome rapide
0234	1183	bombardement par ions
0235	1668	bombardement par particules
0236	2111	boucle (circuit) d'échantillonnage
0237	2522	bras de réactance ajustable
0238	0236	bruit de fond
0239	0196	bruit de fond de rayonnement
0240	0195	bruit de fond du spectre de masse
0241	2267	bruit de fond spectral
0242	0193	bruit de fond, fond continu
0243	0652	brûleur à injection directe
0244	2083	brûleur à injection directe inverse
0245	1839	brûleur à pré-mélange
0246	2464	brûleur à trois fentes
0247	0279	brûleur Bunsen
0248	1429	brûleur Méker
0249	2216	brûleurs à fente
0250	1524	brûleurs multi-fentes
0251	1468	brume, bruine
0252	0288	calorimétrie
0253	1020	calorimétrie de flux thermique à balayage différentiel
0254	0628	calorimétrie différentielle à balayage
0255	1816	calorimétrie différentielle à balayage à compensation de puissance
0256	0713	calorimétrie différentielle dynamique
0257	0293	capacité
0258	1514	capacité de monocouche
0259	0264	capacité de percée
0260	0217	capacité du volume du lit
0261	1821	capacité spécifique pratique
0262	2413	capacité spécifique théorique
0263	2578	capacité volumique
0264	0395	captage
0265	0396	capteur, collecteur
0266	0299	capture
0267	0753	capture d'électrons
0268	1557	capture de neutron
0269	0960	capture de rayonnement gamma
0270	1938	capture radiative
0271	2571	caractéristiques courant-tension
0272	0527	caractéristiques de cristal
0273	2269	caractéristiques spectrales
0274	0957	cascade gamma
0275	0311	cataphorèse
0276	0252	cathode creuse à décharge amplifiée
0277	1619	cathode creuse à extrémités ouvertes
0278	1039	cathode creuse à intensité élevée
0279	1064	cathode creuse chaude
0280	0393	cathode creuse froide
0281	1498	cation moléculaire
0282	2398	cavité conique
0283	1886	cavité de pompage
0284	2068	cavité résonante
0285	1063	cellule chaude
0286	1246	cellule de Kerr
0287	1785	cellule de Pockels
0288	1623	cellule opérationnelle de mesure du pH
0289	0325	cémentation
0290	0569	chaîne de désintégration

0291	1946	chaîne radioactive (chaîne de décomposition)
0292	0402	chaînes de collisions
0293	2244	chambre à étincelles
0294	1203	chambre d'ionisation
0295	2307	chambre de formation d'aérosol
0296	2116	chambre de saturation
0297	0378	chambre propre
0298	1381	champ magnétique
0299	1056	champ magnétique homogène
0300	0524	champs électrique et magnétique croisés
0301	1579	champs magnétiques inhomogènes
0302	1342	charge
0303	2242	charge d'espace
0304	0359	chimi-ionisation
0305	1060	chimie molécule hôte-molécule réceptrice
0306	1589	chimie nucléaire
0307	1966	chimie radioanalytique
0308	1928	chimie sous rayonnement
0309	0360	chimiluminescence
0310	0363	chimisorption
0311	0365	chromatogramme
0312	0619	chromatogramme différentiel
0313	1124	chromatogrammes intégraux
0314	0366	chromatographe
0315	0368	chromatographie
0316	2530	chromatographie à deux dimensions
0317	0906	chromatographie à écoulement programmé
0318	0879	chromatographie à filament
0319	0967	chromatographie à gaz
0320	0985	chromatographie à perméation de gel
0321	2084	chromatographie à phases inversées
0322	2403	chromatographie à température programmée
0323	1621	chromatographie à tube ouvert
0324	0620	chromatographie différentielle
0325	2461	chromatographie en couche mince
0326	0946	chromatographie frontale
0327	0972	chromatographie gaz-liquide
0328	0976	chromatographie gaz-solide
0329	1184	chromatographie ionique
0330	1323	chromatographie liquide
0331	1325	chromatographie liquide-gel
0332	1329	chromatographie liquide-liquide
0333	1339	chromatographie liquide-solide
0334	0046	chromatographie par absorption
0335	0670	chromatographie par déplacement
0336	1188	chromatographie par échange d'ion
0337	0783	chromatographie par élution
0338	1678	chromatographie par partition
0339	1703	chromatographie par perméation
0340	2099	chromatographie par relargage
0341	0409	chromatographie sur colonne
0342	1659	chromatographie sur papier
0343	0367	chromatographier
0344	0369	chronoampérométrie
0345	0697	chronoampérométrie à double saut potentiostatique
0346	0371	chronocoulométrie
0347	0698	chronocoulométrie à double saut potentiostatique
0348	0482	chronocoulométrie par convection
0349	1808	chronocoulométrie potentiostatique
0350	0375	chronopotentiométrie
0351	0543	chronopotentiométrie à cessation de courant
0352	0067	chronopotentiométrie à courant alternatif
0353	0545	chronopotentiométrie à courant inversé
0354	1863	chronopotentiométrie à courant programmé
0355	0070	chronopotentiométrie à tension alternative
0356	0550	chronopotentiométrie cyclique
0357	0551	chronopotentiométrie cyclique à courant variant en fonction échelon
0358	0588	chronopotentiométrie dérivée
0359	0483	chronopotentiométrie par convection
0360	2399	cible
0361	0389	circuit à coincidences
0362	2397	circuit d'accord
0363	1074	circuit d'allumage
0364	1077	circuit d'allumage
0365	0343	circuit de charge
0366	0657	circuit de décharge
0367	2120	circuit démultiplicateur
0368	1377	circuit oscillant , circuit d'accord
0369	1662	circuit paralysant
0370	0449	classification des constituants
0371	2108	classification pondérale d'échantillons
0372	0392	co-ions
0373	0490	co-précipitation
0374	0382	coagulation
0375	0060	coefficient α
0376	0011	coefficient d'absorption
0377	0038	coefficient d'activité
0378	1192	coefficient d'activité ionique
0379	1491	coefficient d'activité molaire
0380	1487	coefficient d'activité molale
0381	0166	coefficient d'atténuation
0382	1390	coefficient d'atténuation de masse
0383	1657	coefficient d'atténuation de production de paires
0384	1302	coefficient d'atténuation linéaire
0385	1726	coefficient d'atténuation photoélectrique
0386	0728	coefficient d'Einstein en émission spontanée
0387	0860	coefficient d'extraction

0388	1204	coefficient d'ionisation
0389	0295	coefficient de capacité
0390	1155	coefficient de conversion interne
0391	0638	coefficient de diffusion
0392	0679	coefficient de distribution
0393	1055	coefficient de distribution homogène
0394	1352	coefficient de distribution logarithmique
0395	2579	coefficient de distribution volumique
0396	0866	coefficient de Fano
0397	1679	coefficient de partition
0398	2008	coefficient de récupération
0399	1126	coefficient de réflexion intégral
0400	2158	coefficient de sélectivité
0401	1810	coefficient de sélectivité potentiométrique
0402	0384	coefficient de variation
0403	1310	coefficient linéaire d'absorption photoélectrique
0404	1315	coefficient linéaire de diffusion
0405	1688	coefficient maximum de diffraction
0406	1408	coefficient moyen d'activité
0407	1410	coefficient moyen d'activité ionique
0408	2580	coefficient volumique de gonflement des échangeurs d'ions
0409	1987	coincidence aléatoire
0410	1866	coincidence instantanée
0411	0556	collecteur cylindrique
0412	0871	collecteur faradique en forme de creuset
0413	0398	collimateur
0414	0397	collimation
0415	0400	collision
0416	0731	collision élastique
0417	1101	collision non élastique
0418	0227	collisions binaires
0419	1358	collisions de faible énergie
0420	1139	collisions interatomiques
0421	1973	colloide radioactif
0422	0408	colonne
0423	0133	colonne d'arc
0424	1801	colonne positive (de lumière)
0425	1655	colonne remplie
0426	2416	colonne thermique
0427	1622	colonne tubulaire ouverte
0428	0949	combustible
0429	1593	combustible nucléaire
0430	0414	combustion
0431	2261	combustion nucléaire spécifique
0432	1845	combustion primaire
0433	2141	combustion secondaire
0434	2113	commutateur à colorant saturable
0435	1892	commutateur Q
0436	2173	commuté semi-Q
0437	0415	comparateur (de colorations)
0438	0420	complexation
0439	0106	complexe anionique
0440	1472	complexes à ligands mixtes
0441	1287	complexes à ligands pontants
0442	1515	complexes binaires mononucléaires
0443	0230	complexes binucléaires
0444	1473	complexes hétérométalliques
0445	1469	complexes mixtes
0446	1797	complexes polynucléaires
0447	1516	complexes ternaires mononucléaires
0448	0423	complexométrie
0449	0418	complexone
0450	0424	complexone
0451	0425	composant
0452	1855	composé primaire
0453	1663	composés paramagnétiques
0454	0390	comptage à coincidence
0455	0002	comptage absolu
0456	2039	comptage relatif
0457	0505	compte
0458	1337	compteur à scintillateur liquide
0459	2131	compteur à scintillation
0460	1929	compteur de rayonnement
0461	1901	compteurs de quanta
0462	0431	concentration
0463	0125	concentration apparente
0464	0334	concentration caractéristique
0465	0039	concentration d'activité
0466	0277	concentration globale (concentration totale initiale)
0467	2384	concentration superficielle
0468	1399	condensateur d'adaptation
0469	1036	condensateur de coupure haute fréquence
0470	0845	condition exoénergétique
0471	0426	conditions (de mesure) moyennées
0472	1636	conditions optima d'étincelle
0473	1626	conductance optique
0474	0442	conductimétrie
0475	0649	conductimétrie à courant continu
0476	1035	conductimétrie à haute fréquence
0477	0440	conductivité
0478	0734	conductivité électrique
0479	2417	conductivité thermique
0480	2462	configuration à trois électrodes
0481	2584	configuration de longueur d'onde
0482	1661	configuration de miroirs parallèle-parallèle
0483	1467	configuration des miroirs
0484	1786	configuration pointe-plan
0485	1787	configuration pointe-pointe
0486	0444	constante
0487	0370	constante chronocoulométrique
0488	0372	constante chronopotentiométrique
0489	0535	constante cumulative de formation
0490	0536	constante cumulative de protonation
0491	0537	constante cumulative de stabilité
0492	0028	constante d'acidité
0493	0558	constante d'amortissement

0494	0817	constante d'équilibre
0495	0434	constante d'équilibre de concentration
0496	0682	constante d'équilibre de distribution
0497	0861	constante d'extraction
0498	0640	constante de courant de diffusion
0499	1665	constante de décomposition partielle
0500	0570	constante de désintégration
0501	0665	constante de désintégration
0502	0672	constante de dissociation
0503	0680	constante de distribution
0504	0870	constante de Faraday
0505	0934	constante de formation
0506	2349	constante de formation partielle
0507	1680	constante de partition
0508	1870	constante de protonation
0509	2069	constante de réponse
0510	2315	constante de stabilité
0511	2468	constante de temps
0512	0346	constante de temps de charge
0513	2493	constante de transformation
0514	1398	constante de vitesse d'électrolyse contrôlée par transport de masse
0515	0614	constante diélectrique
0516	1650	constante globale de formation
0517	2574	constante voltampérométrique
0518	0432	constantes de concentrations
0519	1649	constantes de distribution globales
0520	2316	constantes de stabilité de complexes métal-ion
0521	2434	constantes thermodynamiques
0522	0448	constituant
0523	1466	constituant mineur
0524	2489	constituant trace
0525	1386	constituants principaux
0526	0454	contamination d'un précipité
0527	0453	contamination par piégeage
0528	2385	contamination superficielle
0529	0450	contenu en constituants
0530	0462	continuum
0531	0507	contre-ions
0532	2466	contrôle par thyristor
0533	0923	contrôleur de flux
0534	0481	convection
0535	0213	convention de Bates-Guggenheim
0536	1078	convention de signe de l'équation d'Ilkovič
0537	0317	convention de signe de processus cathodique
0538	2197	conventions de signes
0539	1154	conversion interne
0540	0078	convertisseur analogique/digital
0541	1081	convertisseur d'impédance
0542	0641	convertisseur digital-analogique
0543	1083	corps incandescents
0544	0233	corps noir
0545	0548	correcteur de courbe
0546	1092	correction d'indicateur
0547	0194	correction de bruit de fond
0548	0237	correction de bruit de fond
0549	0501	correction de pesées directes
0550	2062	correction de temps de résolution
0551	0565	correction de temps mort
0552	2615	correction Zeeman de bruit de fond
0553	1130	correspondance d'intensités
0554	0997	couche à gradient
0555	1996	couche de réaction
0556	1375	couche lumineuse
0557	0504	coulométrie
0558	0467	coulométrie à courant contrôlé
0559	0474	coulométrie à potentiel contrôlé
0560	0706	coulométrie sur électrode à gouttes tombantes
0561	0579	coupelle de Delves
0562	2048	couplé à distance
0563	0059	coupure en α
0564	0218	coupure en β
0565	0541	courant
0566	1116	courant (électrique) instantané
0567	2313	courant alternatif à impulsions carrées
0568	0294	courant capacitif
0569	0310	courant catalytique
0570	1248	courant cinétique
0571	0445	courant constant
0572	0646	courant continu
0573	0047	courant d'adsorption
0574	2483	courant d'ions total
0575	0559	courant d'obscurité
0576	0344	courant de charge
0577	1687	courant de crête
0578	1695	courant de crête d'étincelle
0579	0658	courant de décharge
0580	0868	courant de démodulation faradique
0581	0639	courant de diffusion
0582	0696	courant de double couche
0583	1462	courant de migration
0584	0869	courant de rectification faradique
0585	2034	courant de régénération
0586	2369	courant du maximum
0587	0740	courant électrique
0588	0867	courant faradique
0589	1104	courant initial
0590	1185	courant ionique
0591	1253	courant laminaire
0592	1293	courant limite
0593	1292	courant limite catalytique
0594	1295	courant limite cinétique
0595	1291	courant limite d'adsorption
0596	1294	courant limite de diffusion
0597	1296	courant limite de migration
0598	1568	courant linéaire théorique
0599	1550	courant net faradique
0600	1569	courant non additif
0601	1701	courant périodique
0602	2051	courant résiduel
0603	0088	courbe analytique

0604 0547 courbe courant-temps
0605 2470 courbe d'attente
0606 1205 courbe d'efficacité d'ionisation
0607 0784 courbe d'élution
0608 0287 courbe d'étalonnage
0609 0791 courbe d'étalonnage d'émulsion
0610 1007 courbe de croissance d'activité
0611 0571 courbe de désintégration (de décroissance)
0612 1065 courbe de Hurter et Driffield
0613 2280 courbe de réponse spectrale
0614 2475 courbe de titrage
0615 1317 courbe de titrage linéaire
0616 1353 courbe de titrage logarithmique
0617 2441 courbe thermogravimétrique
0618 0092 courbes d'évaluation analytiques
0619 1145 courbes d'interférence
0620 1022 courbes de chauffage
0621 1176 courbes de l'inverse de la vitesse de refroidissement
0622 0488 courbes de refroidissement
0623 0489 courbes de vitesse de refroidissement
0624 0518 cratère
0625 2110 creuset d'échantillonnage
0626 1470 cristal mixte
0627 0540 curie
0628 0711 cycle de charge (cycle de travail)
0629 0950 cycle du combustible nucléaire
0630 0554 cyclotron
0631 1948 datation radioactive
0632 0228 de liaison
0633 1117 débit instantané
0634 0655 décharge
0635 0522 décharge à amortissement critique
0636 1051 décharge à cathode creuse
0637 0853 décharge à étincelle à allumage externe
0638 2246 décharge à étincelles
0639 1382 décharge à luminescence par champ magnétique
0640 1651 décharge amortie en régime supercritique
0641 0181 décharge auxiliaire
0642 2164 décharge d'étincelle auto-allumée
0643 0735 décharge électrique
0644 1363 décharge électrique à basse pression
0645 2190 décharge électrique protégée
0646 2308 décharge en aérosol (décharge diffuse)
0647 0270 décharge en aigrette
0648 2193 décharge en court-circuit
0649 0136 décharge en forme d'arc
0650 0995 décharge luminescente
0651 0451 décharge luminescente à pincement d'enceinte
0652 0254 décharge luminescente à sortie amplifiée
0653 0253 décharge luminescente amplifiée, à champ magnétique
0654 0001 décharge luminescente anormale
0655 2074 décharge luminescente limitée
0656 1584 décharge luminescente normale
0657 2540 décharge oscillante à amortissement sous-critique
0658 0648 décharge oscillante à polarisation par courant continu
0659 1846 décharge primaire
0660 2142 décharge secondaire
0661 0661 décharge sous pression réduite
0662 2374 décharge supplémentaire
0663 2541 décharge unidirectionnelle
0664 2252 décharges de type étincelle
0665 0968 décharges gazeuses
0666 1964 déchet radioactif
0667 1620 décomposition
0668 1442 décomposition métastable
0669 1949 décomposition radioactive
0670 2419 décomposition thermique
0671 1303 décrément linéaire
0672 1351 décrément logarithmique
0673 0850 décroissance exponentielle
0674 1590 décroissance nucléaire, désintégration nucléaire
0675 0575 degré (avancement) de titrage
0676 0574 degré d'ionisation
0677 0573 degré de dissociation
0678 0580 démasquage
0679 1013 demi-épaisseur
0680 0405 demi-largeur de collision
0681 0687 demi-largeur Doppler d'une raie spectrale
0682 1011 demi-vie
0683 1953 demi-vie de radioactivité
0684 2119 démultiplicateur
0685 1919 densité d'énergie rayonnante
0686 1802 densité de courant de la colonne positive
0687 0920 densité de flux
0688 2531 densité de flux à 2200 mètres par seconde
0689 0798 densité de flux d'énergie
0690 1560 densité de flux de neutrons
0691 1671 densité de flux de particules
0692 1751 densité de flux de photons
0693 1558 densité de neutrons
0694 1627 densité optique
0695 2040 densité relative
0696 2278 densité spectrale d'énergie radiante
0697 0583 densitomètre
0698 2333 départ, début, démarrage
0699 0407 déplacement (spectral) dû aux collisions
0700 0214 déplacement bathochrome
0701 0357 déplacement chimique
0702 1321 déplacement de raie
0703 2274 déplacement de raie spectrale

0704	0688	déplacement Doppler
0705	1068	déplacement hyperchromique
0706	1069	déplacement hypochromique
0707	1070	déplacement hypsochromique
0708	1412	déplacement linéaire moyen
0709	1413	déplacement moyen en masse
0710	0247	déplacement vers le bleu
0711	2016	déplacement vers le rouge
0712	0584	dépolarisant
0713	0921	dépression de flux
0714	0599	dérivatographie
0715	0587	dérivé
0716	0702	dérive
0717	0404	desexcitation par collision
0718	1939	désexcitation radiative
0719	0568	désintégration (nucléaire) , décroissance (rayonnement)
0720	0061	désintégration alpha
0721	0219	désintégration bêta
0722	0258	désintégration multiple, désintégration à embranchement
0723	0602	désolvatation
0724	0606	détecteur
0725	0645	détecteur à batterie de diodes
0726	0754	détecteur à capture d'électrons
0727	2418	détecteur à conductivité thermique
0728	0932	détecteur à feuille
0729	1338	détecteur à scintillateur liquide
0730	2132	détecteur à scintillation
0731	2169	détecteur à semi-conducteur
0732	0326	détecteur Cerenkov
0733	1669	détecteur de particules
0734	1930	détecteur de rayonnement
0735	1603	détecteur de trace nucléaire
0736	0621	détecteur différentiel
0737	1125	détecteur intégral
0738	0974	détecteur proportionnel à gaz
0739	2135	détecteur scintillateur
0740	2383	détecteur semi-conducteur à barrière de surface
0741	0636	détecteur semi-conducteur à jonction diffuse
0742	1762	détecteur semiconducteur P.I.N.
0743	2224	détecteur solide
0744	1891	détecteurs pyroélectriques
0745	0603	détection
0746	0073	détection ampérométrique de fin de titrage
0747	0221	détection biampérométrique du point de fin de titrage
0748	0726	détection de gaz effluent(s)
0749	0826	détection du gaz dégagé
0750	2590	détermination de changement de poids
0751	1021	détermination de la courbe de chauffe
0752	1232	détermination de la variation de poids dans des conditions isothermes
0753	1226	détermination isobare des changements de masse
0754	1545	détermination néphélométrique du point de fin de titrage
0755	0610	développer
0756	0611	déviation
0757	1380	déviation magnétique
0758	2043	déviation relative standard
0759	0123	diamètre d'ouverture du diaphragme
0760	0928	diamètre de tache focale
0761	0520	diamètre du cratère
0762	2533	diamètre limite de la tache focale
0763	0613	diatomique
0764	0616	diélectrométrie
0765	0617	différentiel
0766	0529	diffraction cristalline
0767	0755	diffraction d'électron
0768	2155	diffraction d'électrons à surface sélectionnée
0769	1359	diffraction d'électrons de faible énergie
0770	2504	diffraction d'électrons transmis à haute énergie
0771	2601	diffraction des rayons X
0772	2030	diffraction-réflexion d'électrons de haute énergie
0773	0637	diffusion
0774	2125	diffusion (diffraction)
0775	0430	diffusion Compton
0776	1460	diffusion de Mie
0777	2126	diffusion de rayonnement
0778	0732	diffusion élastique
0779	1102	diffusion non élastique
0780	1985	diffusion Raman
0781	1993	diffusion Rayleigh
0782	2095	diffusion rétrograde de Rutherford
0783	0642	dilatométrie
0784	0589	dilatométrie dérivée
0785	0622	dilatométrie différentielle
0786	0643	diluant
0787	2286	diluant spectrochimique
0788	1235	dilution d'isotopes
0789	0664	discriminateur
0790	0667	dispersion
0791	0668	dispersion (en analyse par injection dans le flux)
0792	0101	dispersion angulaire
0793	0796	dispersion d'énergie
0794	0669	dispersion d'un matériau
0795	2585	dispersion de longueur d'onde
0796	0239	dispersion du blanc
0797	1304	dispersion linéaire
0798	2004	dispersion linéaire inverse
0799	1355	dispositif à tube long
0800	1132	dispositif d'étalonnage d'intensité
0801	2256	dispositif de production d'étincelles
0802	2266	dispositif porte spécimen
0803	1079	dispositifs à génération d'images

0804	0671	dissociation
0805	0675	dissolution
0806	2237	distance de migration du solvant
0807	0864	distance extrapolée
0808	0926	distance focale
0809	0676	distillation
0810	0936	distillation fractionnée
0811	0678	distribution
0812	0533	distribution cumulative
0813	1876	distribution d'amplitude d'impulsions
0814	1392	distribution de masse
0815	1677	distribution de taille des particules
0816	1330	distribution liquide-liquide
0817	1988	domaine, gamme (de résultats)
0818	0689	dose
0819	0007	dose absorbée
0820	0350	dosimètre chimique
0821	0470	durée d'étincelle contrôlée
0822	1880	durée d'impulsion
0823	2400	durée de l'échantillonnage
0824	2050	durée de séjour
0825	0491	durées corrigées de déclin de la fluorescence
0826	0492	durées corrigées de déclin de la phosphorescence
0827	0576	eau dé-ionisée
0828	0581	eau déminéralisée
0829	2322	écart type
0830	2323	écart type de comptage
0831	0104	échange d'anion
0832	1187	échange d'ion
0833	1237	échange d'isotope
0834	0323	échange de cation
0835	0340	échange de charge
0836	0356	échange isotopique chimique
0837	0226	échangeur bifonctionnel
0838	0105	échangeur d'anion
0839	1512	échangeur d'ion monofonctionnel
0840	1191	échangeur d'ions
0841	1796	échangeur d'ions polyfonctionnel
0842	0324	échangeur de cation
0843	2013	échangeurs d'ions rédox
0844	2100	échantillon
0845	1814	échantillon de poudre
0846	2222	échantillon de solution solide
0847	2106	échantillon de taille limitée
0848	0278	échantillon en volume
0849	1335	échantillon liquide
0850	2220	échantillon solide
0851	0097	échantillons d'analyse
0852	0801	échelle d'énergie
0853	1710	échelle opérationnelle de pH
0854	1624	échelle opérationnelle des pH
0855	2248	éclateur
0856	2391	éclateur à rotation synchrone
0857	0479	éclateur de contrôle
0858	2395	éclateur tandem
0859	2527	écoulement (courant) turbulent
0860	1864	écran de projection
0861	2192	écrans, protections
0862	2137	écranté
0863	0718	effet
0864	0167	effet Auger
0865	0327	effet Cerenkov
0866	0428	effet Compton
0867	0103	effet d'anion
0868	1024	effet d'atome lourd
0869	0322	effet de cation
0870	0280	effet de combustion différenciée
0871	1761	effet de pincement
0872	1894	effet électro-optique quadratique
0873	1306	effet linéaire électrooptique
0874	1517	effet Mössbauer
0875	1638	effet organique
0876	1383	effet Penning magnétique
0877	1727	effet photoélectrique
0878	2392	effet Szilard-Chalmers
0879	2616	effet Zeeman
0880	0852	effets d'atome lourd externe
0881	2161	effets d'auto-absorption
0882	1401	effets de matrice
0883	1842	effets de pression
0884	2032	effets de réfraction
0885	2234	effets de solvant
0886	2402	effets de température
0887	1142	effets interéléments
0888	1805	effets post-filtre(s)
0889	1837	effets pré-filtres
0890	0724	efficacité
0891	0004	efficacité absolue de comptage dans le pic photoélectrique
0892	2038	efficacité biologique relative
0893	0003	efficacité absolue de comptage dans le pic d'énergie totale absorbée
0894	0509	efficacité de comptage
0895	0954	efficacité de comptage du pic d'énergie totale
0896	0516	efficacité de couplage
0897	0604	efficacité de détection
0898	0910	efficacité de fluorescence
0899	1535	efficacité de nébulisation
0900	2241	efficacité de source
0901	0608	efficacité du détecteur
0902	1173	efficacité intrinsèque
0903	1172	efficacité intrinsèque de détecteur
0904	1174	efficacité intrinsèque du pic d'énergie totale
0905	1175	efficacité intrinsèque du pic photoélectrique
0906	1345	efficacité locale d'atomisation
0907	1372	efficacité lumineuse
0908	1373	efficacité lumineuse
0909	2484	efficacité quantique totale de fluorescence
0910	1950	effluent radioactif
0911	0269	élargissement

0912	1356	élargissement de Lorentz
0913	0686	élargissement Doppler
0914	0401	élargissement dû aux collisions
0915	1529	élargissement naturel
0916	2160	élargissement par auto-absorption
0917	2332	élargissement par effet Stark
0918	2063	élargissement par résonance
0919	0747	électro-érosion
0920	2548	électrode à creuset à succion (à dépression)
0921	0814	électrode à enzyme
0922	2460	électrode à film mince
0923	1067	électrode à gaz hydrogène
0924	0707	électrode à gouttes de mercure tombantes
0925	2085	électrode à matrice rigide
0926	1029	électrode à membrane hétérogène
0927	1057	électrode à membrane homogène
0928	1431	électrode à nappe de mercure
0929	1475	électrode à porteur mobile
0930	2196	électrode à tamis
0931	0182	électrode auxiliaire
0932	0506	électrode auxiliaire
0933	0189	électrode axiale
0934	0413	électrode combinée
0935	1054	électrode creuse
0936	0090	électrode d'analyse
0937	2102	électrode d'échantillon
0938	2018	électrode de référence
0939	1161	électrode de référence interne dans une électrode de verre
0940	2595	électrode de travail
0941	0992	électrode de verre
0942	0539	électrode en forme de cupule
0943	1093	électrode indicatrice
0944	2409	électrode indicatrice
0945	1706	électrode indicatrice de pH
0946	1574	électrode non cristalline
0947	0703	électrode percée
0948	1848	électrode primaire iono-spécifique
0949	0532	électrode spécifique cristalline
0950	0307	électrode vecteur
0951	1222	électrodes à réponse spécifique aux ions
0952	0297	électrodes capillaires
0953	1059	électrodes horizontales
0954	1220	électrodes ion-sélectives
0955	2375	électrodes supplémentaires
0956	1799	électrodes-creusets poreuses
0957	1800	électrodes-creusets poreuses
0958	0748	électrographie
0959	0749	électrogravimétrie
0960	0475	électrogravimétrie à potentiel contrôlé
0961	1156	électrogravimétrie interne
0962	2303	électrogravimétrie spontanée
0963	1157	électrolyse interne
0964	0750	électrolyte
0965	0208	électrolyte de base (de conductibilité)
0966	2377	électrolyte de conductibilité
0967	1094	électrolyte indifférent
0968	2560	électromètre à condensateur vibrant
0969	0168	electron Auger
0970	0429	électron Compton
0971	0484	électron de conversion
0972	0008	électrons absorbés
0973	2551	électrons de valence
0974	0635	électrons diffractés
0975	0940	électrons libres
0976	0198	électrons rétrodiffusés
0977	2144	électrons secondaires
0978	0770	électrophorèse
0979	0771	électrophoretogramme
0980	0772	électroséparation
0981	0476	électroséparation à potentiel contrôlé
0982	0773	électroséparation à potentiel contrôlé
0983	0774	électrospray
0984	0775	élément
0985	0079	élément analysé
0986	0080	élément analysé
0987	0951	élément combustible
0988	2019	élément de référence
0989	2211	élément mono-ionisé
0990	0780	éluant
0991	0779	éluat
0992	0781	élué
0993	2348	élution par étapes
0994	0996	élution par gradient
0995	2156	élution sélective
0996	0149	émission atomique
0997	1917	émissivité radiante
0998	0793	endotherme
0999	0760	énergie cinétique d'un électron
1000	0014	énergie d'absorption
1001	0127	énergie d'apparition
1002	0730	énergie d'éjection
1003	0830	énergie d'excitation
1004	0844	énergie d'excitation
1005	1881	énergie d'impulsion
1006	1206	énergie d'ionisation
1007	0756	énergie d'un électron
1008	0719	énergie de coupure utile du cadmium
1009	0229	énergie de liaison
1010	0171	énergie de pic Auger
1011	1887	énergie de pompage
1012	2064	énergie de résonance
1013	2465	énergie de seuil
1014	1271	énergie de sortie laser
1015	2093	énergie de transition de rotation
1016	1931	énergie du rayonnement
1017	1492	énergie molaire d'ionisation
1018	1918	énergie rayonnante
1019	2006	enregistrement

1020	0808	enrichissement
1021	2517	ensemble de tubes
1022	1015	épaisseur de demi-absorption
1023	1014	épaisseur de la couche de demi-absorption
1024	2458	épaisseur de la couche de réaction
1025	0721	épaisseur efficace infinie
1026	0284	épuisement (du combustible nucléaire)
1027	2138	épuration
1028	0257	équation de Bragg
1029	0567	équation de Debye-Hückel
1030	0685	équation de Doerner et Hoskins
1031	1354	équation de Lomakin-Scheibe
1032	1548	équation de Nernst
1033	1481	équation de Nernst modifiée
1034	2112	équation de Sand
1035	1402	équilibrage de matrice
1036	0351	équilibre chimique
1037	2251	équilibre d'étincelage
1038	0824	équilibre d'évaporation
1039	1755	équilibre physique
1040	1951	équilibre radioactif
1041	2153	équilibre séculaire
1042	2420	équilibre thermique
1043	1350	équilibre thermique local
1044	2435	équilibre thermodynamique
1045	0681	équilibres de distribution
1046	0421	équilibres de formation de complexe
1047	1871	équilibres de protonation
1048	2231	équilibres en solution
1049	1331	équilibres liquide-liquide
1050	0819	équivalent
1051	0352	équivalent chimique
1052	0690	équivalent de dose
1053	0720	équivalent de dose efficace
1054	1464	équivalent en mg de la sensibilité de lecture
1055	1264	érosion par laser
1056	0820	erreur
1057	0993	erreur d'électrode de verre
1058	2476	erreur de titrage
1059	1699	erreur relative
1060	2041	erreur relative
1061	0312	espace cathodique obscur
1062	0112	espace obscur anodique
1063	0143	espace obscur d'Aston
1064	0523	espace obscur de Crookes
1065	0872	espace obscur de Faraday
1066	1048	espace obscur de Hittorf
1067	2405	espacement temporel
1068	1080	essai de réaction immunologique (antigène-anticorps)
1069	1127	étalon interne
1070	1163	étalon interne
1071	1853	étalon primaire
1072	2150	étalon secondaire
1073	0086	étalonnage analytique
1074	1734	étalonnage d'émulsion photographique
1075	0098	étalons d'analyse
1076	1711	étalons de pH
1077	1625	étalons opérationnels de pH
1078	1849	étalons primaires de pH
1079	1851	étalons primaires de référence
1080	2345	étape, marche, vague, pas
1081	1006	état fondamental
1082	1445	état métastable
1083	2344	état stationnaire
1084	0843	états excités
1085	1228	états isomères
1086	1781	états plasmons
1087	0478	étincelle à forme d'onde contrôlée
1088	1427	étincelle à tension moyenne
1089	1076	étincelle d'allumage
1090	2388	étincelle de surface
1091	2212	étincelle glissante
1092	1641	étincelle oscillante
1093	1001	étincelle sur graphite
1094	2542	étincelle unidirectionnelle
1095	1365	étincelles à basse tension
1096	1045	étincelles à haute fréquence
1097	1047	étincelles à haute tension
1098	0739	étincelles électriques
1099	2255	étincelles par semi-période
1100	2550	étincelles sous pression réduite
1101	0823	évaporation
1102	0417	évaporation complète
1103	1666	évaporation partielle
1104	2421	évaporation thermique
1105	0828	excitation
1106	2245	excitation croisée d'étincelle
1107	0736	excitation électrique
1108	1260	excitation et atomisation par laser
1109	1414	excitation mesurée en fluorescence
1110	1416	excitation mesurée en phosphorescence
1111	1940	excitation radiative
1112	0743	excitation sans électrodes
1113	0846	exotherme
1114	2118	expansion d'échelle
1115	0851	exposition
1116	1920	exposition à l'énergie rayonnante
1117	1368	extinction de luminescence
1118	1911	extinction, desactivation
1119	0857	extractabilité
1120	0858	extractant
1121	0859	extraction
1122	0190	extraction en retour
1123	0192	extraction en retour
1124	1324	extraction liquide
1125	1332	extraction liquide-liquide
1126	2235	extraction par solvant
1127	0856	extraire
1128	0191	extraire en retour
1129	0855	extrait
1130	0015	facteur d'absorption

1131	2162	facteur d'auto-absorption
1132	0809	facteur d'enrichissement
1133	0485	facteur de conversion
1134	2479	facteur de conversion titrimétrique
1135	1843	facteur de correction du gradient de pression
1136	1131	facteur de correspondance d'intensités
1137	0988	facteur de géométrie
1138	1897	facteur de qualité
1139	2029	facteur de réflexion
1140	2076	facteur de retardement
1141	2184	facteur de séparation
1142	2503	facteur de transmission
1143	1308	facteur proportionnel dépendant de l'énergie transférée (cf. facteur de qualité)
1144	2588	faible gradient de potentiel
1145	1360	faible perte (par absorption optique)
1146	1133	faisceau à intensité modulée
1147	2017	faisceau de référence
1148	2101	faisceau échantillon
1149	1262	faisceau laser
1150	2213	fente
1151	1629	fibres optiques
1152	2459	film mince
1153	0882	film témoin
1154	0984	filtration à gel
1155	0883	filtre
1156	0884	filtrer
1157	1630	filtres optiques
1158	0916	fin de titrage fluorimétrique
1159	0888	fissible
1160	0885	fissile
1161	0887	fission
1162	1591	fission nucléaire
1163	2305	fission spontanée
1164	2422	fission thermique
1165	0895	flamme
1166	1252	flamme laminaire
1167	2191	flamme protégée
1168	0952	flamme riche en combustible
1169	2182	flamme séparée
1170	2526	flamme turbulente
1171	0353	flammes chimiques
1172	0904	floculation
1173	0908	fluorescence
1174	0653	fluorescence à ligne directe
1175	0122	fluorescence anti-Stokes
1176	0151	fluorescence atomique
1177	2354	fluorescence de Stokes
1178	2350	fluorescence impliquant plus qu'un niveau supérieur
1179	2605	fluorescence par rayons X
1180	0577	fluorescence retardée
1181	0822	fluorescence retardée de type E
1182	1873	fluorescence retardée de type P
1183	2145	fluorescence secondaire
1184	0901	fluorimétrie d'éclair
1185	1912	fluorimétrie d'extinction
1186	1714	fluorimétrie de phase
1187	0919	flux
1188	0925	flux à travers monochromateur
1189	1559	flux de neutrons
1190	1670	flux de particules
1191	1750	flux de photons
1192	1921	flux énergétique de rayonnement
1193	1374	flux lumineux
1194	1017	fonction acidité de Hammett
1195	0792	fonction d'étalonnage d'émulsion
1196	1681	fonction de partition
1197	1319	fonction de profil de raie
1198	2565	fonction de Voigt
1199	1893	fonctionnement à commutateur Q
1200	0461	fonctionnement en onde continue
1201	0029	fonctions acidité
1202	0093	fonctions analytiques
1203	0087	fonctions d'étalonnage analytiques
1204	2270	fond continu spectral
1205	0896	fond de flamme
1206	1643	force de l'oscillateur
1207	0752	force électromotrice
1208	1198	force ionique
1209	0933	formalité
1210	1775	formation du plasma
1211	0557	forme cylindrique
1212	1320	forme de raie
1213	2273	forme de raie spectrale
1214	0521	forme du cratère
1215	0969	forme gazeuse
1216	0304	four à barreau de carbone
1217	1000	four à barreau de graphite
1218	1002	four à tube de graphite
1219	0163	fraction atomisée
1220	0259	fraction d'embranchement
1221	2340	fraction de phase stationnaire
1222	0601	fraction désolvatée
1223	1165	fraction interstitielle
1224	1346	fraction locale atomisée
1225	1347	fraction locale désolvatée
1226	1348	fraction locale volatilisée
1227	0889	fragments de fission
1228	0561	fragments, espèces "filles"
1229	0943	fréquence
1230	2247	fréquence d'étincelles
1231	0945	fréquence de modulation
1232	1483	fréquence de modulation
1233	0893	fréquence fixe (constante)
1234	0903	fréquence flottante
1235	0727	fréquences propres
1236	2236	front du solvant
1237	0956	fusion
1238	1594	fusion nucléaire
1239	1254	gaine laminaire
1240	1910	gaz d'extinction (de désactivation)
1241	1107	gaz d'injection
1242	0659	gaz de décharge

1243	0880	gaz de remplissage
1244	2596	gaz de travail
1245	1566	gaz nobles (gaz rares)
1246	1773	gaz plasma
1247	1856	gaz principal
1248	2052	gaz résiduels
1249	0309	gaz vecteur
1250	0050	gaz vecteur d'aérosol
1251	0472	générateur d'étincelle à haute tension contrôlée
1252	2539	générateur d'étincelle à haute tension non contrôlée
1253	2249	générateur d'étincelles
1254	1428	générateur d'étincelles à tension moyenne
1255	0704	générateur de goutte
1256	2606	génération de rayons X
1257	2607	génération de rayons X par électrons
1258	2609	génération de rayons X par ions positifs
1259	2608	génération de rayons X par photons
1260	0510	géométrie de comptage
1261	1404	géométrie de Mattauch-Herzog
1262	1565	géométrie de Nier-Johnson
1263	0990	getterisation
1264	1005	gray
1265	1217	groupes ionogéniques
1266	0383	guide d'onde coaxial
1267	1027	hauteur d'observation
1268	1612	hauteur d'observation
1269	2214	hauteur de fente
1270	0812	hauteur de la fente d'entrée
1271	2346	hauteur de marche, de vague
1272	1691	hauteur de pic
1273	2583	hauteur de vague
1274	1026	hauteur équivalente à un plateau théorique
1275	1025	hauteur équivalente à un plateau théorique efficace
1276	1071	hystérèse
1277	1774	impédance de plasma
1278	1777	impédance du plasma
1279	1530	impédances naturelles
1280	0333	implantation d'ions dans un cristal par pulvérisation cathodique
1281	1075	impulsion d'allumage
1282	1088	impulsion de rayonnement incohérente
1283	0991	impulsion géante
1284	1274	impulsion laser
1285	2208	impulsion unique
1286	1933	impulsions de rayonnement
1287	1225	impulsions de rayonnement irrégulières
1288	2467	inclinaison de la plaque photographique
1289	1087	inclusion
1290	1090	indicateur
1291	0023	indicateur acide-base
1292	2136	indicateur avec filtre coloré interne
1293	2529	indicateur bicolore
1294	0362	indicateur chimiluminescent
1295	0048	indicateur d'adsorption
1296	0862	indicateur d'extraction
1297	1653	indicateur d'oxydation-réduction
1298	1018	indicateur de Hammett
1299	1707	indicateur de pH
1300	1826	indicateur de précipitation
1301	1436	indicateur métallochromique
1302	1437	indicateur métallofluorescent
1303	1471	indicateur mixte
1304	1615	indicateur monochrome
1305	1954	indicateur radioactif
1306	2012	indicateur rédox
1307	2564	indicateur visuel
1308	1567	indication (valeur) à vide
1309	1120	indication instrumentale
1310	0435	indice de concentration
1311	2033	indice de réfraction
1312	2078	indice de rétention
1313	0276	indice de tampon
1314	1100	inducteur
1315	1106	initier
1316	1110	injecteur
1317	0286	injecteur auxiliaire
1318	2104	injecteur d'échantillon
1319	1336	injection d'échantillon liquide
1320	0261	instabilité de rupture
1321	1122	instrumentation
1322	2290	instruments spectrographiques
1323	0922	intégration de flux
1324	1989	intensimètre
1325	1129	intensité
1326	0495	intensité corrigée
1327	1692	intensité de pic
1328	1136	intensité de raie spectrale
1329	1135	intensité de rayonnement
1330	2610	intensité de rayons X
1331	2022	intensité de référence
1332	1922	intensité énergétique de rayonnement
1333	1415	intensité mesurée
1334	1137	intensité relative au signal du pic de base
1335	2271	intensité spectrale
1336	0024	interaction acide-base
1337	2228	interaction soluté solvant
1338	1143	interface
1339	0746	interface électrode-solution
1340	1146	interférant
1341	1144	interférence
1342	0354	interférence chimique
1343	0897	interférence de géométrie de la flamme
1344	1281	interférence latérale de diffusion
1345	1526	interférence mutuelle
1346	1582	interférence non spectrale

1347	1756	interférence physique
1348	0831	interférences d'excitation
1349	1207	interférences d'ionisation
1350	0673	interférences de dissociation
1351	2257	interférences de distribution spatiale
1352	2508	interférences de transport
1353	2229	interférences de volatilisation du soluté
1354	2556	interférences en phase vapeur
1355	1581	interférences non spécifiques
1356	2262	interférences spécifiques
1357	2272	interférences spectrales
1358	1708	interprétation de pH
1359	2600	interprétation des valeurs obtenues par rayons X
1360	0364	interrupteur rotatif
1361	0964	intervalle
1362	0094	intervalle analytique (entre électrodes)
1363	0113	intervalle anodique
1364	0134	intervalle d'arc (entre électrodes)
1365	2105	intervalle d'échantillonnage
1366	1995	intervalle de réaction
1367	2494	intervalle de transition (d'un indicateur)
1368	2514	intervalle stationnaire déclenchable (entre électrodes)
1369	2362	intervalle stroboscopique
1370	1798	inversion de population
1371	1179	iodimétrie
1372	1181	iodométrie
1373	1182	ion
1374	2002	ion de réarrangement
1375	0937	ion fragment
1376	1242	ion isotope
1377	1501	ion moléculaire
1378	2001	ion moléculaire réarrangé
1379	1540	ion négatif
1380	1664	ion parent
1381	1858	ion parent (ou ion précurseur)
1382	1835	ion précurseur
1383	1847	ion primaire
1384	2146	ion secondaire
1385	1200	ionisation
1386	0355	ionisation chimique
1387	1202	ionisation par bombardement (par pulvérisation)
1388	0878	ionisation par champ électrique
1389	2254	ionisation par étincelles
1390	1263	ionisation par faisceau laser
1391	0759	ionisation par impact électronique
1392	2053	ionisation résiduelle
1393	2263	ionisation spécifique
1394	2423	ionisation thermique
1395	0083	ions d'élément analysé
1396	1803	ions positifs
1397	0894	ions stationnaires
1398	1223	irradiance
1399	1224	irradiation
1400	1596	isobares nucléaires
1401	1227	isoformation
1402	1597	isomères nucléaires
1403	1231	isotherme
1404	1189	isotherme d'échange d'ions
1405	2239	isotherme de sorption
1406	1233	isotones
1407	1836	isotope précurseur
1408	1955	isotope radioactif
1409	1234	isotopes
1410	1238	isotopes d'atomes
1411	1776	jet de plasma
1412	2555	jet de vapeur
1413	1326	jonction liquide
1414	2054	jonction liquide résiduelle
1415	1333	jonction liquide-liquide
1416	0464	lacune de contrôle
1417	1255	lampe
1418	0135	lampe à arc
1419	1439	lampe à arc à vapeur de métal
1420	1052	lampe à cathode creuse
1421	1134	lampe à cathode creuse à intensité modulée
1422	0487	lampe à cathode creuse refroidie
1423	2140	lampe à cathode creuse scellée
1424	1878	lampe à décharge pulsée
1425	0742	lampe à décharge sans électrodes
1426	0902	lampe à éclair
1427	2205	lampe à élément unique
1428	2520	lampe à filament de tungstène
1429	2521	lampe à tungstène-halogène
1430	2597	lampe à xénon
1431	1042	lampe à xénon à haute pression
1432	1888	lampe de pompage
1433	1361	lampes à arc à basse pression
1434	1362	lampes à décharge à basse pression
1435	0983	lampes de Geissler
1436	0205	largeur de bande
1437	2268	largeur de bande spectrale
1438	1009	largeur de demi-intensité de la bande d'absorption
1439	1010	largeur de demi-intensité de la bande d'émission
1440	2594	largeur de faisceau lumineux
1441	2215	largeur de fente
1442	0813	largeur de la fente d'entrée
1443	1696	largeur de pic
1444	1697	largeur de pic à mi-hauteur
1445	1322	largeur de raie
1446	1465	largeur minimum de raie
1447	0955	largeur totale à mi-maximum
1448	2593	largueur
1449	1256	laser
1450	0712	lasers à colorant
1451	0971	lasers à gaz
1452	1328	lasers liquides
1453	2225	lasers solides
1454	1998	lecture

1455	0929	lentille convergente
1456	1286	ligand, coordinat
1457	0209	ligne de base
1458	2334	ligne de départ
1459	2335	ligne de départ
1460	1908	lignes quart d'onde
1461	0605	limite de détection
1462	1297	limite de détection
1463	1298	limite de détermination
1464	0710	limite de Duane-Hunt des courtes longueurs d'onde
1465	1299	limite de quantification
1466	1997	lisibilité
1467	1282	loi d'action de masse
1468	1405	loi de Maxwell-Boltzmann
1469	1768	loi de Planck
1470	2096	loi de Saha-Eggert
1471	2351	loi de Stern-Volmer
1472	0683	lois de distribution
1473	0013	longueur d'onde de discontinuité d'absorption
1474	0243	longueur d'onde échelette
1475	1284	longueur de base efficace du prisme
1476	1195	longueur du parcours ionique
1477	0114	lueur anodique
1478	2356	lumière parasite ambiante
1479	2276	luminance énergétique spectrale
1480	1366	luminescence
1481	0315	luminescence cathodique
1482	0827	luminescence d'excimère
1483	1539	luminescence négative, lueur négative
1484	2181	luminescence sensibilisée
1485	1384	magnétron
1486	1956	marquage radioactif
1487	1250	marque
1488	1251	marquer
1489	1387	marqueur
1490	1388	masquage (d'un élément interférant)
1491	2187	masquage (d'un élément interférant)
1492	0154	masse atomique
1493	2037	masse atomique relative
1494	0335	masse caractéristique
1495	1610	masse du nucléide
1496	2129	matériau à scintillation
1497	0031	matériau activable
1498	2024	matériau de référence
1499	0329	matériau de référence certifié
1500	2327	matériau étalon de référence
1501	0876	matériau fertile
1502	1580	matériau inhomogène
1503	1061	matériaux hôte, d'accueil, support
1504	1957	matériaux radioactifs
1505	1400	matrice
1506	2056	matrice en résine
1507	1646	maturation d'Ostwald
1508	2368	maximum (courbe)
1509	1693	maximum de pic
1510	1422	mécanisation
1511	1423	mécaniser
1512	1421	mécanisme
1513	0833	mécanisme d'excitation
1514	2046	mécanisme de libération
1515	1698	mélange de gaz de Penning
1516	1430	membrane
1517	1190	membrane échangeuse d'ions
1518	0973	membrane perméable au gaz
1519	1432	méso
1520	1433	mésoéchantillon
1521	1418	mesure
1522	1709	mesure de pH
1523	1819	mesure pratique du pH
1524	0238	mesure témoin (de contrôle, à blanc)
1525	1735	mesures d'intensité photographique
1526	2611	mesures de rayons X
1527	1118	mesures instantanées
1528	1440	métastable
1529	1441	métastable
1530	1521	méthode à passages multiples
1531	2321	méthode d'additions étalonnées
1532	1474	méthode de la solution mixte
1533	2183	méthode de la solution séparée
1534	2331	méthode de soustractions étalonnées
1535	1522	méthodes de multiplication
1536	1503	méthodes moléculaires
1537	1446	microanalyse
1538	0764	microanalyse électronique
1539	0765	microanalyse électronique à rayons X
1540	1114	microanalyse in situ
1541	1218	microanalyse par sonde ionique
1542	1275	microanalyse Raman par laser
1543	1447	microbalance
1544	1449	microconstituant
1545	1451	microcoulométrie
1546	1115	microdiffraction in situ des rayons X
1547	1454	microéchantillon
1548	1457	microonde
1549	1450	microordinateur
1550	1452	microphotomètre
1551	1453	microprocesseur
1552	0761	microscopie électronique
1553	2121	microscopie électronique à balayage
1554	2123	microscopie électronique à balayage par transmission
1555	0091	microscopie électronique d'analyse
1556	1215	microscopie ionique
1557	2502	microscopie par transmission d'électrons
1558	1267	microspectrométrie de masse à laser
1559	1266	microspectroscopie d'émission par laser

1560	1456	microtrace
1561	1461	migration
1562	1424	milieu
1563	0035	milieu actif
1564	1194	milieu ionique
1565	1463	millicoulométrie
1566	0930	miroir convergent
1567	1269	miroir laser
1568	2174	miroir semi-transparent
1569	1283	mise en équilibre de la couche (interface)
1570	0834	mode d'excitation
1571	0942	mode de fonctionnement libre
1572	2509	mode transversal électromagnétique
1573	1478	modérateur
1574	1477	modération
1575	1479	modes
1576	1482	modificateur
1577	1403	modificateurs de matrice
1578	2586	modulation de longueur d'onde
1579	1289	modulation de lumière
1580	1484	modulation de rayonnement polarisé linéairement
1581	1634	modulation du signal optique
1582	1488	molalité
1583	1493	molarité
1584	1507	molécule
1585	1509	mono-alternance
1586	1511	monochromateur
1587	0788	monochromateur d'émission
1588	0835	monochromateur d'excitation
1589	1513	monocouche
1590	1407	moyen
1591	0184	moyenne
1592	1520	multicouche
1593	2143	multiplicateur d'électrons secondaires
1594	0762	multiplication d'électrons
1595	0944	multiplication de fréquence
1596	1561	multiplication de neutrons
1597	2086	mûrissement
1598	2109	nacelle d'échantillonnage
1599	1527	nanotrace
1600	1534	nébulisation
1601	1536	nébuliseur
1602	1003	nébuliseur à alimentation gravifique
1603	0332	nébuliseur à chambre
1604	0471	nébuliseur à courant de fluide contrôlé
1605	2367	nébuliseur à succion
1606	0102	nébuliseur angulaire
1607	0437	nébuliseur concentrique
1608	0705	nébuliseur de goutte
1609	2528	nébuliseur jumelé
1610	1784	nébuliseur pneumatique
1611	2535	nébuliseur ultrasonique
1612	2031	nébuliseurs à reflux
1613	1543	négatron (= électron)
1614	1547	néphélométrie
1615	0380	nettoyage
1616	1553	neutre
1617	1554	neutron
1618	2066	neutron de résonance
1619	1867	neutron instantané
1620	1149	neutron intermédiaire
1621	0890	neutrons de fission
1622	0815	neutrons épicadmiques
1623	0816	neutrons épithermiques
1624	0394	neutrons froids
1625	2217	neutrons lents
1626	0874	neutrons rapides
1627	0578	neutrons retardés
1628	2425	neutrons thermiques
1629	1285	niveau de titrage
1630	0799	niveaux d'énergie
1631	0832	niveaux d'excitation
1632	2587	nombre d'onde
1633	1288	nombre de ligands
1634	1394	nombre de masse
1635	1562	nombre de neutrons
1636	1608	nombre de nucléons
1637	2411	nombre de plateaux théoriques
1638	1872	nombre de protons (= numéro atomique)
1639	0722	nombre efficace de plateaux théoriques
1640	0188	nombre moyen de ligands
1641	2543	non polarisé(e)
1642	1585	normalité
1643	1959	noyaux radioactifs
1644	2554	nuage de vapeur
1645	1273	nuage de vapeur produit par laser
1646	1606	nucléation
1647	1609	nucléide
1648	0886	nucléide fissile
1649	0877	nucléides fertiles
1650	1960	nucléides radioactifs
1651	1607	nucléon
1652	0156	numéro atomique
1653	2326	observateur standard (normal)
1654	1611	observation
1655	0769	obturateur électro-optique
1656	2090	obturateurs à disque rotatif
1657	2091	obturateurs à miroir rotatif
1658	1420	obturateurs mécaniques
1659	1614	occlusion
1660	1583	oeil normal
1661	0460	onde continue
1662	1637	ordre du spectre
1663	1639	orifice
1664	0975	orifice de limitation du débit de gaz
1665	1642	oscillateur
1666	0528	oscillateur à cristal
1667	2519	oscillateur à ligne accordée (résonante)
1668	1644	oscillographie
1669	1645	oscillopolarographie

1670	1652	oxydant
1671	0377	paillasse propre
1672	2117	palier de saturation
1673	0062	paramètre α
1674	1318	paramètre d'élargissement de raie
1675	1926	paramètre de rayonnement
1676	0738	paramètre de source électrique
1677	0737	paramètre électrique
1678	1736	paramètre photographique
1679	1367	paramètres de luminescence
1680	1934	paramètres de rayonnement
1681	2279	paramètres de rayonnement spectral
1682	2079	paramètres de rétention
1683	1778	paramètres du plasma
1684	0018	parcours d'absorption
1685	0063	particule alpha
1686	0220	particule bêta
1687	0777	particule élémentaire
1688	1212	particule ionisante
1689	1598	particule nucléaire
1690	0339	particules chargées
1691	1923	particules rayonnantes
1692	2070	pente de réponse
1693	0410	performance de colonne
1694	1822	période de pré-arc
1695	1889	période du pompage
1696	1840	période pré-étincelle
1697	2275	période spectrale
1698	1704	permsélectivité
1699	0511	perte de comptage
1700	0924	perturbation de flux
1701	1705	pH
1702	0126	pH apparent
1703	1334	phase liquide
1704	1476	phase mobile
1705	2339	phase stationnaire
1706	1340	phase stationnaire liquide
1707	0455	phénomènes de contamination en précipitation
1708	1718	phosphorescence
1709	1722	phosphorimétrie
1710	1723	phosphoroscopes
1711	1724	photocourant
1712	1725	photodétecteur
1713	0979	photodétecteurs à déclenchement périodique
1714	2198	photodiodes au silicium
1715	1729	photoélectron
1716	1739	photoionisation
1717	1743	photométrie
1718	1028	photométrie hétérochromatique
1719	1737	photométrie photographique
1720	1909	photométrie quasi-monochromatique
1721	2218	photomultiplicateur non sensible au spectre visible
1722	1745	photon
1723	1082	photon incident
1724	2613	photons X
1725	1753	photopic, pic photoélectrique
1726	2549	phototube à vide
1727	1758	physiosorption
1728	1682	pic
1729	2482	pic d'absorption (photoélectrique) totale
1730	0785	pic d'élution
1731	0953	pic d'énergie totale
1732	1443	pic d'ion métastable
1733	0211	pic de base (spectroscopie de masse)
1734	0694	pic de double libération (de photons)
1735	0055	pic de l'air
1736	0794	pic endothermique
1737	0847	pic exothermique
1738	1444	pic métastable
1739	1728	pic photoélectrique
1740	2604	pic photoélectrique associé à l'émission d'un photon X
1741	2206	pic unique photoélectrique associé à l'émission d'un photon X
1742	1759	picotrace
1743	2127	piège
1744	2128	piégeage
1745	2510	piégeage
1746	1419	piégeage mécanique
1747	0452	pincement de plasma d'arc
1748	1766	pixel
1749	1771	plan polarisé
1750	0443	plaque de confinement
1751	2378	plaque support
1752	1772	plasma
1753	0290	plasma (obtenu) par couplage capacitif
1754	1364	plasma à basse pression induit par micro-onde
1755	0544	plasma à courant nul
1756	2463	plasma à trois électrodes
1757	0542	plasma d'arc à transport de courant
1758	0291	plasma de microonde (obtenu) sur électrode à cavité résonante
1759	1616	plasma de microonde à une électrode
1760	0745	plasma de microondes à cavité résonante avec électrodes
1761	0744	plasma de microondes à cavité résonante sans électrodes
1762	1150	plasma gazeux intermédiaire
1763	0145	plasma induit par microonde à pression atmosphérique
1764	1458	plasma induit par microondes
1765	1631	plasma optiquement mince
1766	1099	plasmas à induction couplée
1767	1459	plasmas de microonde
1768	1974	plasmas de radiofréquence
1769	1575	plasmas sans transport de courant
1770	2491	plasmas transférés

1771	2092	plate-forme rotative
1772	1783	plate-forme tournante
1773	1782	plateau, palier
1774	2557	plume de vapeur
1775	1272	plume laser
1776	1924	plume rayonnante
1777	0160	poids atomique
1778	0865	poids d'échantillon à facteur de conversion
1779	2343	poids statistique
1780	0249	point d'ébullition
1781	0818	point d'équivalence
1782	0281	point de combustion
1783	2336	point de départ
1784	0563	point de fin de titrage bi-ampérométrique
1785	0373	point de fin de titrage chronopotentiométrique
1786	1033	point de fin de titrage conductimétrique à haute fréquence
1787	0590	point de fin de titrage dérivé
1788	0810	point de fin de titrage enthalpimétrique
1789	1741	point de fin de titrage photométrique
1790	1809	point de fin de titrage potentiométrique
1791	1978	point de fin de titrage radiométrique
1792	2447	point de fin de titrage thermométrique
1793	2523	point de fin de titrage turbidimétrique
1794	0863	point extrapolé de début de croissance
1795	0795	point final (de titrage)
1796	1230	point isopotentiel
1797	2352	point stoechiométrique de fin de titrage
1798	2410	point théorique de fin de titrage
1799	2563	point visuel de fin de titrage
1800	2618	point zéro d'une échelle
1801	2617	point zéro d'une électrode de verre
1802	2301	pointes d'impulsions
1803	1279	pointes laser
1804	0224	polarisation
1805	1789	polarisation
1806	0376	polarisation circulaire
1807	0778	polarisation élliptique
1808	1311	polarisation linéaire de luminescence
1809	1770	polarisation plane
1810	1790	polarisé(e)
1811	1791	polariseur
1812	1312	polariseur linéaire
1813	1788	polarité
1814	1792	polarographie
1815	2314	polarographie (en courant alternatif) à impulsions carrées
1816	2512	polarographie à balayage triangulaire
1817	0068	polarographie à courant alternatif
1818	1031	polarographie à courant alternatif d'harmonique supérieure
1819	1032	polarographie à courant alternatif d'harmonique supérieure avec redressement sensible aux variations de phase
1820	0650	polarographie à courant continu
1821	0660	polarographie à décharge
1822	0582	polarographie à démodulation
1823	0701	polarographie à doubles potentiels alternatifs surimposés
1824	2320	polarographie à fonction échelon
1825	1883	polarographie à impulsions
1826	1089	polarographie à incréments de charge
1827	1485	polarographie à modulation
1828	1525	polarographie à multibalayage
1829	0477	polarographie à potentiel contrôlé
1830	1975	polarographie à radiofréquence
1831	0341	polarographie à sauts de charge
1832	0071	polarographie à tension alternative
1833	0546	polarographie à variation contrôlée du courant
1834	0552	polarographie cyclique à balayage triangulaire
1835	1245	polarographie de Kalousek
1836	0591	polarographie dérivée
1837	0623	polarographie différentielle
1838	0593	polarographie impulsionnelle dérivée
1839	0626	polarographie impulsionnelle différentielle
1840	0318	polarographie oscillographique
1841	2209	polarographie oscillographique à balayage unique
1842	2170	polarographie semi-intégrale
1843	1794	polyatomique
1844	1795	polychromateur
1845	2014	polymères rédox
1846	0627	pompage différentiel
1847	1632	pompage optique
1848	0183	pont d'éclatement auxiliaire
1849	2103	porte échantillon
1850	2021	porte-référence
1851	2073	position de repos
1852	1804	positron
1853	1806	post-précipitation
1854	1807	potentiel
1855	1012	potentiel à mi-pic
1856	0130	potentiel appliqué
1857	1907	potentiel au quart du temps de transition
1858	0336	potentiel caractéristique
1859	0128	potentiel d'apparition
1860	0144	potentiel d'asymétrie
1861	0836	potentiel d'excitation

1862	1209	potentiel d'ionisation
1863	1016	potentiel de demi-vague
1864	0674	potentiel de dissociation
1865	1327	potentiel de jonction liquide
1866	2055	potentiel de jonction liquide résiduelle
1867	2370	potentiel du maximum (courbe)
1868	1813	potentiométrie
1869	0469	potentiométrie à courant contrôlé
1870	1829	potentiométrie de précision à point zéro
1871	0625	potentiométrie différentielle
1872	1573	poudre non conductrice
1873	2355	pouvoir d'arrêt (de particules)
1874	2060	pouvoir de résolution
1875	0274	pouvoir tampon
1876	1044	pré-étincellage à haute fréquence
1877	1838	pré-ionisation
1878	1825	précipitation à partir d'une solution homogène
1879	1823	précipité
1880	1824	précipiter
1881	0022	précision
1882	1828	précision
1883	1832	précision d'indication
1884	1830	précision d'une balance
1885	1831	précision d'une pesée
1886	0512	précision de comptage
1887	2281	précision spectrochimique
1888	1834	précurseur
1889	2285	précurseur de gaz vecteur pour spectrochimie
1890	2486	premier coefficient de Townsend
1891	1841	pression
1892	0763	pression d'électrons
1893	0186	pression moyenne de gaz
1894	1210	probabilité d'ionisation
1895	2495	probabilité de transition
1896	0019	probabilité de transition d'absorption
1897	2496	probabilité de transition en absorption
1898	2497	probabilité de transition en émission spontanée
1899	2498	probabilité de transition en émission stimulée
1900	2304	probabilité de transition par émission spontanée (de photon)
1901	0729	probabilité de transmission d'Einstein
1902	0457	procédure continue
1903	0255	procédure d'encadrement
1904	2589	procédure de pesée
1905	0246	procédures d'insufflation
1906	0662	procédures discontinues
1907	0758	processus à transfert d'énergie électronique
1908	0406	processus de collision
1909	0562	processus de désactivation
1910	2185	processus de séparation
1911	1941	processus radiatifs
1912	2426	processus thermiques
1913	1658	production de paires
1914	2227	produit de solubilité
1915	0891	produits de fission
1916	0585	profil (de concentration) en profondeur
1917	0020	profil d'absorption
1918	0789	profil d'émission
1919	2566	profil de Voigt
1920	0821	profondeur de fuite
1921	0519	profondeur du cratère
1922	1861	programme
1923	1859	programme (informatique)
1924	1862	programmer
1925	1860	programmer (informatique)
1926	0947	prolongement du front
1927	2394	prolongement en queue
1928	2124	propagation par diffusion
1929	2287	propriétés spectrochimiques
1930	1869	proton
1931	1884	puissance d'impulsion
1932	2300	puissance d'impulsion à pointe
1933	1818	puissance de détection
1934	1890	puissance de pompage
1935	1925	puissance de rayonnement par unité d'angle solide
1936	1085	puissance incidente
1937	1426	puissance moyenne
1938	2027	puissance réfléchie
1939	2310	pulvérisation
1940	0319	pulvérisation cathodique
1941	1981	pureté en radionucléide
1942	1968	pureté radiochimique
1943	1967	purification radioanalytique
1944	1900	quantité
1945	1376	quantité lumineuse
1946	1757	quantité physique
1947	0986	quantités géométriques
1948	1633	quantités optiques
1949	0958	quantum gamma
1950	0914	quantum total de fluorescence
1951	0012	quotient des coefficients d'absorption de part et d'autre de la discontinuité d'adsorption
1952	1913	rad
1953	2277	radiance spectrale du corps noir
1954	1915	radiance, luminance
1955	1982	radio-réactif
1956	1944	radioactif
1957	1965	radioactivité
1958	0138	radioactivité artificielle
1959	1096	radioactivité induite
1960	1533	radioactivité naturelle
1961	1971	radiochimie
1962	0095	radiochimie analytique
1963	1972	radiochromatographe
1964	1976	radioisotope

1965	1977	radiolyse
1966	1980	radionucléide
1967	1983	raffinat
1968	1300	raie
1969	1193	raie (spectrale) d'ion
1970	0153	raie atomique
1971	0137	raie d'arc
1972	2023	raie de référence
1973	2065	raie de résonance
1974	1058	raie homologue
1975	1162	raie interne de référence
1976	0158	raies du spectre atomique
1977	0989	raies homologues de Gerlach
1978	1147	raies interférantes
1979	1197	raies spectrales ioniques
1980	0436	rapport de concentration
1981	0684	rapport de distribution
1982	1393	rapport de distribution de masse
1983	0433	rapport de distribution des concentrations
1984	1715	rapport de phase
1985	2505	rapport de transmission
1986	1717	rapport de volumes des phases
1987	1618	rapport des durées de fonctionnement et de non-fonctionnement
1988	1391	rapport masse/charge
1989	0961	rayon gamma
1990	0119	rayon gamma anti-Compton
1991	1927	rayonnement
1992	0328	rayonnement Cerenkov
1993	0386	rayonnement cohérent
1994	0458	rayonnement continu
1995	1577	rayonnement continu (à fond continu)
1996	0502	rayonnement cosmique
1997	0110	rayonnement d'annihilation
1998	1084	rayonnement d'incandescence
1999	0234	rayonnement de corps noir
2000	0266	rayonnement de freinage
2001	1563	rayonnement de neutrons
2002	1674	rayonnement de particules
2003	1752	rayonnement de photons
2004	0751	rayonnement électromagnétique
2005	0385	rayonnement électromagnétique cohérent
2006	1571	rayonnement électromagnétique non cohérent
2007	0959	rayonnement gamma
2008	0301	rayonnement gamma de capture
2009	1086	rayonnement incident
2010	1112	rayonnement interne de freinage
2011	1213	rayonnement ionisant
2012	1504	rayonnement moléculaire
2013	1510	rayonnement monochromatique
2014	1532	rayonnement naturel
2015	2357	rayonnement parasite ambiant
2016	1850	rayonnement primaire
2017	1984	rayonnement Raman
2018	2149	rayonnement secondaire
2019	2342	rayonnement stationnaire
2020	2427	rayonnement thermique
2021	2598	rayonnement X
2022	0337	rayonnement X caractéristique
2023	1600	réacteur nucléaire
2024	1833	réacteur pré-colonne
2025	0077	réaction d'amplification
2026	1592	réaction de fission nucléaire
2027	1595	réaction de fusion nucléaire
2028	1216	réaction ion-molécule
2029	1599	réaction nucléaire
2030	1640	réaction oscillante
2031	2387	réaction superficielle
2032	0848	réactions exothermiques
2033	1097	réactions induites
2034	2432	réactions thermochimiques
2035	1406	réarrangement de McLafferty
2036	1942	recombinaisons radiatives
2037	1019	recouvrement d'harmoniques
2038	2386	recouvrement de surface
2039	1040	rectification faradique à niveau élevé
2040	2005	recul
2041	2007	récupération
2042	1700	récupération relative
2043	2359	redissolution (à partir d'un extrait)
2044	0115	redissolution anodique par chronoampérométrie à variation linéaire de potentiel
2045	0116	redissolution anodique par coulométrie à potentiel contrôlé
2046	0320	redissolution cathodique
2047	1043	réflectivité élevée
2048	1947	refroidissement radioactif (= atténuation de radioactivité)
2049	0837	région d'excitation
2050	1170	région interzones
2051	2302	règle de conservation du spin
2052	2098	relargage, diminution de solubilité par addition d'un sel
2053	1844	relation pression-tension-courant
2054	2572	relation tension-temps d'un arc
2055	2047	rem
2056	0381	remontée de l'arc
2057	1656	remplissage
2058	0998	remplissage à gradient
2059	0173	rendement Auger
2060	0358	rendement chimique
2061	0534	rendement cumulatif de fission
2062	0162	rendement d'atomisation
2063	0892	rendement de fission
2064	0330	rendement de fission en chaîne
2065	0915	rendement de fluorescence
2066	1732	rendement de photoélectrons
2067	1749	rendement de photoémission
2068	2312	rendement de pulvérisation superficielle
2069	0651	rendement direct de fission

2070	0170	rendement en électrons Auger
2071	2148	rendement en ions secondaires
2072	0911	rendement énergétique de fluorescence
2073	1817	rendement énergétique de fluorescence (cf. 911)
2074	0805	rendement énergétique de luminescence
2075	1904	rendement quantique
2076	0498	rendement quantique corrigé de luminescence
2077	0912	rendement quantique de fluorescence
2078	1902	rendement quantique de fluorescence
2079	1905	rendement quantique de fluorescence
2080	1906	rendement quantique de luminescence
2081	1417	rendement quantique de luminescence mesuré
2082	1903	rendement quantique de luminiscence
2083	1970	rendement radiochimique
2084	1549	réponse nernstienne
2085	2049	reprécipitation
2086	0531	réseau cristallin
2087	2223	réseau cristallin de l'état solide
2088	2057	résistance
2089	0965	résistance à lacune
2090	0345	résistance de charge
2091	1343	résistance de charge
2092	2058	résolution
2093	2059	résolution à 10 % interpics (spectrométrie de masse)
2094	0800	résolution d'énergie
2095	0399	résolution de collimateur
2096	1694	résolution de pic
2097	0586	résolution de profondeur
2098	0609	résolution du détecteur
2099	2293	résolution du spectromètre
2100	1820	résolution pratique
2101	2258	résolution spatiale
2102	2404	résolution temporelle
2103	1276	résonateur laser
2104	2075	résultat
2105	0096	résultat d'analyse
2106	2077	rétention
2107	2042	rétention relative
2108	1952	retombée radioactive
2109	0900	retour de flamme
2110	0197	rétrodiffuser, rétrodiffusion
2111	2089	Roentgen
2112	0308	sans vecteur
2113	2115	saturation
2114	0331	saturation de la chambre
2115	0572	schéma de désintégration
2116	2134	scintillateur
2117	2130	scintillation
2118	2347	secteur à gradins, secteur échelonné
2119	0525	section efficace
2120	0034	section efficace d'activation
2121	0526	section efficace d'ionisation
2122	0300	section efficace de capture
2123	0403	section efficace de collision
2124	2592	section efficace de Westcott
2125	1378	section efficace macroscopique
2126	1455	section efficace microscopique
2127	0723	section efficace thermique effective
2128	1877	sélecteur d'amplitude d'impulsions
2129	2168	semi-conducteur
2130	2171	semi-micro
2131	2172	semi-microéchantillon
2132	2072	sensibilité
2133	2176	sensibilité
2134	2177	sensibilité (pour une charge donnée du plateau de la balance)
2135	2179	sensibilité d'un système d'admission
2136	2178	sensibilité d'une sonde directe
2137	2180	sensibilité d'une source d'ion
2138	1969	séparation chimique des radioisotopes d'un élément
2139	1243	séparation isotopique
2140	2188	série
2141	1961	séries radioactives
2142	0802	seuil d'énergie
2143	0982	seuil de Geiger-Müller
2144	1280	seuil laser
2145	2195	sievert
2146	0036	solide actif
2147	1480	solide actif modifié
2148	2230	solution
2149	0416	solution de comparaison
2150	2361	solution de redissolution
2151	0881	solution de remplissage
2152	0267	solution du pont (de jonction)
2153	0268	solution du pont (de jonction) pour une électrode de référence à double jonction
2154	2107	solution échantillon
2155	2329	solution étalon
2156	2328	solution étalon de référence
2157	1854	solution étalon primaire
2158	2151	solution étalon secondaire
2159	2324	solution étalonnée
2160	1158	solution interne de remplissage
2161	1159	solution interne de remplissage d'une électrode de verre
2162	1586	solution normale
2163	2114	solution saturée
2164	2221	solution solide
2165	2372	solution sursaturée
2166	0240	solution témoin (de contrôle)
2167	2473	solution titrante
2168	2325	solution titrante étalonnée
2169	2139	solutions d'épuration
2170	2025	solutions de référence
2171	2232	solvant

2172	0076	solvant amphotère
2173	0538	somme cumulative
2174	0124	sommet
2175	0654	sonde directe
2176	2238	sorption
2177	1270	sortie laser
2178	2240	source
2179	0838	source d'excitation
2180	1617	source d'excitation en une étape
2181	1221	source d'ion
2182	1769	source de décharge luminescente à cathode plane
2183	1278	source laser
2184	1852	source primaire
2185	2011	source rectifiée
2186	2428	source thermique
2187	1053	sources à cathode creuse
2188	0387	sources cohérentes
2189	1523	sources d'étincelles multifonctions
2190	1935	sources de rayonnement
2191	1879	sources laser pulsées
2192	1572	sources optiques non cohérentes
2193	2265	spécimen
2194	2298	spectre
2195	0157	spectre atomique
2196	0493	spectre d'émission corrigé
2197	1747	spectre d'émission de photons
2198	0494	spectre d'excitation corrigé
2199	0841	spectre d'excitation de fluorescence
2200	0842	spectre d'excitation de phosphorescence
2201	1720	spectre d'excitation-émission en phosphorescence
2202	2389	spectre de fluorescence à excitation synchrone
2203	1397	spectre de masse
2204	1541	spectre de masse d'ion négatif
2205	2390	spectre de phosphorescence à excitation synchrone
2206	0496	spectre de polarisation corrigé en émission de luminescence
2207	0497	spectre de polarisation corrigé en excitation de luminescence
2208	1937	spectre de rayonnement
2209	2299	spectre de rayonnement
2210	2614	spectre de rayons X
2211	0790	spectres d'émission
2212	0839	spectres d'excitation
2213	0829	spectres d'excitation-émission
2214	0204	spectres de bandes
2215	2544	spectres de bandes non résolubles
2216	2545	spectres de bandes non résolus
2217	1730	spectres de photoélectrons
2218	0963	spectres de rayons gammas
2219	2194	spectres de Shpol'skii
2220	0766	spectres électroniques
2221	1196	spectres ioniques
2222	2288	spectrogramme
2223	2289	spectrographe
2224	2291	spectrographie
2225	2292	spectromètre
2226	0530	spectromètre à diffraction cristalline
2227	1369	spectromètre à luminescence
2228	0120	spectromètre à rayon gamma anti-Compton
2229	0962	spectromètre à rayons gammas
2230	2133	spectromètre à scintillation
2231	0913	spectromètre de fluorescence
2232	1395	spectromètre de masse
2233	1301	spectromètre de masse à accélération linéaire
2234	0714	spectromètre de masse à champ dynamique
2235	2337	spectromètre de masse à champ statique
2236	0695	spectromètre de masse à double focalisation
2237	2207	spectromètre de masse à focalisation unique
2238	1865	spectromètre de masse à parcours trochoide allongé
2239	1186	spectromètre de masse à résonance cyclotron ionique
2240	2469	spectromètre de masse à temps de vol
2241	0935	spectromètre de masse à transformée de Fourier
2242	1721	spectromètre de phosphorescence
2243	1936	spectromètre de rayonnement
2244	2067	spectromètre de résonance
2245	2203	spectromètre monofaisceau
2246	2294	spectrométrie
2247	1219	spectrométrie à diffusion d'ions
2248	0692	spectrométrie à double faisceau
2249	0699	spectrométrie à double faisceau spectral
2250	0700	spectrométrie à double faisceau synchrone
2251	0147	spectrométrie d'absorption atomique
2252	1494	spectrométrie d'absorption moléculaire
2253	0150	spectrométrie d'émission atomique
2254	1499	spectrométrie d'émission moléculaire
2255	1628	spectrométrie d'émission optique
2256	1672	spectrométrie d'émission X induite par bombardement de particules
2257	0898	spectrométrie de flamme
2258	0152	spectrométrie de fluorescence atomique
2259	1370	spectrométrie de luminescence
2260	1396	spectrométrie de masse
2261	2147	spectrométrie de masse d'ions secondaires
2262	1505	spectrométrie moléculaire

2263	1502	spectrométrie moléculaire de luminescence
2264	2602	spectrométrie par émission de rayons X
2265	0594	spectrophotométrie dérivée
2266	2296	spectroscope
2267	2297	spectroscopie
2268	0757	spectroscopie à perte d'énergie d'électrons
2269	2501	spectroscopie à perte d'énergie par transmission d'électrons
2270	2471	spectroscopie à résolution temporelle
2271	0159	spectroscopie atomique
2272	0172	spectroscopie Auger
2273	1495	spectroscopie d'absorption moléculaire
2274	0169	spectroscopie d'électrons Auger
2275	0767	spectroscopie d'électrons pour analyse chimique (ESCA)
2276	1748	spectroscopie d'émission de photons (= de photoémission)
2277	1500	spectroscopie d'émission moléculaire
2278	1673	spectroscopie d'émission X induite par bombardement de particules
2279	1767	spectroscopie d'émission X par bombardement de particules
2280	1371	spectroscopie de luminescence
2281	1046	spectroscopie de perte d'énergie à haute résolution
2282	1731	spectroscopie de photoélectrons
2283	2612	spectroscopie de photoélectrons par rayons X
2284	2537	spectroscopie de photoélectrons U.V.
2285	1733	spectroscopie de photoémission
2286	1740	spectroscopie de photoionisation
2287	2028	spectroscopie de réflexion à perte d'énergie d'électrons
2288	0199	spectroscopie de rétrodiffusion
2289	1506	spectroscopie moléculaire
2290	2603	spectroscopie par émission de rayons X
2291	1986	spectroscopie Raman
2292	1128	sphère d'intégration
2293	2532	sphère d'Ulbricht
2294	2259	stabilisation spatiale
2295	2406	stabilisation temporelle
2296	0854	standard externe
2297	2353	stoechiométrie
2298	2358	structure en stries
2299	2363	sublimation
2300	2364	submicroanalyse
2301	1258	substance active pour effet laser
2302	0244	substance décolorable
2303	0677	substance distribuée
2304	0741	substance électroactive
2305	1148	substance interférante
2306	2330	substances étalons
2307	1962	substances radioactives
2308	0438	substances simultanément présentes
2309	2371	superposition
2310	2376	support (matériau)
2311	2226	support solide
2312	2379	supresseurs (d'interférences)
2313	2381	surface
2314	1684	surface de pic
2315	0849	surface expérimentale
2316	0948	surface frontale
2317	0379	surface propre
2318	2167	surface semi-conductrice
2319	2373	sursaturation
2320	0555	synchrométrie, spectrométrie de masse cyclotron
2321	0693	système à double faisceau
2322	0875	système à rétroaction
2323	0388	système cohérent d'unités
2324	0030	système cribleur (tamis) à onde de choc acoustique
2325	1004	système cribleur à alimentation gravifique
2326	1111	système d'admission
2327	2518	système de mesure à amplificateur accordé
2328	0715	système dynamique
2329	1601	système nucléaire
2330	1885	système pompé
2331	2396	système tandem de spectrométrie de masse
2332	0099	systèmes d'analyse
2333	2200	systèmes d'analyse simultanée
2334	1537	systèmes nébuliseur-flamme
2335	1635	systèmes optiques
2336	2306	tache
2337	1675	taille de particule(s)
2338	0927	tache focale
2339	0271	tampon
2340	1199	tampon d'ajustement de la force ionique
2341	2284	tampon spectrochimique
2342	0272	tamponner
2343	1201	tampons d'ionisation
2344	2401	tast-polarographie
2345	0260	taux d'embranchement
2346	0285	taux d'épuisement (du combustible nucléaire)
2347	0084	technique analytique d'addition
2348	0081	technique d'addition de l'élément analysé
2349	1249	technique de Kossel
2350	2020	technique de l'élément de référence
2351	0089	technique de la courbe analytique
2352	0210	technique de la ligne de base en spectroscopie atomique
2353	2199	technique de simulation
2354	2243	techniques à résolution spatiale

2355	2472	techniques à résolution temporelle
2356	1815	techniques de poudre(s)
2357	0273	techniques par addition de tampon
2358	2202	techniques simultanées
2359	0515	techniques simultanées couplées
2360	0663	techniques simultanées discontinues
2361	2430	techniques thermoanalytiques
2362	2233	témoin du solvant
2363	1091	témoin indicateur
2364	0005	température absolue
2365	0129	température d'apparition
2366	0250	température d'ébullition
2367	0768	température d'électrons
2368	0840	température d'excitation
2369	1108	température d'injection
2370	1211	température d'ionisation
2371	0411	température de colonne
2372	0899	température de flamme
2373	0978	température de gaz
2374	1349	température de gaz locale
2375	1564	température de neutrons
2376	1916	température de radiance
2377	2080	température de rétention
2378	2094	température de rotation
2379	2186	température de séparation
2380	2562	température de vibration
2381	1779	température du plasma
2382	1105	température initiale
2383	1587	température normel
2384	2436	température thermodynamique
2385	1138	temps d'interaction
2386	2511	temps de déplacement
2387	0708	temps de goutte
2388	2087	temps de montée (temps de croissance)
2389	2003	temps de réception
2390	2071	temps de réponse
2391	2061	temps de résolution
2392	0391	temps de résolution des coincidences
2393	2081	temps de rétention
2394	0499	temps de rétention corrigé
2395	1551	temps de rétention net
2396	0262	temps de rupture, temps d'amorçage
2397	2500	temps de transit
2398	2499	temps de transition
2399	0564	temps mort
2400	0607	temps mort du détecteur
2401	2000	temps réel
2402	2407	tensammétrie
2403	2570	tension
2404	0069	tension alternative
2405	0292	tension capacitive
2406	0447	tension constante
2407	1385	tension d'alimentation secteur
2408	0263	tension d'amorçage
2409	1214	tension d'ionisation
2410	0314	tension de chute cathodique
2411	1686	tension de crête du condensateur
2412	2035	tension de réallumage
2413	0283	tension de régime (d'arc)
2414	1702	tension périodique
2415	0644	test de dilution
2416	2009	test de récupération
2417	2429	thermoacoustimétrie
2418	2431	thermobalance
2419	2433	thermodilatométrie
2420	2437	thermoélectrométrie
2421	2438	thermogramme
2422	0631	thermogramme différentiel
2423	2439	thermographie
2424	2442	thermogravimétrie
2425	0596	thermogravimétrie dérivée
2426	0633	thermogravimétrie différentielle
2427	2443	thermoluminescence
2428	2444	thermomagnétométrie
2429	2446	thermomécanométrie
2430	0717	thermomécanométrie dynamique
2431	2449	thermomicroscopie
2432	2451	thermophotométrie
2433	2452	thermopile
2434	2453	thermoptométrie
2435	2454	thermoréfractométrie
2436	2455	thermosonimétrie
2437	2456	thermospectrométrie
2438	1277	tir de faisceau laser
2439	2474	titrage
2440	1038	titrage à haute fréquence
2441	0025	titrage acide-base
2442	0026	titrage acidimétrique
2443	0056	titrage alcalimétrique
2444	0074	titrage ampérométrique
2445	0222	titrage biampérométrique
2446	0231	titrage bipotentiométrique
2447	0374	titrage chronopotentiométrique
2448	0347	titrage complexométrique
2449	0422	titrage complexométrique
2450	0441	titrage conductimétrique
2451	1034	titrage conductimétrique à haute fréquence
2452	0503	titrage coulométrique
2453	0473	titrage coulométrique à potentiel contrôlé
2454	0480	titrage de contrôle
2455	1716	titrage de phase
2456	0615	titrage diélectrométrique
2457	1570	titrage en milieu non aqueux
2458	0201	titrage en retour
2459	0811	titrage enthalpimétrique
2460	0917	titrage fluorimétrique
2461	1095	titrage indirect
2462	1546	titrage néphélométrique
2463	1827	titrage par précipitation
2464	1742	titrage photométrique
2465	1793	titrage polarométrique
2466	2591	titrage pondéral

2467	1811	titrage potentiométrique
2468	0468	titrage potentiométrique à courant contrôlé
2469	2152	titrage potentiométrique à dérivée seconde
2470	0592	titrage potentiométrique dérivé
2471	0624	titrage potentiométrique différentiel
2472	1177	titrage potentiométrique par dérivée inverse
2473	1812	titrage potentiométrique pondéral
2474	1979	titrage radiométrique
2475	1654	titrage rédox
2476	2295	titrage spectrophotométrique
2477	0241	titrage témoin (de contrôle) à blanc
2478	2448	titrage thermométrique
2479	2524	titrage turbidimétrique
2480	1178	titrages iodimétriques
2481	1180	titrages iodométriques
2482	2015	titrages rédox
2483	1712	titration pH-métrique
2484	2477	titre
2485	0200	titrer en retour
2486	2480	titrimétrie
2487	1780	torche à plasma
2488	2487	trace
2489	1602	trace nucléaire
2490	2490	traceur
2491	1244	traceur isotopique
2492	1963	traceur radioactif
2493	0803	transfert d'énergie
2494	0342	transfert de charge
2495	1508	transfert de moment (d'énergie)
2496	1307	transfert linéaire d'énergie
2497	1037	transformateur progressif haute fréquence
2498	2492	transformation
2499	0202	transformation de Baker-Sampson-Seidel
2500	1604	transformation nucléaire
2501	0804	transition d'énergie
2502	2561	transition d'énergie de vibration
2503	1140	transition interchromophore non radiative
2504	1169	transition intersystèmes
2505	1229	transition isomère
2506	0941	transition libre-libre (c.à.d. entre niveaux non quantifiés)
2507	0938	transition libre-liée
2508	1932	transition non radiative
2509	1152	transition non radiative intermoléculaire
2510	1171	transition non radiative intrachromophore
2511	1605	transition nucléaire
2512	1943	transitions radiatives
2513	2506	transmittance
2514	1738	transmittance photographique
2515	2507	transport
2516	1049	transporteur (vecteur) sélectif
2517	1241	transporteur isotopique
2518	0298	tube capillaire
2519	1109	tube d'injection
2520	0508	tube de comptage
2521	0980	tube de comptage Geiger-Müller
2522	1868	tube de comptage proportionnel
2523	1151	tube intermédiaire
2524	1744	tube photomultiplicateur
2525	2393	tubes T
2526	2525	turbidimétrie
2527	2534	ultramicroanalyse
2528	0517	unité de couplage
2529	0155	unités de masse atomique
2530	2546	Untergrund
2531	2552	valeur d'une division
2532	1994	valeur de R_B
2533	1008	valeur de G
2534	2026	valeur étalon de référence de pH
2535	0560	valeurs
2536	1538	valve à aiguille
2537	2309	vaporisateur
2538	2553	vaporisation
2539	0118	vaporisation anodique
2540	0321	vaporisation cathodique
2541	2559	variable aléatoire
2542	2558	variance
2543	1313	variation linéaire de potentiel
2544	0305	vecteur
2545	1544	verre au néodyme
2546	0187	vie moyenne
2547	1411	vie moyenne
2548	1958	vie moyenne radioactive
2549	0054	vieillissement
2550	0349	vieillissement chimique
2551	1754	vieillissement physique
2552	2414	vieillissement thermique
2553	2547	vitesse d'apport
2554	0141	vitesse d'aspiration
2555	1990	vitesse d'aspiration de liquide
2556	2581	vitesse d'écoulement (débit) volumétrique
2557	1648	vitesse d'écoulement de sortie
2558	0907	vitesse d'écoulement, débit
2559	1023	vitesse de chauffage
2560	0513	vitesse de comptage, taux de comptage
2561	0282	vitesse de combustion
2562	1991	vitesse de consommation de liquide
2563	0666	vitesse de désintégration
2564	2088	vitesse de montée
2565	1992	vitesse de nucléation
2566	2311	vitesse de pulvérisation superficielle
2567	2568	vitesse de volatilisation
2568	1166	vitesse interstitielle
2569	1167	vitesse interstitielle à la pression de sortie
2570	1309	vitesse linéaire d'écoulement
2571	0185	vitesse moyenne d'écoulement

2572	1409	vitesse moyenne interstitielle du gaz vecteur
2573	0514	vitesses de comptage
2574	2253	vitesses de répétition d'étincelles
2575	0931	voile
2576	2567	volatilisation
2577	2157	volatilisation sélective
2578	2576	voltampérogramme
2579	2577	voltampérogramme
2580	2575	voltampérométrie
2581	1316	voltampérométrie à balayage linéaire
2582	0117	voltampérométrie à redissolution anodique
2583	2513	voltampérométrie à variation triangulaire du potentiel
2584	0553	voltampérométrie cyclique à balayage triangulaire
2585	0597	voltampérométrie dérivée
2586	0634	voltampérométrie différentielle
2587	0486	voltampérométrie en balayage linéaire à intégrale de convolution
2588	1066	voltampérométrie hydrodynamique
2589	2338	voltampérométrie sur électrode stationnaire
2590	0786	volume d'élution
2591	1689	volume d'élution (Gel Perm. Chrom.) ou volume de rétention (CPG et HPLC)
2592	0412	volume de colonne
2593	0265	volume de percée
2594	2341	volume de phase stationnaire
2595	2082	volume de rétention
2596	0042	volume de rétention ajusté
2597	0500	volume de rétention corrigé
2598	0216	volume du lit
2599	0600	volume identifié
2600	1168	volume interstitiel
2601	1290	volume limité
2602	1341	volume liquide
2603	0566	volume mort
2604	1050	volume mort (de l'appareil)
2605	0970	volume mort d'un chromatographe à gaz
2606	1552	volume net de rétention
2607	2175	volume sensible d'un détecteur
2608	2264	volume spécifique de rétention
2609	2412	volume théorique de rétention
2610	2485	volume total de rétention
2611	2516	vraie coincidence
2612	2619	zone
2613	1613	zone d'observation de plasma
2614	0313	zone de chute cathodique (de tension)
2615	0981	zone de Geiger-Müller
2616	1647	zone externe
2617	1141	zone intercônes (de combustion)
2618	1113	zone interne
2619	1542	zones négatives

ADDENDA

A01	A07	électrode de référence interne
A02	A05	noircissement
A03	A01	obturateurs acoustico-optiques
A04	A02	spectroscopie d'absorption atomique
A05	A03	spectroscopie d'émission atomique
A06	A09	spectroscopie d'émission optique
A07	A06	spectroscopie de flamme
A08	A04	spectroscopie de fluorescence atomique
A09	A10	spectroscopie de fluorescence par rayons X
A10	A08	spectroscopie de luminescence moleculaire

German

0001	1442	Abbau eines metastabilen Ions
0002	1079	abbildende Systeme
0003	0284	Abbrand
0004	2510	Abfangen
0005	0884	abfiltrieren
0006	0635	abgelenkte Elektronen
0007	2137	abgeschirmt
0008	2140	abgeschmolzene Hohlkathodenlampe
0009	2182	abgetrennte Flamme, getrennte Flamme
0010	1947	Abklingen der Radioaktivität
0011	0489	Abkühlungsgeschwindigkeitskurven
0012	0488	Abkühlungskurven
0013	1997	Ablesbarkeit
0014	1998	Ablesung
0015	0001	abnormale Glimmentladung
0016	1164	Abreißbögen
0017	0165	Abschwächung, Schwächung
0018	0005	absolute Temperatur
0019	0002	absolute Zählung
0020	0004	absoluter Photopeakwirkungsgrad
0021	0003	absoluter Vollenergiepeakwirkungsgrad
0022	0501	Absolutwägungskorrektur
0023	0009	Absorber
0024	0007	absorbierte Dosis
0025	0008	absorbierte Elektronen
0026	0010	Absorption
0027	0014	Absorptionsenergie
0028	0015	Absorptionsfaktor
0029	0012	Absorptionskantensprungverhältnis
0030	0011	Absorptionskoeffizient
0031	0019	Absorptionsprofil
0032	0018	Absorptionsweglänge
0033	0611	Abweichung
0034	0027	Acidimetrie
0035	0026	acidimetrische Titration
0036	0041	Additivität
0037	0043	Adsorbat, Adsorbend
0038	0044	Adsorbens, Adsorptionsmittel
0039	0045	Adsorption
0040	0046	Adsorptionschromatographie
0041	1291	Adsorptionsgrenzstrom
0042	0048	Adsorptionsindikator
0043	0047	Adsorptionsstrom
0044	0049	Aerosol
0045	0050	Aerosolträgergas
0046	0051	Agglomeration
0047	0052	Aggregat
0048	0053	Aggregation, Zusammenballung
0049	0036	aktiver Feststoff
0050	0035	aktives Medium
0051	0031	aktivierbares Material
0052	0032	Aktivierung
0053	0338	Aktivierung mit geladenen Teilchen
0054	1746	Aktivierung mit Photonen
0055	0033	Aktivierungsanalyse
0056	0034	Aktivierungsquerschnitt
0057	0037	Aktivität
0058	0038	Aktivitätskoeffizient
0059	0039	Aktivitätskonzentration
0060	1007	Aktivitätswachstumskurve
0061	0030	akustisches Stoßwellen-Siebsystem
0062	0057	Alkalimetrie
0063	0056	alkalimetrische Titration
0064	0058	alkalisieren
0065	0060	Alphakoeffizient
0066	0062	Alphaparameter
0067	0059	Alphaspaltung
0068	0063	Alphateilchen
0069	0061	Alphazerfall
0070	0054	Alterung
0071	0072	Aminopolycarbonsäure
0072	0075	Amperometrie
0073	0073	amperometrische Endpunkterkennung
0074	0074	amperometrische Titration
0075	0076	amphiprotisches Lösungsmittel
0076	2583	Amplitude
0077	0078	Analog-Digital-Wandler
0078	1763	Analysator mit magnetischem π-Sektorfeld
0079	1764	Analysator mit magnetischem π/2-Sektorfeld
0080	1765	Analysator mit magnetischem π/3-Sektorfeld
0081	1914	Analysator mit radialem elektrostatischen Feld
0082	0806	Analyse mit verstärkter Phosphoreszenz
0083	0096	Analysenergebnis
0084	0093	Analysenfunktionen
0085	0088	Analysenkurve
0086	0089	Analysenkurvenverfahren
0087	0097	Analysenproben
0088	0098	Analysenstandards, analytische Standards
0089	0099	Analysensysteme, analytische Systeme
0090	0085	Analysenwaage
0091	0079	Analyt
0092	0080	Analyt
0093	0082	Analytatome
0094	0083	Analytionen
0095	0092	analytische Auswertungskurven

0096	0087	analytische Eichfunktionen, Kalibrationsfunktionen
0097	0090	analytische Elektrode
0098	0091	analytische Elektronenmikroskopie
0099	0086	analytische Kalibration
0100	0095	analytische Radiochemie
0101	1966	analytische Radiochemie
0102	0094	analytische Strecke
0103	1104	Anfangsstrom
0104	1105	Anfangstemperatur
0105	0130	angelegte Spannung
0106	0843	angeregte Zustände
0107	0104	Anionenaustausch
0108	0105	Anionenaustauscher
0109	0103	Anioneneffekt
0110	0106	Anionenkomplex
0111	0107	Anisotropie
0112	0109	Annihilation, Vernichtung
0113	0110	Annihilationsstrahlung, Zerstrahlung
0114	0112	Anodendunkelraum
0115	0114	Anodenglimmhaut
0116	0113	Anodenhohlraum, Anodenlücke
0117	0118	anodische Verdampfung
0118	1399	Anpassungskondensator
0119	1106	anregen (einen Bogen), zünden (einen Bogen)
0120	0828	Anregung
0121	1940	Anregung durch Strahlung
0122	0829	Anregungs-Emissionsspektrum
0123	1720	Anregungs-Emissionsspektrum bei der Phosphoreszenz
0124	0841	Anregungs-Fluoreszenzspektrum
0125	0834	Anregungsart
0126	0830	Anregungsenergie
0127	0844	Anregungsenergie
0128	0833	Anregungsmechanismus
0129	0835	Anregungsmonochromator
0130	0832	Anregungsniveaus
0131	0836	Anregungspotential
0132	0838	Anregungsquelle
0133	0839	Anregungsspektren
0134	0840	Anregungstemperatur
0135	0837	Anregungszone, Anregungsgebiet
0136	0808	Anreicherung
0137	0809	Anreicherungsfaktor
0138	0139	ansaugen
0139	0140	Ansaugen, Aspiration
0140	0141	Ansaugrate
0141	2367	Ansaugzerstäuber
0142	2072	Ansprechempfindlichkeit
0143	2069	Ansprechkonstante
0144	2070	Ansprechneigung
0145	2071	Ansprechzeit
0146	2088	Anstiegsgeschwindigkeit
0147	2087	Anstiegszeit
0148	2340	Anteil der stationären Phase, Fraktion der stationären Phase
0149	0119	Anti-Compton-Gammastrahlung
0150	0120	Anti-Compton-Gammastrahlungs-spektrometer
0151	0122	anti-Stokes Fluoreszenz
0152	0121	Antipartikel, Antiteilchen
0153	0123	Aperturblende (Durchmesser)
0154	0124	Apex
0155	0819	Äquivalent
0156	0818	Äquivalenzpunkt
0157	2596	Arbeitsgas
0158	0142	Aspirator
0159	0143	Astonscher Dunkelraum
0160	0144	Asymmetriepotential
0161	0146	Atomabsorption
0162	0147	Atomabsorptionsspektrometrie
0163	0149	Atomemission
0164	0150	Atomemissionsspektrometrie
0165	0151	Atomfluoreszenz
0166	0152	Atomfluoreszenzspektrometrie
0167	0160	Atomgewicht, relative Atommasse
0168	0163	atomisierte Fraktion
0169	0161	Atomisierung
0170	1260	Atomisierung und Anregung durch Laserstrahlung
0171	0162	Atomisierungswirkungsgrad
0172	0153	Atomlinie
0173	0154	Atommasse
0174	0155	Atommasseneinheiten
0175	0158	Atomspektrallinien
0176	0157	Atomspektren
0177	0159	Atomspektroskopie
0178	2050	Aufenthaltzeit
0179	1023	Aufheizgeschwindigkeit
0180	1022	Aufheizkurven
0181	1021	Aufheizkurvenbestimmung
0182	0675	Auflösung (*s.* Anmerkung nach 'receiving time')
0183	2058	Auflösung
0184	0609	Auflösung eines Detektors
0185	2293	Auflösung eines Spektrometers
0186	2060	Auflösungsvermögen
0187	2003	Auflösungszeit
0188	2061	Auflösungszeit
0189	2062	Auflösungszeitkorrektur
0190	2547	Aufnahmerate
0191	0081	Aufstockverfahren, Analytadditionsverfahren
0192	0084	Aufstockverfahren
0193	1082	auftreffendes Photon
0194	0127	Auftrittsenergie
0195	0128	Auftrittspotential
0196	0129	Auftrittstemperatur
0197	2006	Aufzeichnung
0198	0173	Auger-Ausbeute
0199	0167	Auger-Effekt
0200	0168	Auger-Elektron
0201	0170	Auger-Elektronenausbeute
0202	0169	Auger-Elektronenspektroskopie
0203	0171	Auger-Peakenergie
0204	0172	Auger-Spektroskopie
0205	0280	Ausbrenneffekt
0206	0904	Ausflockung
0207	1885	ausgepumptes System
0208	2098	Aussalzen, Aussalzung
0209	2099	Aussalzschromatographie
0210	1647	äußere Zone
0211	0852	äußerer Schweratomeffekt
0212	0854	äußerer Standard
0213	1648	Ausströmgeschwindigkeit, Ausflußgeschwindigkeit
0214	0821	Austrittstiefe
0215	0176	Automation
0216	0175	automatisieren

0217	0178	automatisieren
0218	0177	Automatisierung
0219	0179	Autoradiograph
0220	0180	Autoxidation
0221	0189	axiale Elektrode
0222	0202	Baker-Sampson-Seidel-Transformation
0223	0205	Bandbreite
0224	0204	Bandenspektren
0225	0206	Barn
0226	0207	Base
0227	1857	Basis eines Prismas
0228	0209	Basislinie
0229	0210	Basislinienverfahren in der Atomspektroskopie
0230	0211	Basispeak
0231	0212	Basizität
0232	0213	Bates-Guggenheim-Konvention
0233	0214	bathochrome Verschiebung
0234	0539	Becherelektrode
0235	0215	Becquerel
0236	0438	Begleitbestandteile, Begleitstoffe
0237	0443	Begrenzerplatte
0238	0451	begrenzte Glimmentladung
0239	1290	begrenztes Volumen
0240	0293	Belastbarkeit (einer Waage)
0241	1342	Belastung
0242	1611	Beobachtung
0243	1027	Beobachtungshöhe
0244	1612	Beobachtungshöhe
0245	1613	Beobachtungszone des Plasmas
0246	0148	Beschuß mit Atomen
0247	0873	Beschuß mit schnellen Atomen
0248	0380	Beseitigung unerwünschter Gase, Reinigung
0249	1798	Besetzungsinversion
0250	0425	Bestandteil, Komponente
0251	0448	Bestandteil, Komponente
0252	0449	Bestandteileinteilung, Bestandteilklassifizierung
0253	2590	Bestimmung der Massenänderung
0254	1298	Bestimmungsgrenze
0255	1224	Bestrahlung
0256	1915	Bestrahlungsdichte, Strahlungsdichte, spezifische Strahlungsleistung
0257	1223	Bestrahlungsstärken
0258	0218	Betaspaltung
0259	0220	Betateilchen
0260	0219	Betazerfall
0261	0216	Bettvolumen
0262	0529	Beugung am Kristall
0263	1359	Beugung langsamer (niederenerget-ischer) Elektronen (LEED)
0264	2018	Bezugselektrode, Referenzelektrode
0265	2019	Bezugselement, Referenzelement
0266	2020	Bezugselementtechnik, Referenzelementtechnik
0267	2022	Bezugsintensität, Referenzintensität
0268	2023	Bezugslinie, Referenzlinie
0269	2025	Bezugslösungen, Referenzlösungen
0270	2026	Bezugswert pH-Standardlösung
0271	0223	Biamperometrie
0272	0221	biamperometrische Endpunkterkennung
0273	0222	biamperometrische Titration
0274	0226	bifunktioneller Ionenaustauscher
0275	1766	Bildpunkt
0276	0934	Bildungskonstante
0277	0228	Bindung
0278	0229	Bindungsenergie
0279	0231	bipotentiometrische Titration
0280	0232	Bipotentiometrie
0281	0247	Blauverschiebung
0282	0243	Blazewellenlänge
0283	0242	Blazewinkel
0284	0244	bleichbare Substanz
0285	0240	Blindlösung
0286	0235	Blindprobe
0287	0238	Blindwert
0288	0237	Blindwertkorrektur
0289	0239	Blindwertstreuung
0290	0902	Blitzlampe, Blitzlichtlampe
0291	0901	Blitzlichtfluorimetrie
0292	0245	Block (Probenträgeranordnung bei der DTA)
0293	0248	Boersma-Anordnung
0294	0131	Bogen (siehe auch Lichtbogen)
0295	2010	Bogen mit gleichgerichtetem Wechselstrom
0296	0136	bogenartige Entladung
0297	0132	Bogenatmosphäre
0298	0135	Bogenlampe
0299	0137	Bogenlinie
0300	0134	Bogenstrecke
0301	0251	Bolometer
0302	0257	Braggsche Gleichung
0303	0256	Braggscher Winkel, Glanzwinkel
0304	2033	Brechungsindex
0305	2593	Breite
0306	0266	Bremsstrahlung
0307	1477	Bremsung, Moderierung
0308	2355	Bremsvermögen
0309	1839	Brenner für vorgemischte Gase
0310	0281	Brennfleck
0311	0952	brenngasreiche Flamme
0312	0282	Brenngeschwindigkeit
0313	0927	Brennpunkt
0314	0928	Brennpunktdurchmesser
0315	0283	Brennspannung
0316	0949	Brennstoff
0317	0951	Brennstoffelement
0318	0926	Brennweite
0319	0876	Brutmaterial, Brutstoff
0320	0877	Brutnuklide
0321	1650	Bruttokomplexbildungskonstante
0322	1649	Bruttoverteilungskonstante
0323	0279	Bunsenbrenner
0324	0270	Büschelentladung
0325	0286	Bypass-Probengeber
0326	0289	Calvet-Anordnung
0327	0334	charakteristische Konzentration
0328	0335	charakteristische Masse
0329	0337	charakteristische Röntgenstrahlung
0330	0336	charakteristisches Potential
0331	0347	chelatometrische Titration
0332	0359	Chemiionisation
0333	0360	Chemilumineszenz
0334	0361	Chemilumineszenzanalyse
0335	0362	Chemilumineszenzindikator
0336	0349	chemische Alterung

0337	0358	chemische Ausbeute
0338	0353	chemische Flammen
0339	0355	chemische Ionisation
0340	0354	chemische Störung
0341	0357	chemische Verschiebung
0342	0356	chemischer Isotopenaustausch
0343	0348	chemisches Actinometer
0344	0352	chemisches Äquivalent
0345	0350	chemisches Dosimeter
0346	0351	chemisches Gleichgewicht
0347	0363	Chemisorption
0348	0365	Chromatogramm
0349	0366	Chromatograph
0350	0368	Chromatographie
0351	0367	chromatographieren
0352	0369	Chronoamperometrie
0353	0115	Chronoamperometrie mit anodischer Auflösung durch linearen Potentialanstieg
0354	0697	Chronoamperometrie mit doppelter Potentialstufe
0355	0371	Chronocoulometrie
0356	0698	Chronocoulometrie mit doppelter Potentialstufe
0357	0370	chronocoulometrische Konstante
0358	0375	Chronopotentiometrie
0359	0543	Chronopotentiometrie mit unterbrochenem Strom
0360	0373	chronopotentiometrische Endpunkterkennung
0361	0372	chronopotentiometrische Konstante
0362	0374	chronopotentiometrische Titration
0363	0392	Co-Ionen
0364	0428	Compton-Effekt
0365	0429	Compton-Elektron
0366	0430	Compton-Streuung
0367	0427	Compton-Winkel
0368	0504	Coulometrie
0369	0503	coulometrische Titration
0370	0304	CRA-Ofen
0371	0523	Crookesscher Dunkelraum
0372	0540	Curie
0373	0550	cyclische Chronopotentiometrie
0374	0552	cyclische Dreieckswellenpolarography
0375	0553	cyclische Dreieckswellenvoltammetrie
0376	0551	cyclische Stromstufenchrono-potentiometrie
0377	0555	Cyclotronresonanzmassen-spektrometer
0378	2555	Dampfstrahl
0379	2557	Dampfstrahl
0380	0558	Dämpfungskonstante
0381	2554	Dampfwolke
0382	0560	Daten
0383	0460	Dauerstrich
0384	0461	Dauerstrichbetrieb
0385	0563	Dead-stop-Endpunkt
0386	0567	Debye-Hückel-Gleichung
0387	0579	Delves-Cup
0388	0580	Demaskierung
0389	0581	demineralisiertes Wasser
0390	0582	Demodulationspolarographie
0391	0583	Densitometer
0392	0584	Depolarisator
0393	0587	derivativ
0394	0592	derivativ-potentiometrische Titration
0395	0588	Derivativchronopotentiometrie, derivative Chronopotentiometrie
0396	0589	Derivativdilatometerie, derivative Dilatometrie
0397	0590	derivativer Endpunkt
0398	0591	Derivativpolarographie, derivative Polarographie
0399	0593	Derivativpulspolarographie, derivative Pulspolarographie
0400	0594	Derivativspektrophotometrie, derivative Spektrophotometrie
0401	0595	Derivativthermoanalyse, derivative thermische Analyse
0402	0596	Derivativthermogravimetrie, derivative Thermogravimetrie
0403	0597	Derivativvoltammetrie, derivative Voltammetrie
0404	0599	Derivatographie
0405	0598	derivatographische Analyse
0406	0562	Desaktivierungsprozeß
0407	0602	Desolvatation
0408	0601	desolvatisierte Fraktion
0409	0676	Destillation
0410	0604	Detektionsausbeute, Detektionswirkugsgrad
0411	0606	Detektor
0412	0608	Detektorwirkungsgrad
0413	0549	Dicyanmolekülbanden
0414	0614	Dielektrizitätskonstante
0415	0616	Dielektrometrie
0416	0615	dielektrometrische Titration
0417	0617	differential
0418	0633	Differential-Thermogravimetrie
0419	0619	Differentialchromatogramm
0420	0620	Differentialchromatographie
0421	0621	Differentialdetektor
0422	0622	Differentialdilatometrie
0423	0627	Differentialpumpen
0424	0628	Differentialscanningkalorimetrie
0425	0629	Differentialthermoanalyse
0426	0630	Differentialthermoanalyse in einer isothermen Umgebung
0427	0631	Differentialthermogramm
0428	0632	differentielle thermogravimetrische Analyse
0429	0618	Differenz-Amperometrie
0430	0626	Differenz-Pulspolarographie
0431	0623	Differenzpolarographie
0432	0625	Differenzpotentiometrie
0433	0624	differenzpotentiometrische Titration
0434	0634	Differenzvoltammetrie
0435	0637	Diffusion
0436	1294	Diffusionsgrenzstrom
0437	0638	Diffusionskoeffizient
0438	1327	Diffusionspotential
0439	0636	Diffusionsschicht-Halbleiterdetektor
0440	0639	Diffusionsstrom
0441	0640	Diffusionsstromkonstante
0442	0641	Digital-Analog-Wandler
0443	0642	Dilatometrie
0444	0645	Diodenarraydetektor
0445	0651	direkte Spaltausbeute

0446	0654	Direkteinlaßsystem
0447	0652	Direktinjektionsbrenner
0448	0653	Direktlinienfluoreszenz
0449	0663	diskontinuierliche simultane Methoden
0450	0662	diskontinuierliche Verfahren
0451	0664	Diskriminator
0452	0667	Dispersion
0453	0668	Dispersion bei der Fließinjektionsanalyse
0454	0669	Dispersion eines Materials
0455	0671	Dissoziation
0456	0573	Dissoziationsgrad
0457	0672	Dissoziationskonstante
0458	0674	Dissoziationspotential
0459	0677	Distribuent, verteilte Substanz
0460	0100	Divergenzwinkel
0461	0685	Doerner-Hoskins-Gleichung
0462	0691	Doppelbogen
0463	0694	Doppelescapepeak
0464	0695	doppelfokussierendes Massenspektrometer
0465	0696	Doppelschichtstrom
0466	0701	Doppeltonpolarographie
0467	0687	Doppler-Halbwertsbreite einer Spektrallinie, Doppler-Halbintensitätsbreite
0468	0686	Doppler-Verbreiterung
0469	0688	Doppler-Verschiebung
0470	0689	Dosis
0471	0690	Dosisäquivalent
0472	2512	Dreieckwellenpolarographie
0473	2513	Dreieckwellenvoltammetrie
0474	2462	Dreielektrodenanordnung
0475	2463	Dreielektrodenplasma
0476	2464	Dreischlitzbrenner
0477	0702	Drift
0478	1841	Druck
0479	1844	Druck-Spannung-Strom-Verhältnis
0480	1842	Druckeinfluß
0481	1843	Druckgradient-Korrekturfaktor
0482	0710	Duane-Hunt-Grenzwellenlänge
0483	0559	Dunkelstrom
0484	2459	Dünnfilm
0485	2460	Dünnfilmelektrode
0486	2461	Dünnschichtchromatographie
0487	0703	durchbohrte Elektrode
0488	0264	Durchbruchskapazität
0489	0265	Durchbruchsvolumen
0490	0456	Durchflußanalyse
0491	2500	Durchgangszeit
0492	0713	dynamische Differentialcalorimetrie
0493	0716	dynamische thermogravimetrische Analyse
0494	0717	dynamische Thermomechanometrie
0495	0714	dynamisches Massenspektrometer
0496	0715	dynamisches System
0497	2516	echte Koinzidenz
0498	2000	Echtzeit
0499	1566	Edelgase
0500	0718	Effekt
0501	1284	effektive Basislänge eines Prismas
0502	0719	effektive Cadmiumabscheideenergie
0503	1025	effektive theoretische Trennstufenhöhe
0504	0722	effektive Trennstufenzahl
0505	0721	effektive unendliche Dicke
0506	0723	effektiver thermischer Wirkungsquerschnitt
0507	0720	effektives Dosisäquivalent
0508	0727	Eigenfrequenzen
0509	1618	ein-aus Verhältnis
0510	0246	Einblas-Verfahren
0511	1616	Einelektrodenmikrowellenplasma
0512	2205	Einelementlampe
0513	2207	einfachfokussierendes Massenspektrometer
0514	2211	einfachionisiertes Element
0515	1085	einfallende Leistung
0516	1086	einfallende Strahlung
0517	0299	Einfang
0518	0301	Einfanggammastrahlung
0519	0300	Einfangquerschnitt
0520	1615	einfarbiger Indikator
0521	2324	eingestellte Lösung, Standardlösung
0522	2325	eingestellte Maßlösung
0523	2204	Einkanalimpulshöhenanalysator
0524	1515	einkernige binäre Komplexe
0525	1516	einkernige ternäre Komplexe
0526	1111	Einlaßsystem
0527	2097	Einsalzen
0528	1087	Einschluß
0529	0452	Einschnürung des Bogenplasmas
0530	2541	einseitig gerichtete Entladung
0531	2542	einseitig gerichteter Funken, gerichtete Funkenentladung
0532	0728	Einstein-Koeffizient für spontane Emission
0533	0729	Einstein-Übergangswahrscheinlichkeit
0534	2522	einstellbare Stichleitung
0535	2203	Einstrahlspektrometer
0536	1617	einstufige Anregungsquelle
0537	0813	Eintrittsspaltbreite
0538	0812	Eintrittsspalthöhe
0539	2208	Einzelimpuls
0540	0732	elastische Streuung
0541	0731	elastischer Stoß
0542	0741	electroactive Substanz
0543	0736	elektrische Anregung
0544	0735	elektrische Entladung
0545	1363	elektrische Entladung bei geringem Druck
0546	2190	elektrische Entladung in Schutzgasatmosphäre
0547	0739	elektrische Funken
0548	0734	elektrische Leitfähigkeit
0549	0733	elektrischer Lichtbogen
0550	0737	elektrischer Parameter
0551	0740	elektrischer Strom
0552	1475	Elektrode mit mobilem Träger
0553	0745	Elektrode Resonanzhohlraumplasma
0554	0746	Elektroden-Lösungs-Grenzfläche
0555	0743	elektrodenlose Anregung
0556	0742	elektrodenlose Entladungslampe
0557	0747	Elektroerosion
0558	0748	Elektrographie
0559	0749	Elektrogravimetrie
0560	0750	Elektrolyt

0561	0268	Elektrolyt für eine Doppelstrom-schlüssel-Bezugselektrode
0562	0751	elektromagnetische Strahlung
0563	0752	elektromotorische Kraft
0564	0755	Elektronenbeugung
0565	2155	Elektronenbeugung an einer selektierten Fläche
0566	0763	Elektronendruck
0567	0753	Elektroneneinfang
0568	0754	Elektroneneinfangdetektor
0569	0756	Elektronenenergie
0570	0758	Elektronenenergieübertragungs-prozess
0571	0757	Elektronenenergieverlust-spektroskopie
0572	0761	Elektronenmikroskopie
0573	0766	Elektronenspektren
0574	0767	Elektronenspektroskopie für chemische Analyse
0575	0759	Elektronenstoßionisierung
0576	0765	Elektronenstrahl-Röntgen-Mikroanalyse
0577	0764	Elektronenstrahlmikroanalyse
0578	0768	Elektronentemperatur
0579	0762	Elektronenvervielfachung
0580	0769	elektrooptische Verschlüsse
0581	0771	Elektropherogramm, Elektrophoretogramm
0582	0770	Elektrophorese
0583	0772	Elektroseparation
0584	0773	Elektroseparation mit kontrolliertem Potential
0585	0774	Elektrospray
0586	0775	Element
0587	0776	Elementanalyse
0588	0777	Elementarteilchen
0589	0778	elliptische Polarisation
0590	0779	Eluat
0591	0780	Eluent
0592	0781	eluieren
0593	0782	Elutionsbande
0594	0783	Elutionschromatographie
0595	0784	Elutionskurve
0596	0785	Elutionspeak
0597	0786	Elutionsvolumen
0598	0787	Emanationsthermoanalyse
0599	0730	Emissionsenergie
0600	0788	Emissionsmonochromator
0601	0789	Emissionsprofil
0602	0790	Emissionsspektren
0603	2175	empfindliches Detektorvolumen
0604	2176	Empfindlichkeit
0605	2180	Empfindlichkeit einer Ionenquelle
0606	2178	Empfindlichkeit eines Direkteinlaßsystems
0607	2179	Empfindlichkeit eines Einlaßsystems
0608	2177	Empfindlichkeit für eine festgelegte Masse
0609	0793	Endotherm
0610	0794	endothermer Peak
0611	0795	Endpunkt
0612	0800	Energieauflösung
0613	1817	Energieausbeute bei der Fluoreszenz, Leistungsausbeute
0614	0805	Energieausbeute bei der Lumineszenz
0615	0797	energiedispersive Röntgenfluor-eszenzanalyse
0616	0798	Energieflußdichte
0617	0799	Energieniveaus
0618	0802	Energieschwelle
0619	0801	Energieskala
0620	0804	Energieübergang
0621	0803	Energieübertragung
0622	0796	Energieverteilung
0623	0810	enthalpimetrische Endpunkterkennung
0624	0811	enthalpimetrische Titration
0625	0576	entionisiertes Wasser, deionisiertes Wasser
0626	0655	Entladung
0627	0661	Entladungen im Vakuum
0628	0656	Entladungsatmosphäre
0629	0659	Entladungsgas
0630	0660	Entladungspolarographie
0631	0658	Entladungsstrom
0632	0657	Entladungsstromkreis
0633	0610	entwickeln
0634	0814	Enzymelektrode, Enzym-Substrat-Electrode
0635	0815	epicadmische Neutronen
0636	0816	epithermische Neutronen
0637	2075	Ergebnis, Resultat
0638	1264	Erosion durch Laserstrahlung
0639	2486	erster Townsend-Koeffizient
0640	2206	Escapepeak
0641	0827	Excimer-Lumineszenz
0642	0845	exoenergetische Bedingung
0643	0846	exotherm
0644	0848	exotherme Reaktionen
0645	0847	exothermer Peak
0646	0849	experimentell erfaßte Oberfläche
0647	0850	exponentieller Zerfall
0648	0851	Exposition
0649	0006	Extinktion, Absorbanz
0650	0021	Extinktionskoeffizient
0651	0857	Extrahierbarkeit
0652	0856	extrahieren, ausschütteln
0653	0855	Extrakt
0654	0859	Extraktion
0655	0862	Extraktionsindikator
0656	0860	Extraktionskoeffizient
0657	0861	Extraktionskonstante
0658	0858	Extraktionsmittel
0659	0864	extrapolierte Bereich
0660	0863	extrapolierter Anfang
0661	0879	Fadenchromatographie
0662	0865	Faktorgewicht, Faktor-Einwaage
0663	1824	fällen
0664	1823	Fällung, Niederschlag
0665	1825	Fällung aus homogener Lösung
0666	1826	Fällungsindikator
0667	1827	Fällungstitration
0668	0486	Faltungsintegral-Voltammetrie mit linearem Spannungsanstieg
0669	0866	Fanofaktor
0670	0871	Faraday-Becher-Kollektor
0671	0870	Faraday-Konstante
0672	0868	Faradayscher Demodulationsstrom
0673	0872	Faradayscher Dunkelraum
0674	0869	Faradayscher Gleichrichtungsstrom

0675	1550	Faradayscher Nettostrom
0676	0867	Faradayscher Strom
0677	2564	Farbindikator, visueller Indikator
0678	0712	Farbstofflasern
0679	0820	Fehler
0680	0878	Feldionisierung
0681	2048	ferngekoppelt
0682	2221	feste Lösung
0683	2222	feste Lösung als Meßprobe
0684	2220	feste Probe
0685	2226	fester Träger
0686	0893	Festfrequenz
0687	2223	Festkörper-Kristallgitter
0688	2224	Festkörperdetektor
0689	2225	Festkörperlaser
0690	2085	Festmatrixelektroden
0691	0882	Filmdosimeter
0692	0883	Filter
0693	0895	Flamme
0694	0898	Flammenspektrometrie
0695	0899	Flammentemperatur
0696	0896	Flammenuntergrund
0697	2306	Fleck, Tüpfel
0698	0907	Fließgeschwindigkeit
0699	0905	Fließinjektionsanalyse
0700	2237	Fließmittelwanderungsabstand
0701	2469	Flugzeitmassenspektrometer
0702	0908	Fluoreszenz
0703	0909	Fluoreszenzanalyse
0704	0915	Fluoreszenzausbeute
0705	0911	Fluoreszenzleistungswirkungsgrad
0706	1437	Fluoreszenzmetallindikator
0707	0912	Fluoreszenzquantenwirkungsgrad
0708	0913	Fluoreszenzspektrometer
0709	0910	Fluoreszenzwirkungsgrad
0710	0916	fluorimetrische Endpunkterkennung
0711	0917	fluorimetrische Titration
0712	0919	Fluß
0713	2531	Flußdichte thermischer Neutronen
0714	0923	Flußmonitor
0715	0906	flußprogrammierte Chromatographie
0716	0924	Flußstörung
0717	0921	Flußverringerung
0718	1324	Flüssig-Extraktion
0719	1339	Flüssig-Fest-Chromatographie
0720	1329	Flüssig-Flüssig-Chromatographie
0721	1332	Flüssig-Flüssig-Extraktion
0722	1331	Flüssig-Flüssig-Gleichgewichte
0723	1333	Flüssig-Flüssig-Grenzfläche
0724	1330	Flüssig-Flüssig-Verteilung
0725	1325	Flüssig-Gel-Chromatographie
0726	1334	flüssige Phase
0727	1335	flüssige Probe
0728	1340	flüssige stationäre Phase
0729	1326	Flüssigkeits-Grenzfläche
0730	1338	Flüssigkeits-Szintillationsdetektor
0731	1990	Flüssigkeitsansauggeschwindigkeit
0732	1323	Flüssigkeitschromatographie
0733	1328	Flüssigkeitslaser
0734	1991	Flüssigkeitsverbrauchsrate, Flüssigkeitsverbrauch pro Zeiteinheit
0735	1341	Flüssigkeitsvolumen
0736	0932	Foliendetektor
0737	0933	Formalität
0738	0935	Fourier-Transform-Massenspektrometer
0739	0937	Fragmention
0740	0936	fraktionierte Destillation
0741	0938	Free-bound-Übergang, Frei-Gebunden-Übergang
0742	0941	Free-free-Übergang, Frei-Frei-Übergang
0743	0939	freibrennender Bogen
0744	0940	freie Elektronen
0745	0942	freilaufender Betrieb
0746	2044	Freisetzungmittel
0747	2046	Freisetzungsmechanismus
0748	2045	Freisetzungsmittel
0749	0853	fremdgezündete Funkenentladung
0750	0943	Frequenz
0751	0944	Frequenzvervielfachung
0752	0946	Frontalchromatographie
0753	0948	Frontflächen(geometrie)
0754	0947	Fronting
0755	0880	Füllgas
0756	1045	Funken mit großer Wiederholungsrate
0757	2255	Funken pro Halbwelle
0758	2252	Funkenartige Entladungen
0759	2246	Funkenentladung
0760	2550	Funkenentladung im Vakuum, Vakuumfunken
0761	2247	Funkenfrequenz
0762	2249	Funkengenerator
0763	2251	Funkengleichgewicht
0764	2244	Funkenkammer
0765	2254	Funkenquellenionisation
0766	2256	Funkenstative
0767	2248	Funkenstrecke
0768	2253	Funkenwiederholungs-geschwindigkeiten, Funkenwiederholungsrate
0769	2250	Funkenzündung
0770	1008	G-Wert
0771	0957	Gammakaskade
0772	0958	Gammaquant
0773	0963	Gammaspektrum
0774	0961	Gammastrahl
0775	0016	Gammastrahlenabsorption
0776	0962	Gammastrahlenspektrometer
0777	0959	Gammastrahlenspektrum, Gammaspektrum
0778	0960	Gammastrahlungseinfang
0779	0976	Gas-Fest-Chromatographie
0780	0972	Gas-Flüssig-Chromatographie
0781	0974	Gas-Proportionaldetektor
0782	0967	Gaschromatographie
0783	0973	gasdurchlässige Membran
0784	0968	Gasentladungen
0785	0725	Gasentwicklungsanalyse
0786	0825	Gasentwicklungsanalyse
0787	0726	Gasentwicklungsnachweis
0788	0826	Gasentwicklungsnachweis
0789	0969	gasförmig
0790	0971	Gaslaser
0791	0977	gasstabilisierte Bögen
0792	0978	Gastemperatur
0793	0970	Gastotvolumen
0794	0966	Gasverstärkung

0795	0313	Gebiet des Kathodenfalls, Bereich des Kathodenfalls
0796	0894	gebundene Ionen, Festionen
0797	0506	Gegenelektrode
0798	0507	Gegenionen
0799	1526	gegenseitige Störung
0800	0450	Gehalt an einem Bestandteil
0801	0981	Geiger-Müller-Bereich
0802	0982	Geiger-Müller-Schwelle
0803	0980	Geiger-Müller-Zähler (Zählrohr)
0804	0983	Geißler-Rohr
0805	2154	geimpfte Lichtbögen
0806	0515	gekoppelte simultane Methoden
0807	0524	gekreuzte elektrische und magnetische Felder
0808	0487	gekühlte Hohlkathodenlampe
0809	0339	geladene Teilchen
0810	0984	Gelfiltration
0811	0985	Gelpermeationschromatographie
0812	1414	gemessene Fluoreszenzanregung
0813	1415	gemessene Intensität
0814	1416	gemessene Phosphoreszenzanregung
0815	1417	gemessene Quantenausbeute der Lumineszenz
0816	1469	gemischte Komplexe
0817	1472	Gemischtligandkomplexe
0818	1473	Gemischtmetallkomplexe
0819	0988	Geometriefaktor
0820	0986	geometrische Größen
0821	0987	geometrische Schwächung
0822	1655	gepackte Säule
0823	0275	gepufferte Bögen
0824	1878	gepulste Entladungsröhre
0825	1879	gepulste Laserquellen
0826	2588	geringer Potentialgradient
0827	1360	geringer Verlust
0828	0989	Gerlachsches homologes Linienpaar
0829	2482	Gesamtabsorptionspeak
0830	0277	Gesamtkonzentration
0831	0278	Gesamtprobe
0832	0914	Gesamtquantenausbeute der Fluoreszenz
0833	2484	Gesamtquantenausbeute der Fluoreszenz
0834	2485	Gesamtretentionsvolumen
0835	2114	gesättigte Lösung
0836	2191	geschützte Flamme
0837	1166	Geschwindigkeit in Hohlraumbereich
0838	1167	Geschwindigkeit in Hohlraumbereich bei Ausgangsdruck
0839	0918	gespült
0840	0472	gesteuerter Hochspannungs-funkenerzeuger
0841	0465	gesteuerter Wechselstrombogen
0842	0990	Gettern, Getterung
0843	2007	Gewinnung, Wiederfindung
0844	1072	gezündeter Wechselstrombogen
0845	0992	Glaselektrode
0846	0993	Glaselektrodenfehler
0847	0817	Gleichgewichtskonstante
0848	0646	Gleichstrom
0849	0647	Gleichstrombogen
0850	0649	Gleichstromkonduktometrie
0851	0650	Gleichstrompolarographie
0852	0903	gleitende Frequenz
0853	2212	Gleitfunken
0854	0995	Glimmentladung
0855	2074	Glimmentladung mit begrenzter Kathodenfläche
0856	0254	Glimmentladung mit erhöhter Leistung
0857	1769	Glimmentladungsquelle mit ebener Kathode
0858	0996	Gradientenelution
0859	0998	Gradientenpackung
0860	0997	Gradientenschicht
0861	0999	Graphitbecheratomisator
0862	1001	Graphitfunken
0863	1002	Graphitrohrofen
0864	1000	Graphitstabofen
0865	1005	Gray (Gy)
0866	1143	Grenzfläche
0867	1293	Grenzstrom
0868	2117	Grenzwertplateau, Sättigungsplateau
0869	1006	Grundzustand
0870	2174	halbdurchlässiger Spiegel
0871	2173	halbgütegeschaltete Technik, teilgütegeschaltete Technik
0872	2167	halbleitende Oberfläche
0873	2168	Halbleiter
0874	2169	Halbleiterdetektor
0875	2171	halbmikro
0876	2172	Halbmikroprobe
0877	1016	Halbstufenpotential
0878	0955	Halbwertsbreite
0879	1697	Halbwertsbreite
0880	1009	Halbwertsbreite einer Absorptionslinie
0881	1010	Halbwertsbreite einer Emissionslinie
0882	1013	Halbwertsdicke
0883	1015	Halbwertsdicke
0884	1012	Halbwertspotential, Potential in halber Höhe
0885	1014	Halbwertsschicht
0886	1011	Halbwertszeit
0887	1017	Hammett-Funktion
0888	1018	Hammett-Indikator
0889	1019	harmonische Überlappung
0890	2056	Harzmatrix
0891	1386	Hauptbestandteile
0892	1856	Hauptgas
0893	1064	heiße Hohlkathode
0894	1063	heiße Zelle
0895	1062	heißes Atom
0896	1028	heterochromatische Photometrie
0897	1029	heterogene Membranelektrode
0898	0182	Hilfselektrode
0899	0181	Hilfsentladung
0900	0183	Hilfsfunkenstrecke
0901	1048	Hittorfscher Dunkelraum
0902	1046	hochauflösende Elektronenenergie-verlustspektroskopie
0903	1042	Hochdruck-Xenonlampe
0904	1041	Hochdrucklichtbogen
0905	1037	Hochfrequenz-Aufwärtstransformator
0906	1036	Hochfrequenz-Kurzschlußkondensator
0907	1035	Hochfrequenzkonduktometrie

0908	1034	hochfrequenzkonduktometrische Titration
0909	1033	hochfrequenzkonduktometrischer Endpunkt
0910	1974	Hochfrequenzplasmen, Radiofrequenzplasmen
0911	1038	Hochfrequenztitration
0912	1039	Hochintensitätshohlkathode
0913	1040	Hochniveau-Faradaysche Gleichrichtung
0914	1047	Hochspannungsfunken
0915	1030	Hochstrombogen
0916	1043	hoher Reflexionsgrad
0917	1054	Hohlelektrode, Hohlraumelektrode
0918	1051	Hohlkathodenentladung
0919	1052	Hohlkathodenlampe
0920	1053	Hohlkathodenlichtquellen
0921	1165	Hohlraumanteil
0922	1168	Hohlraumvolumen
0923	1057	homogene Membranelektrode
0924	1055	homogener Verteilungskoeffizient
0925	1056	homogenes Magnetfeld
0926	1058	homologe Linie
0927	1059	horizontale Elektrode
0928	1065	Hurter-Driffield-Kurve
0929	1066	hydrodynamische Voltammetrie
0930	1068	hyperchrome Verschiebung
0931	1069	hypochrome Verschiebung
0932	1070	hypsochrome Verschiebung
0933	1071	Hysterese (Hysteresis)
0934	1078	Ilkovič-Gleichungs-Vorzeichen-Regeln
0935	1080	Immunoassay
0936	1081	Impedanzkonverter
0937	1875	Impulsamplitudenanalysator
0938	1877	Impulsamplitudenselektor
0939	1876	Impulsamplitudenverteilung
0940	1880	Impulsdauer
0941	1989	Impulsdichtemesser, Impulsmeter
0942	1881	Impulsenergie
0943	1882	Impulshöhenanalysator
0944	1508	Impulsübertragung
0945	1874	Impulsverstärker
0946	1115	in-situ-Mikro-Röntgen-Beugung
0947	1114	in-situ-Mikroanalyse
0948	1090	Indikator
0949	1091	Indikatorblindwert
0950	1093	Indikatorelektrode
0951	1092	Indikatorkorrektur
0952	1095	indirekte Titration
0953	2349	individuelle Bildungskonstante
0954	1098	Induktionsspule
0955	1099	induktiv gekoppelte Plasmen
0956	1100	Induktor
0957	1096	induzierte Radioaktivität
0958	1097	induzierte Reaktionen
0959	1103	inerte Entladungsatmosphäre
0960	1579	inhomogene Magnetfelder
0961	1580	inhomogenes Material
0962	1336	Injektion einer flüssigen Probe
0963	1107	Injektionsgas
0964	1109	Injektionsrohr
0965	1088	inkohärenter Strahlungsimpuls
0966	0881	Innenlösung
0967	1158	Innenlösung
0968	1159	Innenlösung einer Glaselektrode
0969	1161	innere Bezugselektrode einer Glaselektrode
0970	1162	innere Bezugslinie
0971	1112	innere Bremsstrahlung
0972	1157	innere Elektrolyse
0973	1153	innere Extinktion
0974	1154	innere Konversion
0975	1160	innere Schwingungsamplitude
0976	1113	innere Verbrennungszone
0977	1172	innerer Detektorwirkungsgrad
0978	1155	innerer Konversionskoeffizient
0979	1175	innerer Photopeakwirkungsgrad
0980	1127	innerer Standard
0981	1163	innerer Standard
0982	1174	innerer Vollenergiepeakwirkungsgrad
0983	1173	innerer Wirkungsgrad
0984	1119	instrumentelle Aktivierungsanalyse
0985	1121	instrumentelle Neutronenaktivierungsanalyse
0986	1120	Instrumentenanzeige
0987	1122	Instrumentierung
0988	1124	Integralchromatogramm
0989	1125	Integraldetektor
0990	1123	integrale Absorption
0991	1126	integraler Reflexionskoeffizient
0992	0922	Integration des Flusses, Flußintegration
0993	1129	Intensität
0994	1136	Intensität einer Spektrallinie
0995	1137	Intensität relativ zum Basispeak
0996	1132	Intensitäts-(Schwärzungs-)Eichgerät
0997	1134	intensitätsmodulierter Hohlkathodenlampe
0998	1133	intensitätsmodulierter Lichtstrahl
0999	1130	Intensitätsübertragung
1000	1131	Intensitätsübertragungsverhältnis
1001	1139	Interatomare Stöße
1002	1140	interchromophorer strahlungsloser Übergang
1003	1142	Interelementeffekte
1004	1141	interkonische Zone
1005	1152	intermolekularer strahlungsloser Übergang
1006	1156	interne Elektrogravimetrie
1007	1170	Interzonengebiet
1008	1171	intrachromophorer strahlungsloser Übergang
1009	1176	inverse Abkühlungsgeschwindigkeitskurven
1010	1177	inverse derivativpotentiometrische Titration
1011	0117	inverse Voltammetrie,Voltammetrie mit anodischer Auflösung
1012	2360	inverse Voltammetrie
1013	2083	inverser Direktinjektionsbrenner
1014	0545	Inversstrom-Chronopotentiometrie
1015	1179	Iodimetrie
1016	1178	iodimetrische Titrationen
1017	1181	Iodometrie
1018	1180	iodometrische Titrationen
1019	1182	Ion
1020	1216	Ion-Molekül-Reaktion
1021	1192	Ionenaktivitätskoeffizient
1022	1187	Ionenaustausch

1023	1188	Ionenaustauschchromatographie
1024	1191	Ionenaustauscher
1025	1190	Ionenaustauschermembran
1026	1189	Ionenaustauschisotherme
1027	1183	Ionenbeschuß
1028	1184	Ionenchromatographie
1029	1193	Ionenlinie
1030	1215	Ionenmikroskopie
1031	1221	Ionenquelle
1032	1220	ionenselektive Elektroden
1033	1848	ionenselektive Primärelektrode
1034	1197	Ionenspektrallinien
1035	1196	Ionenspektren
1036	1222	ionenspezifische Elektroden
1037	1202	Ionensputtern
1038	1198	Ionenstärke
1039	1218	Ionenstrahl-Mikroanalyse
1040	1219	Ionenstreuungsspektrometrie
1041	1185	Ionenstrom
1042	1195	Ionenweglänge
1043	1205	Ionisationsausbeutekurve
1044	1206	Ionisationsenergie
1045	0574	Ionisationsgrad
1046	1203	Ionisationskammer
1047	1208	Ionisationsnadel
1048	1209	Ionisationspotential
1049	1201	Ionisationspuffer
1050	1211	Ionisationstemperatur
1051	1210	Ionisationswahrscheinlichkeit
1052	1194	ionisches Medium
1053	1214	ionisierende Spannung
1054	1213	ionisierende Strahlung
1055	1212	ionisierendes Teilchen
1056	1200	Ionisierung, Ionisation
1057	1204	Ionisierungskoeffizient
1058	0526	Ionisierungsquerschnitt
1059	1217	ionogene Gruppen
1060	1225	irreguläre Strahlungspulse
1061	1226	isobare Bestimmung der Massenänderung
1062	1227	Isoformation
1063	1228	isomere Zustände
1064	1229	isomerer Übergang
1065	1230	Isopotentialpunkt
1066	1231	Isotherm
1067	1232	isotherme thermogravimetrische Analyse
1068	1233	Isotone
1069	1234	Isotop
1070	1238	Isotope von Atomen
1071	1240	Isotopenanalyse
1072	1237	Isotopenaustausch
1073	1239	Isotopenhäufigkeit
1074	1242	Isotopenion
1075	1244	Isotopentracer
1076	1243	Isotopentrennung
1077	1235	Isotopenverdünnung
1078	1236	Isotopenverdünnungsanalyse
1079	1241	isotoper Träger
1080	1734	Kalibration der photographischen Emulsion
1081	2199	Kalibration mit Lösungen vergleichbarer Zusammensetzung, Simulationstechnik
1082	0287	Kalibrationskurve
1083	0255	Kalibrierung durch Interpolation
1084	0288	Kalorimetrie
1085	1245	Kalousek-Polarographie
1086	0393	kalte Hohlkathode
1087	0394	kalte Neutronen
1088	0331	Kammersättigung
1089	0332	Kammerzerstäuber
1090	0333	Kanalbildung
1091	0295	Kapazitätsfaktor
1092	0294	Kapazitätsstrom
1093	0291	kapazitiv gekoppeltes Mikrowellenplasma
1094	0290	kapazitiv gekoppeltes Plasma
1095	0296	Kapillarbogen
1096	1621	Kapillarchromatographie
1097	0297	Kapillarelektrode
1098	0298	Kapillarrohr
1099	1622	Kapillarsäule
1100	1292	katalytischer Grenzstrom
1101	0310	katalytischer Strom
1102	0311	Kataphorese
1103	0312	Kathodendunkelraum
1104	0315	Kathodenglimmlicht, negatives Glimmlicht
1105	0316	Kathodenschichtlichtbogen
1106	0318	Kathodenstrahlpolarographie
1107	0319	Kathodenzerstäubung
1108	0320	kathodische Auflösung
1109	0321	kathodische Verdampfung
1110	0323	Kationenaustausch
1111	0324	Kationenaustauscher
1112	0322	Kationeneffekt
1113	1606	Keimbildung
1114	1992	Keimbildungsgeschwindigkeit
1115	1588	Kernaktivierungsanalyse
1116	1593	Kernbrennstoff
1117	0950	Kernbrennstoffzyklus
1118	1589	Kernchemie
1119	1594	Kernfusion
1120	1595	Kernfusionsreaktion
1121	1596	Kernisobaren
1122	1597	Kernisomere
1123	1599	Kernreaktion
1124	1600	Kernreaktor
1125	1591	Kernspaltung
1126	1592	Kernspaltungsreaktion
1127	1602	Kernspur
1128	1603	Kernspurdetektor
1129	1601	Kernsystem
1130	1598	Kernteilchen
1131	1605	Kernübergang
1132	1604	Kernumwandlung
1133	1590	Kernzerfall
1134	1246	Kerr-Zelle
1135	1247	kinetische Analyse
1136	0760	kinetische Energie eines Elektrons
1137	1295	kinetischer Grenzstrom
1138	1248	kinetischer Strom
1139	2533	kleinster Brennfleckdurchmesser, kleinster Fokus
1140	0382	Koagulation
1141	0383	koaxialer Hohlleiter, Wellenleiter
1142	0385	kohärente elektromagnetische Strahlung
1143	0386	kohärente Strahlung

1144	0387	kohärente Strahlungsquellen
1145	0388	kohärentes System von Einheiten
1146	0302	Kohlebecheratomisator
1147	0303	Kohlefadenatomisator
1148	0391	Koinzidenz-Auflösungszeit
1149	0390	Koinzidenz-Zählung
1150	0389	Koinzidenzschaltung
1151	0397	Kollimation
1152	0398	Kollimator
1153	0399	Kollimatorauflösung
1154	0413	Kombinationselektrode
1155	0415	Komparator
1156	0420	Komplexbildung
1157	0421	Komplexbildungsgleichgewichte
1158	0419	komplexes Anion
1159	0423	Komplexometrie
1160	0422	komplexometrische Titration
1161	0418	Komplexon
1162	0424	Komplexon
1163	0426	Kompromißbedingungen
1164	0292	Kondensatorspannung
1165	0439	kondensierter Bogen
1166	0442	Konduktometrie
1167	0441	konduktometrische Titration
1168	2398	konischer Hohlraum
1169	0930	Konkavspiegel, Hohlspiegel, fokussierender Spiegel
1170	0446	Konstant-Temperatur-Lichtbogen
1171	0444	Konstante
1172	0447	konstante Spannung
1173	0445	konstanter Strom
1174	0455	Kontaminationserscheinungen bei einer Fällung
1175	0458	kontinuierliche Strahlung
1176	0459	kontinuierliche Zugabe
1177	0457	kontinuierliches Verfahren
1178	0462	Kontinuum
1179	0463	Kontrollbande
1180	0466	kontrollierte Atmosphäre
1181	0480	Kontrolltitration
1182	0481	Konvektion
1183	0482	konvektive Chronoamperometrie
1184	0483	konvektive Chronopotentiometrie
1185	0484	Konversionselektron
1186	0431	Konzentration
1187	0434	Konzentrations-Gleichgewichts-konstante
1188	1285	Konzentrationsbereich einer Titration
1189	0435	Konzentrationsindex
1190	0432	Konzentrationskonstanten
1191	0436	Konzentrationsverhältnis
1192	0433	Konzentrationsverteilungsverhältnis
1193	0437	konzentrischer Zerstäuber
1194	1288	Koordinationszahl, Ligandenzahl
1195	0517	Kopplungsanlage
1196	0516	Kopplungswirkungsgrad
1197	0491	korrigierte Abklingzeiten der Fluoreszenz
1198	0492	korrigierte Abklingzeiten der Phosphoreszenz
1199	0495	korrigierte Intensität
1200	0498	korrigierte Quantenausbeute der Lumineszenz
1201	0499	korrigierte Retentionszeit
1202	0494	korrigiertes Anregungsspektrum
1203	0493	korrigiertes Emissionsspektrum
1204	0497	korrigiertes Lumineszenz-Anregungspolarisationsspektrum
1205	0496	korrigiertes Lumineszenz-Emissionspolarisationsspektrum
1206	0500	korrigiertes Retentionsvolumen
1207	0502	kosmische Strahlung
1208	1249	Kossel-Technik
1209	0518	Krater
1210	0520	Kraterdurchmesser
1211	0521	Kraterform
1212	0519	Kratertiefe
1213	0530	Kristallbeugungsspektrometer
1214	0527	Kristallcharakteristik
1215	0528	kristallgesteuerter Oszillator
1216	0531	Kristallgitter
1217	0532	kristalline ionenselektive Elektrode
1218	0522	kritisch gedämpfte Entladung
1219	0994	Kugelbogen
1220	0535	kumulative Bildungskonstante
1221	0536	kumulative Protonierungskonstante
1222	0534	kumulative Spaltausbeute
1223	0537	kumulative Stabilitätskonstante
1224	0533	kumulative Verteilung
1225	0538	Kumulativsumme
1226	0138	künstliche Radioaktivität
1227	0548	Kurvenkorrektor, Linearisierungsgerät
1228	2193	Kurzschlußentladung
1229	0343	Ladekreis, Ladestromkreis
1230	0344	Ladestrom
1231	0345	Ladewiderstand
1232	0346	Ladezeitkonstante, Aufladezeitkonstante
1233	0340	Ladungsaustausch
1234	1089	Ladungsinkrementpolarographie
1235	0341	Ladungsstufenpolarographie
1236	0342	Ladungsübergang
1237	1343	Ladungswiderstand
1238	1252	laminare Flamme
1239	1253	laminare Strömung
1240	1254	laminarer Schutzgasstrom
1241	1255	Lampe
1242	1355	Langrohranordnung
1243	2217	langsame Neutronen
1244	1256	Laser
1245	1279	Laser spikes
1246	1268	Laser-Mikrosonden-Analyse
1247	1275	Laser-Raman-Mikroanalyse
1248	1258	laseraktive Substanz
1249	1259	Laseranalyse
1250	1261	Laseratomisatoren
1251	1273	lasererzeugte Dampfwolke
1252	1263	Laserionisierung
1253	1270	Laserleistung
1254	1271	Laserleistung
1255	1265	Laserlokalanalyse
1256	1266	Lasermikroemissionsspektroskopie
1257	1267	Lasermikromassenspektrometrie
1258	1257	Lasern
1259	1272	Laserplasma
1260	1274	Laserpuls
1261	1277	Laserpulse
1262	2405	Laserpulse pro Zeiteinheit
1263	1276	Laserresonator

1264	1280	Laserschwellenwert
1265	1269	Laserspiegel
1266	1262	Laserstrahl
1267	1278	Laserstrahlungsquelle
1268	1567	Leeranzeige
1269	1816	leistungskompensierte Differentialscanningkalorimetrie
1270	0208	Leitelektrolyt
1271	1094	Leitelektrolyt
1272	2377	Leitelektrolyt
1273	0440	Leitfähigkeit
1274	1375	leuchtende Schicht
1275	1373	Lichtausbeute, relative Spektralempfindlichkeit des Normalauges
1276	2032	Lichtbrechungseffekte
1277	1376	Lichtgrößen
1278	1289	Lichtmodulation
1279	2594	Lichtstrahlbreite
1280	1374	Lichtstrom
1281	1286	Ligand
1282	1287	ligandüberbrückte Komplexe
1283	1305	Linearbeschleuniger
1284	1301	Linearbeschleunigungsmassenspektrometer
1285	1303	lineare Abnahme
1286	1304	lineare Dispersion
1287	1307	lineare Energieübertragung
1288	1311	lineare Polarisation der Lumineszenz
1289	1309	lineare Strömungsgeschwindigkeit
1290	1317	lineare Titrationskurve
1291	1306	linearer elektro-optischer Effekt
1292	1308	linearer Energieübertragungsfaktor
1293	1314	linearer Impulsverstärker
1294	1310	linearer photoelektrischer Absorptionskoeffizient
1295	1312	linearer Polarisator
1296	1313	linearer Potentialanstieg
1297	1302	linearer Schwächungskoeffizient
1298	1315	linearer Streukoeffizient
1299	1770	Linearpolarisation
1300	1771	linearpolarisiert
1301	1300	Linie
1302	1322	Linienbreite
1303	1320	Linienform
1304	1319	Linienprofilfunktion
1305	1318	Linienverbreiterungsparameter
1306	1351	logarithmische Abnahme
1307	1353	logarithmische Titrationskurve
1308	1352	logarithmische Verteilungskoeffizient
1309	1347	lokal desolvatisierter Anteil
1310	1344	Lokalanalyse
1311	1349	lokale Gastemperatur
1312	1346	lokaler atomisierter Anteil
1313	1348	lokaler verdampfter Anteil
1314	1345	lokaler Wirkungsgrad der Atomisierung
1315	1350	lokales thermisches Gleichgewicht
1316	1354	Lomakin-Scheibe-Gleichung
1317	1356	Lorentz-Verbreiterung
1318	1910	Löschgas
1319	1911	Löschung
1320	2227	Löslichkeitsprodukt
1321	2230	Lösung
1322	2231	Lösungsgleichgewichte
1323	2232	Lösungsmittel, Solvens
1324	2228	Lösungsmittel-Wechselwirkung
1325	2233	Lösungsmittelblindwert, Blindwert des Lösungsmittels
1326	2234	Lösungsmitteleffekte
1327	2235	Lösungsmittelextraktion, Solventextraktion
1328	2236	Lösungsmittelfront, Fließmittelfront
1329	0964	Lücke
1330	0965	Lückenwiderstand
1331	0055	Luftpeak
1332	1366	Lumineszenz
1333	1368	Lumineszenzlöschung
1334	1367	Lumineszenzparameter
1335	1369	Lumineszenzspektrometer
1336	1370	Lumineszenzspektrometrie
1337	1371	Lumineszenzspektroskopie
1338	1381	Magnetfeld
1339	1382	Magnetfeld-Glimmentladung
1340	1379	magnetisch stabilisierte Lichtbögen
1341	1380	magnetische Ablenkung
1342	1383	magnetischer Penning-Effekt
1343	1384	Magnetron
1344	1378	makroskopischer Wirkungsquerschnitt
1345	1387	Marker, Markierungssubstanz
1346	1251	markieren
1347	1250	markiertes Atom
1348	1388	Maskierung
1349	2187	Maskierung
1350	1389	Maskierungsreagenz
1351	2478	Maßanalyse, Titrimetrie
1352	2552	Masse pro Skalenteil, Massewert pro Skalenteil
1353	1391	Masse-Ladungs-Verhältnis
1354	1390	Massenschwächungskoeffizient
1355	1395	Massenspektrometer
1356	1865	Massenspektrometer mit gestreckter Trochoidenbahn
1357	1396	Massenspektrometrie
1358	1397	Massenspektrum
1359	1541	Massenspektrum der negativen Ionen
1360	1392	Massenverteilung
1361	1393	Massenverteilungsverhältnis
1362	1282	Massenwirkungsgesetz
1363	1394	Massenzahl
1364	2473	Maßlösung, Titrant
1365	1400	Matrix
1366	1402	Matrixanpassung
1367	1401	Matrixeffekte
1368	1403	Matrixmodifizierer
1369	1404	Mattauch-Herzog-Geometrie
1370	1405	Maxwell-Boltzmannsches Gesetz
1371	1406	McLafferty-Umlagerung
1372	1420	mechanische Verschlüsse
1373	1419	mechanischer Einschließung
1374	1423	mechanisieren
1375	1422	Mechanisierung
1376	1421	Mechanismus
1377	1424	Medium
1378	1794	mehratomig, polyatomig, vielatomig
1379	1521	Mehrfachdurchgangsmethode, Mehrfachreflexionsmethode, Multipass-Methode
1380	1518	Mehrkanalanalysator, Vielkanalanalysator

1381	1519	Mehrkanalimpulshöhenanalysator, Vielkanalimpulshöheanalysator
1382	1524	Mehrschlitzbrenner
1383	1429	Méker-Brenner
1384	1430	Membran
1385	1900	Menge
1386	1432	meso
1387	1433	Mesoprobe
1388	1434	Mesospurenanalyse
1389	2518	Meßeinrichtung mit abgestimmten Verstärker
1390	2105	Meßintervall
1391	1418	Messung
1392	1438	Metalldampfbogen
1393	1439	Metalldampfbogenlampe
1394	1435	Metallfadenatomisator
1395	1436	Metallindikator
1396	1440	metastabil
1397	1441	metastabiler Ionenpeak
1398	1444	metastabiler Peak
1399	1445	metastabiler Zustand
1400	2183	Methode der getrennten Lösungen
1401	1474	Methode der Lösungsgemische
1402	1460	Mie-Streuung
1403	1461	Migration
1404	1296	Migrationsgrenzstrom
1405	1462	Migrationsstrom
1406	1446	Mikroanalyse
1407	1449	Mikrobestandteil, Mikrokomponente
1408	1450	Mikrocomputer
1409	1451	Mikrocoulometrie
1410	1452	Mikrophotometer
1411	1454	Mikroprobe
1412	1453	Mikroprozessor
1413	1455	Mikroskopischer Wirkungsquerschnitt
1414	1456	Mikrospur
1415	1447	Mikrowaage
1416	1448	Mikrowaage
1417	1457	Mikrowelle
1418	1458	mikrowelleninduziertes Plasma
1419	1459	Mikrowellenplasmen
1420	1463	Millicoulometrie
1421	1464	Milligrammäquivalent der Ablesbarkeit
1422	1465	minimale Linienbreite
1423	1471	Mischindikator, Indikatorgemisch
1424	2136	Mischindikator
1425	1470	Mischkristall
1426	0490	Mitfällung
1427	1149	mittelschnelles Neutron
1428	1427	Mittelspannungsfunken
1429	1428	Mittelspannungsfunkenerzeuger
1430	1425	Mittelstrombogen
1431	0184	Mittelwert
1432	1407	Mittelwert
1433	0187	mittlere Lebensdauer
1434	1411	mittlere Lebensdauer
1435	1958	mittlere Lebensdauer eines Radionuklids
1436	1426	mittlere Leistung
1437	0188	mittlere Ligandenzahl
1438	1412	mittlere lineare Reichweite
1439	0185	mittlere Strömungsgeschwindigkeit
1440	1409	mittlere Trägergasgeschwindigkeit im Hohlraumbereich
1441	1408	mittlerer Aktivitätskoeffizient
1442	0186	mittlerer Gasdruck
1443	1410	mittlerer Ionenaktivitätskoeffizient
1444	1413	mittlerer Massenbereich
1445	1476	mobile Phase
1446	1479	Mode, Modus, Betriebsart
1447	1478	Moderator, Bremssubstanz, Bremsstoff
1448	1482	Modifier
1449	1481	modifizierte Nernstsche Gleichung
1450	1480	modifizierter aktiver Festkörper
1451	1484	Modulation der linearpolarisierten Strahlung
1452	1634	Modulation eines optischen Signals
1453	0945	Modulationsfrequenz
1454	1483	Modulationsfrequenz
1455	1485	Modulationspolarographie
1456	1486	molale Aktivität
1457	1487	molaler Aktivitätskoeffizient
1458	1488	Molalität
1459	1490	molare Aktivität
1460	1492	molare Ionisierungsenergie
1461	1491	molarer Aktivitätskoeffiezient
1462	1489	molarer Extinktionskoeffizient
1463	1493	Molarität
1464	1507	Molekül
1465	1494	Molekülabsorptionsspektrometrie
1466	1495	Molekülabsorptionsspektroskopie
1467	1496	Molekülanion
1468	1497	Molekülbanden
1469	1499	Molekülemissionsspektrometrie
1470	1500	Molekülemissionsspektroskopie
1471	1501	Molekülion
1472	1498	Molekülkation
1473	1502	Moleküllumineszenzspektrometrie
1474	1503	Molekülmethoden
1475	1505	Molekülspektrometrie
1476	1506	Molekülspektroskopie
1477	1504	Molekülstrahlung
1478	1117	momentane Strömungsgesch-windigkeit
1479	1118	Momentanmessung
1480	1116	Momentanstrom
1481	1509	Mono-alternance
1482	1510	monochromatische Strahlung
1483	1511	Monochromator
1484	1512	monofunktioneller Ionenaustauscher
1485	1514	Monolayerkapazität, Monoschichtkapazität
1486	1513	Monoschicht
1487	1517	Mössbauer-Effekt
1488	1520	Multilayer, Mehrfachschicht
1489	1525	Multisweep-Polarographie
1490	1664	Mutterion
1491	1806	Nachfällung
1492	1805	Nachfiltereffekte
1493	0603	Nachweis
1494	0605	Nachweisgrenze
1495	1297	Nachweisgrenze
1496	1818	Nachweisstärke
1497	1538	Nadelventil
1498	1527	Nanospur
1499	1528	natürliche Häufigkeit

1500	1530	natürliche Impedanzen
1501	1531	natürliche Isotopenhäufigkeit
1502	1532	natürliche radioaktive Strahlung
1503	1533	natürliche Radioaktivität
1504	1529	natürliche Verbreiterung
1505	1468	Nebel
1506	1466	Nebenbestandteil
1507	1542	negative Zonen
1508	1539	negatives Glimmlicht
1509	1540	negatives Ion
1510	1543	Negatron
1511	2467	Neigung der Photoplatte
1512	0600	Nenninhalt
1513	1544	Neodymglas, Neodymiumglas
1514	1547	Nephelometrie
1515	1545	nephelometrische Endpunkterkennung
1516	1546	nephelometrische Titration
1517	1548	Nernstsche Gleichung
1518	1549	Nernstsches Verhalten
1519	1552	Nettoretentionsvolumen
1520	1551	Nettoretentionszeit
1521	1385	Netzspannung
1522	1553	neutral
1523	1554	Neutron
1524	1555	Neutronenaktivierung
1525	1556	Neutronenaktivierungsanalyse
1526	1558	Neutronendichte
1527	1557	Neutroneneinfang
1528	1559	Neutronenfluß, Neutronenflux
1529	1560	Neutronenflußdichte
1530	1561	Neutronenmultiplikation, Neutronenvervielfachung
1531	1563	Neutronenstrahlung
1532	1564	Neutronentemperatur
1533	1562	Neutronenzahl
1534	2543	nicht polarisiert, unpolarisiert
1535	1578	nicht-Faradaysche Admittanz
1536	1569	nichtadditiver Strom
1537	2545	nichtaufgelöste Bandenspektren
1538	2544	nichtauflösbare Bandenspektren, unauflösbare Bandenspektren
1539	1577	nichtdiskrete kontinuierliche Strahlung
1540	1571	nichtkohärente elektromagnetische Strahlung
1541	1572	nichtkohärente optische Quellen
1542	1574	nichtkristalline Elektrode
1543	1573	nichtleitendes Pulver
1544	1582	nichtspektrale Störung
1545	1361	Niederdruckbogenlampen
1546	1362	Niederdruckentladungslampen
1547	1364	Niederdruckmikrowellenplasma
1548	1358	niederenergetische Zusammenstöße
1549	1365	Niederspannungsfunken
1550	1357	Niederstrombogen
1551	1565	Nier-Johnson-Geometrie
1552	1568	nominelle lineare Strömung
1553	1583	Normalauge
1554	2326	Normalbeobachter
1555	0145	Normaldruck-mikrowelleninduziertes Plasma
1556	1584	normale Glimmentladung
1557	1585	Normalität
1558	1586	Normallösung
1559	1587	Normtemperatur
1560	1607	Nukleon
1561	1608	Nukleonenzahl
1562	1609	Nuklid
1563	1610	Nuklidmasse
1564	2618	Nullpunkt der Zeigerskala
1565	2617	Nullpunkt einer Glaselektrode
1566	2381	Oberfläche
1567	2382	Oberflächenanalyse
1568	2386	Oberflächenbedeckung
1569	2388	Oberflächenfunken
1570	2384	Oberflächenkonzentration
1571	2387	Oberflächenreaktion
1572	2383	Oberflächensperrschicht-Halbleiterdetektor, Sperrschicht-Halbleiterdetektor
1573	2385	Oberflächenverunreinigung
1574	1031	Oberwellenwechselstrom-polarographie
1575	1032	Oberwellenwechselstrom-polarographie mit phasenempfindlicher Gleichrichtung
1576	1619	offene Hohlkathode
1577	1639	Öffnung
1578	1614	Okklusion
1579	1624	operative pH-Skala, pH-Arbeitsskala
1580	1710	operative pH-Skala, pH-Arbeitsskala
1581	1623	operative pH-Zelle, pH-Arbeitszelle
1582	1625	operativer pH-Standard, pH-Arbeitsstandard
1583	1636	optimale Bedingungen für einen Funken
1584	1631	optisch dünnes Plasma
1585	1627	optische Dichte
1586	1628	optische Emissionsspektrometrie
1587	1629	optische Fasern
1588	1630	optische Filter
1589	1633	optische Größen
1590	1626	optische Leitfähigkeit
1591	1635	optische Systeme
1592	1632	optisches Pumpen
1593	0156	Ordnungszahl
1594	1872	Ordnungszahl, Protonenzahl
1595	1646	Ostwald-Reifung
1596	1642	Oszillator
1597	2519	Oszillator mit abgestimmter Leitung
1598	1643	Oszillatorstärke
1599	0648	oszillierende Entladung mit Gleichstromvorspannung
1600	1640	oszillierende Reaktion
1601	1641	oszillierender Funken, oszillierende Funkentladung
1602	1644	Oszillographie
1603	2209	oszillographische Single-sweep-Polarographie
1604	1645	Oszillopolarographie
1605	1652	Oxidans, Oxidationsmittel
1606	1653	Oxidations-Reduktions-Indikator, Redox-Indikator
1607	1654	Oxidations-Reduktions-Titration, Redox-Titration
1608	1762	P.I.N.-Halbleiterdetektor
1609	1658	Paarbildung
1610	1657	Paarschwächungskoeffizient
1611	1656	Packung

1612	1659	Papierchromatographie
1613	1661	Parallel-parallel-Spiegelanordnung
1614	1660	parallelgeschaltete Zündung
1615	1377	Parallelschwingkreis, Schwingkreis
1616	2397	Parallelschwingkreis, Schwingkreis, Tankkreis
1617	1663	paramagnetische Verbindungen
1618	1778	Parameter des Plasmas
1619	1666	partielle Verdampfung
1620	1665	partielle Zerfallskonstante
1621	1672	partikelinduzierte Röntgen-Emissionsspektrometrie
1622	1673	partikelinduzierte Röntgen-Emissionsspektroskopie (PIXES)
1623	1682	Peak
1624	1443	Peak eines metastabilen Ions
1625	1683	Peakanalyse
1626	1690	Peakanpassung, Peakfitting
1627	1694	Peakauflösung
1628	1685	Peakbasis
1629	1688	Peakbeugungskoeffizient
1630	1696	Peakbreite
1631	1689	Peakelutionsvolumen
1632	1684	Peakfläche
1633	1695	Peakfunkenstrom, Spitzenstrom der Funkenentladung
1634	1691	Peakhöhe
1635	1692	Peakintensität
1636	1686	Peakkondensatorspannung
1637	1693	Peakmaximum
1638	1687	Peakstrom
1639	1698	Penning-Gasgemisch
1640	1703	Permeationschromatographie
1641	1704	Permselektivität
1642	1706	pH-Elektrode
1643	1707	pH-Indikator
1644	1708	pH-Interpretation
1645	1709	pH-Messung
1646	1711	pH-Standards
1647	1712	pH-Titration
1648	1705	pH-Wert, pH
1649	1713	Phasenanalyse
1650	1714	Phasenfluorimetrie
1651	1716	Phasentitration
1652	1715	Phasenverhältnis
1653	1717	Phasenvolumenverhältnis
1654	1718	Phosphoreszenz
1655	1719	Phosphoreszenz-Analyse
1656	0842	Phosphoreszenz-Anregungsspektrum, angeregtes Phosphoreszenz-Spektrum
1657	1721	Phosphoreszenz-Spektrometer
1658	1722	Phosphorimetrie
1659	1723	Phosphoroskope
1660	1725	Photodetektor
1661	1727	photoelektrischer Effekt
1662	1728	photoelektrischer Peak
1663	1726	photoelektrischer Schwächungskoeffizient
1664	1729	Photoelektron
1665	1732	Photoelektronenausbeute
1666	1730	Photoelektronenspektren
1667	1731	Photoelektronenspektroskopie
1668	1744	Photoelektronenvervielfacher, Photomultiplier
1669	1733	Photoemissionsspektroskopie
1670	0792	Photoemulsionseichfunktion
1671	1735	photographische Intensitätsmessungen
1672	1737	photographische Photometrie
1673	1736	photographischer Parameter
1674	1739	Photoionisation
1675	1740	Photoionisationsspektroskopie
1676	1128	Photometerkugel, Ulbricht-Kugel
1677	1743	Photometrie
1678	1741	photometrische Endpunkterkennung
1679	1742	photometrische Titration
1680	1745	Photon
1681	1749	Photonenemissionsausbeute
1682	1748	Photonenemissionsspektroskopie
1683	1747	Photonenemissionsspektrum
1684	1750	Photonenfluß
1685	1751	Photonenflußdichte
1686	1752	Photonenstrahlung
1687	1753	Photopeak
1688	1724	Photostrom
1689	1754	physikalische Alterung
1690	1757	physikalische Größe
1691	1756	physikalische Störung
1692	1755	physikalisches Gleichgewicht
1693	1758	Physisorption
1694	1759	Picospur
1695	1760	Pile-up
1696	1761	Pincheffekt
1697	1767	PIXES
1698	1768	Plancksches Strahlungsgesetz
1699	1772	Plasma
1700	1780	Plasmabrenner, Plasmafackel
1701	1773	Plasmagas
1702	1774	Plasmaimpedanz
1703	1777	Plasmaladung
1704	1776	Plasmastrahl
1705	1779	Plasmatemperatur
1706	1775	Plasmazündung
1707	1781	Plasmonenzustände
1708	1782	Plateau
1709	1784	pneumatischer Zerstäuber
1710	1785	Pockels-Zelle
1711	1786	Point-to-plane Anordnung, Point-to-plane Konfiguration
1712	1787	Point-to-point Anordnung, Point-to-point Konfiguration
1713	1789	Polarisation, Polarisierung
1714	1791	Polarisator
1715	1790	polarisiert
1716	1788	Polarität
1717	1792	Polarographie
1718	1793	polarometrische Titration
1719	1795	Polychromator
1720	1796	polyfunktioneller Ionenaustauscher
1721	1797	polynukleare Komplexe
1722	1799	Porous-cup-Elektroden
1723	1800	Porous-cup-Elektroden
1724	1803	positive Ionen
1725	1801	positive Säule
1726	0133	positive Säule des Lichtbogens, Lichtbogensäule
1727	1804	Positron
1728	1807	Potential
1729	0474	potentialkontrollierte Coulometrie

1730	0116	potentialkontrollierte Coulometrie mit anodischer Auflösung
1731	0473	potentialkontrollierte coulometrische Titration
1732	0475	potentialkontrollierte Elektrogravimetrie
1733	0476	potentialkontrollierte Elektroseparation
1734	0477	potentialkontrollierte Polarographie
1735	1808	Potentialstufenchronocoulometrie
1736	1813	Potentiometrie
1737	1809	potentiometrische Endpunkterkennung
1738	1811	potentiometrische Titration
1739	2152	potentiometrische Titration mit zweiter Ableitung
1740	1812	potentiometrische Wägetitration
1741	1810	potentiometrischer Selektivitätskoeffizient
1742	1838	Präionisation
1743	1820	praktische Auflösung
1744	1819	praktische pH-Messung
1745	1821	praktische spezifische Kapazität
1746	1828	Präzision
1747	1832	Präzision der Anzeige
1748	1830	Präzision einer Waage
1749	1831	Präzision einer Wägung
1750	1829	Präzisions-Nullstrompotentiometrie
1751	1851	primäre Bezugsstandards, primäre Referenzstandards
1752	1846	primäre Entladung
1753	1849	primäre pH-Standards
1754	1852	primäre Strahlungsquelle
1755	1845	primäre Verbrennung
1756	1853	primärer Standard, Urtitresubstanz
1757	1847	Primärion
1758	1850	Primärstrahlung
1759	1855	Primärsubstanz, Urtitresubstanz
1760	2100	Probe
1761	2265	Probe, Probestück
1762	2106	Probe von begrenzter Größe
1763	1108	Probeneingabetemperatur, Injektionstemperatur
1764	2102	Probenelektrode
1765	1110	Probengeber, Injektor
1766	2104	Probengeber
1767	2103	Probenhalter
1768	2266	Probenhalter
1769	2107	Probenlösung
1770	2108	Probenmasseneinteilung
1771	2109	Probenschiffchen
1772	2111	Probenschleife
1773	2101	Probenstrahl
1774	2110	Probentiegel
1775	1859	Programm
1776	1861	Programm
1777	1860	programmieren
1778	1862	programmieren
1779	1864	Projektionsschirm
1780	1867	promptes Neutron
1781	1866	Promptkoinzidenz
1782	1868	Proportionalzählrohr
1783	1869	Proton
1784	1871	Protonierungsgleichgewicht
1785	1870	Protonierungskonstante
1786	1699	prozentualer Fehler
1787	0271	Puffer
1788	1199	Puffer zur Einstellung der Ionenstärke, Ionenstärkeneinstellungspuffer
1789	0274	Pufferkapazität
1790	0272	puffern
1791	0276	Pufferwert
1792	0273	Pufferzusatztechniken
1793	1884	Pulsleistung
1794	1883	Pulspolarographie
1795	1814	pulverförmige Probe, Pulverprobe
1796	1004	Pulverschüttsystem
1797	1815	Pulvertechniken
1798	1887	Pumpenenergie
1799	1886	Pumpenhohlraum
1800	1890	Pumpleistung
1801	1888	Pumplichtquelle
1802	1889	Pumpperiode
1803	1891	pyroelektrische Detektoren
1804	1892	Q-switch, Güteschalter
1805	1893	Q-switch Betrieb, gütegeschalteter Betrieb
1806	1894	quadratischer elektrooptischer Effekt
1807	1895	Quadrupolmassenspektrometer
1808	1896	qualitative Analyse
1809	1897	Qualitätsfaktor
1810	1904	Quantenausbeute
1811	1905	Quantenausbeute der Fluoreszenz
1812	1906	Quantenausbeute der Lumineszenz
1813	1902	Quantenwirkungsgrad der Fluoreszenz
1814	1903	Quantenwirkungsgrad der Lumineszenz
1815	1901	Quantenzähler
1816	1299	Quantifizierungsgrenze
1817	1898	quantitative Analyse
1818	1899	quantitative Differentialthermoanalyse
1819	1909	quasimonochromatische Photometrie
1820	1431	Quecksilberpoolelektrode
1821	0707	Quecksilbertropfelektrode
1822	2240	Quelle
1823	2011	Quelle mit gleichgerichtetem Strom, gleichgerichtete Quelle
1824	2241	Quellenwirkungsgrad
1825	1912	Quenchfluorimetrie, Auslöschungsfluorimetrie
1826	2245	Querfunkenanregung
1827	0525	Querschnitt, Wirkungsquerschnitt
1828	1913	Rad
1829	1944	radioaktiv
1830	1982	radioaktiv markiertes Reagens, Radioreagens
1831	1948	radioaktive Altersbestimmung, radiometrische Altersbestimmung
1832	1953	radioaktive Halbwertszeit
1833	1959	radioaktive Kerne
1834	1956	radioaktive Markierung, radioaktiv markiertes Atom
1835	1957	radioaktive Materialien
1836	1962	radioaktive Substanzen
1837	1946	radioaktive Zerfallsreihe
1838	1961	radioaktive Zerfallsreihen
1839	1950	radioaktiver Abfall

1840	1964	radioaktiver Abfall
1841	1952	radioaktiver Fallout
1842	1954	radioaktiver Indikator
1843	1963	radioaktiver Tracer, Radiotracer, radioaktiver Indikator
1844	1949	radioaktiver Zerfall
1845	1945	radioaktives Alter
1846	1951	radioaktives Gleichgewicht
1847	1955	radioaktives Isotop
1848	1960	radioaktives Nuklid
1849	1965	Radioaktivität
1850	1967	radioanalytische Reinigung
1851	1971	Radiochemie
1852	1970	radiochemische Ausbeute
1853	1968	radiochemische Reinheit
1854	1969	radiochemische Trennung
1855	1972	Radiochromatograph
1856	1975	Radiofrequenzpolarographie
1857	1976	Radioisotop
1858	1973	Radiokolloid
1859	1977	Radiolyse
1860	1979	radiometrische Titration
1861	1978	radiometrischer Endpunkt
1862	1980	Radionuklid
1863	1981	Radionuklidreinheit
1864	1983	Raffinat
1865	1986	Raman-Spektroskopie
1866	1984	Raman-Strahlung
1867	1985	Raman-Streuung
1868	2123	Raster-Transmissionselektronen-mikroskopie
1869	2121	Rasterelektronenmikroskopie
1870	2243	raumauflösende Techniken
1871	2242	Raumladung
1872	2258	räumliche Auflösung
1873	2259	räumliche Stabilisierung
1874	2219	Raumwinkel
1875	1993	Rayleigh-Streuung
1876	1994	R_B-Wert
1877	1995	Reaktionsintervall
1878	1996	Reaktionsschicht
1879	2458	Reaktionsschichtdicke
1880	2314	Rechteckwellenpolarographie, Square-wave-Polarographie
1881	2313	Rechteckwellenstrom, Square-wave-Strom
1882	2013	Redoxaustauscher
1883	2012	Redoxindikator
1884	2014	Redoxpolymere
1885	2015	Redoxtitrationen
1886	0042	reduziertes Retentionsvolumen
1887	2021	Referenzhalter, Referenzprobenhalter
1888	2024	Referenzmaterial
1889	2027	reflektierte Leistung
1890	2030	Reflexionsbeugung schneller Elektronen (RHEED)
1891	2028	Reflexionselektronenenergie-verlustspektroskopie (REELS)
1892	2029	Reflexionsvermögen
1893	0875	Regelkreis
1894	2034	Regenerationsstrom
1895	2086	Reifung
1896	2188	Reihen, Serie
1897	0379	reine Oberfläche
1898	0377	reiner Arbeitstisch
1899	0378	Reinraum
1900	2036	relative Aktivität
1901	2039	relative Aktivitätsmessung
1902	2037	relative Atommasse
1903	2038	relative biologische Wirksamkeit
1904	2040	relative Dichte
1905	0711	relative Einschaltdauer, Einschaltdauer
1906	2042	relative Retention
1907	2043	relative Standardabweichung
1908	2580	relative Volumenzunahme der Ionenaustauscher durch Quellung
1909	0285	relativer Abbrand, Abbrandsfraktion
1910	2041	Relativfehler
1911	1939	Relaxation durch Strahlung
1912	2047	rem
1913	2560	Resonanzblättchen-Elektrometer
1914	2064	Resonanzenergie
1915	2068	Resonanzhohlraum
1916	0744	Resonanzhohlraumplasma
1917	2065	Resonanzlinie
1918	2066	Resonanzneutron
1919	2067	Resonanzspektrometer
1920	2063	Resonanzverbreiterung
1921	2054	Rest-Flüssig-Flüssig-Grenzfläche
1922	2055	Restdiffusionspotential
1923	2052	Restgase
1924	2053	Restionisation
1925	2051	Reststrom
1926	2077	Retention
1927	2078	Retentionsindex
1928	2079	Retentionsparameter
1929	2080	Retentionstemperatur
1930	2082	Retentionsvolumen
1931	2081	Retentionszeit
1932	2004	reziproke lineare Dispersion, reziproke Empfindlichkeit
1933	2076	R_f-Wert
1934	0022	Richtigkeit
1935	2281	Richtigkeit spektrochemischer Analyse
1936	0991	Riesenpuls
1937	0111	Ringanode, ringförmige Anode
1938	2517	Röhrenzusammenbau
1939	2089	Röntgen
1940	2599	Röntgen-Analyse
1941	2601	Röntgen-Beugung
1942	2600	Röntgen-Datenauswertung
1943	2602	Röntgen-Emissionsspektrometrie
1944	2603	Röntgen-Emissionsspektroskopie
1945	2604	Röntgen-Escapepeak
1946	2605	Röntgen-Fluoreszenz
1947	2611	Röntgen-Messungen
1948	2612	Röntgen-Photoelektronenspektro-skopie
1949	2613	Röntgen-Photonen
1950	2614	Röntgen-Spektrum
1951	2610	Röntgen-Strahlenintensität
1952	2598	Röntgen-Strahlung
1953	2606	Röntgen-Strahlungserzeugung
1954	2607	Röntgen-Strahlungserzeugung durch Elektronen
1955	2608	Röntgen-Strahlungserzeugung durch Photonen

1956 2609 Röntgen-Strahlungserzeugung durch positive Ionen
1957 2093 Rotationsenergie-Übergang
1958 2094 Rotationstemperatur
1959 1783 rotierende Plattenelektrode
1960 2092 rotierende Plattenelektrode
1961 2090 rotierende Scheibenverschlüsse
1962 0364 rotierende Sektorscheibe, Chopper
1963 2091 rotierende Spiegelverschlüsse
1964 2016 Rotverschiebung
1965 0191 rückextrahieren
1966 0190 Rückextrakt
1967 0192 Rückextraktion
1968 2359 Rückextraktion
1969 2361 Rückextraktionslösung, Stripping-Lösung
1970 2031 Rückflußzerstäuber
1971 0198 rückgestreute Elektronen
1972 1049 Rückhalteträger
1973 2005 Rückstoß
1974 0199 Rückstreuspektroskopie
1975 0197 Rückstreuung
1976 0201 Rücktitration, Zurücktitrieren
1977 0200 rücktitrieren
1978 2073 Ruhepunkt
1979 2095 Rutherford-Rückstreuung
1980 2096 Saha-Eggert-Gesetz
1981 2153 Säkulargleichgewicht
1982 0929 Sammellinse, fokussierende Linse
1983 0396 Sammler, Kollektor
1984 0395 Sammlung, Kollektion
1985 2112 Sand-Gleichung
1986 2113 sättigbarer Farbstoffschalter
1987 2115 Sättigung
1988 2116 Sättigungskammer
1989 0408 Säule
1990 0409 Säulenchromatographie
1991 0410 Säuleneffektivität, Trennleistung einer Säule
1992 0411 Säulentemperatur
1993 0412 Säulenvolumen
1994 0023 Säure-Base-Indikator
1995 0025 Säure-Base-Titration
1996 0024 Säure-Base-Wechselwirkung
1997 0029 Säurefunktionen
1998 0028 Säurekonstante
1999 2122 Scanning-Lasermikroanalyse
2000 0125 scheinbare Konzentration
2001 0126 scheinbarer pH-Wert
2002 2368 Scheitel
2003 2370 Scheitelpotential
2004 2369 Scheitelstrom
2005 1283 Schichtäquilibrierung
2006 0931 Schleier
2007 2216 Schlitzbrenner
2008 0956 Schmelze
2009 0874 schnelle Neutronen
2010 2192 Schutzschilde, Blenden
2011 0166 Schwächungskoeffizient
2012 0233 schwarzer Körper
2013 0234 Schwarzkörper-Strahlung, Hohlraumstrahlung
2014 0791 Schwärzungskurve
2015 2465 Schwellenergie
2016 1024 Schweratomeffekt
2017 1003 schwerkraftbetriebener Zerstäuber
2018 2275 Schwingungsdauer
2019 2561 Schwingungsenergie-Übergang
2020 2562 Schwingungstemperatur
2021 2144 sekundäre Elektronen
2022 2142 sekundäre Entladung
2023 2145 sekundäre Fluoreszenz
2024 2151 sekundäre Maßlösung, sekundäre Standardlösung
2025 2141 sekundäre Verbrennung
2026 2143 Sekundärelektronenvervielfacher
2027 2150 sekundärer Standard
2028 2146 Sekundärion
2029 2148 Sekundärionenausbeute
2030 2147 Sekundärionenmassenspektrometrie
2031 2149 Sekundärstrahlung
2032 2166 Selbstabschirmung
2033 2159 Selbstabsorption, Eigenabsorption
2034 2161 Selbstabsorptionseffekte
2035 2162 Selbstabsorptionsfaktor
2036 0174 Selbstionisation
2037 2163 selbstleitende Elektrode
2038 2165 Selbstumkehr
2039 2164 selbstzündende Funkenentladung
2040 2156 selektive Elution
2041 2157 selektive Verdampfung
2042 2158 Selektivitätskoeffizient
2043 2170 semiintegrale Polarographie
2044 2181 sensibilisierte Lumineszenz
2045 2189 seriengeschalteter Zündstromkreis
2046 2194 Shpol'skii-Spektren
2047 2196 Siebelelektrode
2048 0249 Siedepunkt
2049 0250 Siedetemperatur
2050 2195 Sievert
2051 2198 Siliciumphotodioden
2052 2200 Simultanbestimmungssysteme
2053 2201 simultane Differentialthermoanalyse
2054 2202 simultane Techniken
2055 2210 Singulett-Singulett-Absorption
2056 2118 Skalendehnung
2057 2218 Sonnenblind-Photovervielfacher
2058 2238 Sorption
2059 2239 Sorptionsisotherme
2060 2213 Spalt
2061 0892 Spaltausbeute
2062 0885 spaltbar
2063 0888 spaltbar
2064 0886 spaltbares Nuklid
2065 2215 Spaltbreite
2066 0889 Spaltbruchstücke, Spaltstücke
2067 2214 Spalthöhe
2068 0330 Spaltkettenausbeute
2069 0890 Spaltneutronen
2070 0891 Spaltprodukte
2071 0887 Spaltung
2072 2422 Spaltung durch thermische Neutronen
2073 2570 Spannung
2074 0314 Spannung im Bereich des Kathodenfalls
2075 2572 Spannungs-Zeit-Beziehung eines Lichtbogens
2076 1988 Spannweite
2077 2280 spektrale Ansprechkurve, Empfindlichkeitskurve

2078	2268	spektrale Bandbreite
2079	2271	spektrale Intensität
2080	2272	spektrale Störungen
2081	2278	spektrale Strahlungsenergiedichte
2082	2279	spektrale Strahlungsgrößen
2083	1372	Spektralempfindichkeit des Normalauges, Lichtwirksamkeit
2084	2267	spektraler Untergrund
2085	2269	spektrales Verhalten, spektrale Charakteristik
2086	2270	Spektralkontinuum
2087	2273	Spektrallinienform
2088	2274	Spektrallinienverschiebung
2089	1321	Spektrallinienverschiebung, Linienverschiebung
2090	1637	Spektrenordnung
2091	2282	spektrochemische Analyse
2092	2283	spektrochemische Anwendungen
2093	2287	spektrochemische Eigenschaften
2094	2284	spektrochemischer Puffer
2095	2285	spektrochemischer Träger
2096	2286	spektrochemisches Verdünnungsmittel
2097	2288	Spektrogramm
2098	2289	Spektrograph
2099	2291	Spektrographie
2100	2290	spektrographische Instrumente
2101	2292	Spektrometer
2102	2294	Spektrometrie
2103	2295	spektrophotometrische Titration
2104	2296	Spektroskop
2105	2297	Spektroskopie
2106	2298	Spektrum
2107	1662	Sperrkreis für Zähler
2108	2260	spezifische Aktivität
2109	2263	spezifische Ionisation
2110	2276	spezifische spektrale Strahlungsleistung, spektrale Strahlungsleistung
2111	2277	spezifische spektrale Strahlungsleistung des schwarzen Körpers
2112	2262	spezifische Störungen
2113	2261	spezifischer Abbrand
2114	2264	spezifisches Retentionsvolumen
2115	1467	Spiegelanordnung, Spiegelkonfiguration
2116	2300	Spikeenergie
2117	2301	Spikes
2118	2302	Spinerhaltungs-Regel
2119	2303	spontane Elektrogravimetrie
2120	2305	Spontanspaltung, spontane Kernspaltung
2121	2308	Sprühentladung
2122	2127	Spülmittel, Spurenfänger
2123	2128	Spülung
2124	2487	Spur
2125	2488	Spurenanalyse
2126	2489	Spurenbestandteil
2127	2312	Sputterausbeute
2128	2310	Sputtern
2129	2311	Sputterrate
2130	2318	stabilisierte Bögen
2131	2319	stabilisierte Stromquellen
2132	2317	Stabilisierungsringe
2133	2315	Stabilitätskonstante
2134	2316	Stabilitätskonstanten von Metallionenkomplexen
2135	2322	Standardabweichung
2136	2323	Standardabweichung für die Impulszählung
2137	2321	Standardadditionsmethode, Standardzusatzmethode
2138	2328	Standardbezugslösung
2139	2327	Standardbezugsmaterial
2140	2329	Standardlösung, Maßlösung
2141	2330	Standardsubstanzen
2142	2331	Standardsubtraktionsmethode
2143	2332	Stark-Verbreiterung
2144	2333	Start
2145	2334	Startlinie
2146	2335	Startlinie
2147	2336	Startpunkt
2148	2339	stationäre Phase
2149	2342	stationäre Strahlung
2150	2344	stationärer Zustand
2151	2337	statisches Massenspektrometer
2152	2343	statistisches Gewicht
2153	2351	Stern-Vollmersches-Gesetz
2154	2514	steuerbare stationäre Funkenstrecke, triggerbare stationäre Funkenstrecke
2155	0479	Steuerfunkenstrecke
2156	0464	Steuerstrecke
2157	2353	Stöchiometrie
2158	2352	stöchiometrischer Endpunkt
2159	1398	stoffübergangskontrollierte Elektrolysen-Geschwindigkeitskonstante
2160	2354	Stokes-Fluoreszenz
2161	1147	störende Linien
2162	1146	störende Substanz
2163	1148	störende Substanz
2164	1145	Störkurven
2165	1144	Störung
2166	0897	Störung durch Änderung der Flammengeometrie
2167	1281	Störung durch laterale Diffusion
2168	2229	Störungen bei der Verflüchtigung des Gelösten
2169	2257	Störungen durch Änderung der räumlichen Verteilung
2170	0673	Störungen durch Änderung des Dissoziation
2171	0831	Störungen durch Anregung
2172	1207	Störungen durch Ionisation
2173	2556	Störungen in der Dampfphase
2174	1638	Störwirkung organischer Stoffe, organischer Effekt
2175	0400	Stoß, Zusammenstoß
2176	0405	Stoßhalbwertsbreite
2177	0402	Stoßketten
2178	0406	Stoßprozesse
2179	0404	Stoßrelaxation
2180	0401	Stoßverbreiterung
2181	0407	Stoßverschiebung
2182	0403	Stoßwirkungsquerschnitt
2183	1928	Strahlenchemie
2184	1942	strahlende Rekombinationen
2185	1923	strahlende Teilchen
2186	1924	strahlendes Plasma
2187	1920	Strahlenexposition

2188	1927	Strahlung
2189	1930	Strahlungsdetektor
2190	0920	Strahlungsdichte
2191	1938	Strahlungseinfang
2192	1917	Strahlungsemissionsvermögen
2193	1918	Strahlungsenergie
2194	1931	Strahlungsenergie
2195	1919	Strahlungsenergiedichte
2196	1921	Strahlungsfluß
2197	0925	Strahlungsfluß durch einen Monochromator
2198	1926	Strahlungsgröße
2199	1934	Strahlungsgrößen
2200	1933	Strahlungsimpulse
2201	1135	Strahlungsintensität
2202	1925	Strahlungsleistung pro Raumwinkel
2203	1932	strahlungsloser Übergang
2204	1941	Strahlungsprozesse
2205	1935	Strahlungsquellen
2206	1936	Strahlungsspektrometer
2207	1937	Strahlungsspektrum
2208	2299	Strahlungsspektrum
2209	1922	Strahlungsstärke
2210	1916	Strahlungstemperatur
2211	1943	Strahlungsübergänge
2212	1929	Strahlungszähler
2213	2358	Streifen
2214	2356	Streulicht
2215	2357	Streustrahlung
2216	2124	Streuung
2217	2125	Streuung
2218	2126	Streuung der Strahlung
2219	2362	Strobe-Intervall
2220	0541	Strom
2221	2576	Strom-Spannungs-Kurve, Voltammogramm
2222	2571	Strom-Spannungscharakteristik (Verhalten)
2223	0547	Strom-Zeit-Kurve
2224	1802	Stromdichte der positiven Säule
2225	0544	stromfreies Plasma
2226	0542	stromführendes Bogenplasma
2227	0467	stromkontrollierte Coulometrie
2228	0469	stromkontrollierte Potentiometrie
2229	0468	stromkontrollierte potentiometrische Titration
2230	1575	stromlose Gleichstromplasmen, nichtstromführende Plasmen
2231	1863	stromprogrammierte Chronopotentiometrie
2232	0738	Stromquellenparameter
2233	0546	Stromscanningpolarographie
2234	0267	Stromschlüssellösung, Salzbrückelösung
2235	2345	Stufe, Schritt
2236	2346	Stufenhöhe
2237	2347	Stufensektor
2238	2348	stufenweise Elution
2239	2350	stufenweise Linienfluoreszenz
2240	2363	Sublimation
2241	2364	Submikroanalyse
2242	2366	Subspurenanalyse
2243	2365	substöchiometrische Isotopenverdünnungsanalyse
2244	2379	Suppressoren, Unterdrücker
2245	2380	suprathermale Lumineszenz
2246	2389	synchron angeregtes Fluoreszenzspektrum
2247	2390	synchron angeregtes Phosphoreszenzspektrum
2248	2391	synchron rotierende Funkenstrecke, synchron rotierender Unterbrecher
2249	0224	systematischer Fehler, Verzerrung
2250	2392	Szilard-Chalmers-Effekt
2251	2130	Szintillation
2252	2132	Szintillationsdetektor
2253	2135	Szintillationsdetektor
2254	1337	Szintillationsdetektor mit flüssigem Szintillator
2255	2133	Szintillationsspektrometer
2256	2131	Szintillationszähler
2257	2134	Szintillator
2258	2129	szintillierender Stoff
2259	2393	T-Rohre, T-Stücke
2260	2394	Tailing
2261	2395	Tandemfunkenstrecke
2262	2396	Tandenmassenspektrometrie
2263	2399	Target
2264	2400	Tast-Intervall
2265	2401	Tast-Polarographie
2266	0017	Teilchenabsorption
2267	1668	Teilchenbeschuß
2268	1669	Teilchendetektor
2269	1670	Teilchenfluß
2270	1671	Teilchenflußdichte
2271	1675	Teilchengröße
2272	1676	Teilchengrößenanalyse
2273	1677	Teilchengrößenverteilung
2274	1674	Teilchenstrahlung
2275	1667	Teilchenvernichtung, Teilchenannihilation
2276	2402	Temperatureffekte
2277	2403	temperaturprogrammierte Chromatographie
2278	2407	Tensametrie
2279	2408	Tesla-Spule
2280	2411	theoretische Bodenzahl
2281	2413	theoretische spezifische Kapazität
2282	1026	theoretische Trennstufenhöhe
2283	2410	theoretischer Endpunkt
2284	2412	theoretisches Retentionsvolumen
2285	2424	thermisch gezündeter Lichtbogen
2286	2414	thermische Alterung
2287	2415	thermische Analyse
2288	2423	thermische Ionisation, thermische Ionisierung
2289	2425	thermische Neutronen
2290	2426	thermische Prozesse, thermische Vorgänge
2291	2416	thermische Säule
2292	2421	thermische Verdampfung
2293	2419	thermische Zersetzung
2294	2420	thermisches Gleichgewicht
2295	2429	Thermoakustometrie
2296	2430	thermoanalytische Methoden, thermoanalytische Techniken
2297	2432	thermochemische Reaktionen
2298	2433	Thermodilatometrie
2299	2434	thermodynamische Konstanten
2300	2436	thermodynamische Temperatur

2301	2435	thermodynamisches Gleichgewicht
2302	2437	Thermoelektrometrie
2303	2438	Thermogramm
2304	2439	Thermographie
2305	2442	Thermogravimetrie
2306	2440	thermogravimetrische Analyse
2307	2441	thermogravimetrische Kurve
2308	2443	Thermolumineszenz
2309	2444	Thermomagnetometrie
2310	2445	thermomechanische Analyse
2311	2446	Thermomechanometrie
2312	2447	thermometrische Endpunkterkennung
2313	2448	thermometrische Titration
2314	2449	Thermomikroskopie
2315	2453	Thermooptometrie
2316	2450	Thermopartikelanalyse
2317	2451	Thermophotometrie
2318	2454	Thermorefraktometrie
2319	2452	Thermosäule
2320	2455	Thermosonometrie
2321	2456	Thermospektrometrie
2322	2457	Thermoverdunstungsanalyse
2323	2431	Thermowaage
2324	2466	Thyristorkontrolle
2325	0586	Tiefenauflösung
2326	0585	Tiefenprofil
2327	2477	Titer
2328	2474	Titration
2329	0241	Titration einer Blindprobe
2330	1570	Titration in nichwäßrigem Medium
2331	2476	Titrationsfehler
2332	0575	Titrationsgrad
2333	2475	Titrationskurve
2334	2480	Titrimetrie, Volumetrie, Maßanalyse
2335	2479	titrimetrischer Umrechnungsfaktor, volumetrischer Umrechnungsfaktor
2336	0561	Tochternuklid
2337	0979	torgesteuerte Photodetektoren, torgeschaltete Photodetektoren
2338	2481	Torsionslitzenanalysen
2339	2483	Totalionenstrom
2340	0566	Totvolumen
2341	1050	Totvolumen
2342	0564	Totzeit
2343	0607	Totzeit eines Detektors
2344	0565	Totzeitkorrektur
2345	2490	Tracer
2346	0305	Träger
2347	2376	Träger
2348	0306	Trägerdestillationsbogen
2349	0307	Trägerelektrode
2350	0308	trägerfrei
2351	0309	Trägergas
2352	2378	Trägerplatte
2353	2493	Transformationskonstante
2354	2501	Transmissionselektronen-Energieverlustspektroskopie
2355	2502	Transmissionselektronen-Mikroskopie
2356	2503	Transmissionsgrad, Durchlässigkeitsverhältnis, Durchlässigkeitsgrad
2357	2505	Transmissionsgrad, Durchlässigkeitsverhältnis, Durchlässigkeitsgrad
2358	2506	Transmissionsgrad Durchlässigkeitsgrad, Durchlässigkeitsverhältnis
2359	1738	Transmissionsgrad (Photoplatte), Durchlässigkeitsgrad (Photoplatte)
2360	2504	Transmissionshochenergie-Elektronenbeugung
2361	2507	Transport
2362	2508	Transportstörungen
2363	2511	Transportzeit
2364	2509	transversaler elektromagnetischer Modus
2365	2184	Trennfaktor, Separationsfaktor
2366	2185	Trennprozeß
2367	2186	Trenntemperatur
2368	2320	Treppenstufenpolarographie
2369	2515	Triplet-Triplet-Absorption
2370	0709	trockenes Aerosol
2371	0706	Tropfelektrodencoulometrie
2372	0704	Tropfenerzeuger
2373	0708	Tropfzeit
2374	0326	Tscherenkow-Detektor
2375	0327	Tscherenkow-Effekt
2376	0328	Tscherenkow-Strahlung
2377	2525	Turbidimetrie
2378	2523	turbidimetrische Endpunkterkennung
2379	2524	turbidimetrische Titration
2380	2526	turbulente Flamme
2381	2527	turbulente Strömung
2382	2495	Übergangswahrscheinlichkeit
2383	0020	Übergangswahrscheinlichkeit für Absorption
2384	2496	Übergangswahrscheinlichkeit für Absorption
2385	2498	Übergangswahrscheinlichkeit für angeregte Emission
2386	2304	Übergangswahrscheinlichkeit für spontane Emission
2387	2497	Übergangswahrscheinlichkeit für spontane Emission
2388	2499	Übergangszeit
2389	1651	überkritisch gedämpfte Entladung
2390	2371	Überlagerung
2391	2372	übersättigte Lösung
2392	2373	Übersättigung
2393	0261	Überschlagsinstabilität
2394	0263	Überschlagsspannung, Durchschlagsspannung
2395	0262	Überschlagszeit
2396	2491	übertragene Plasmen
2397	2532	Ulbricht-Kugel
2398	2534	Ultramikroanalyse
2399	2535	Ultraschallzerstäuber
2400	2536	Ultraspurenanalyse
2401	2537	Ultraviolett-Photoelektronen-spektroskopie
2402	2049	Umfällung
2403	2001	umgelagertes Molekülion
2404	2084	Umkehrphasen-Chromatographie, reversed-phase-Chromatographie
2405	2002	Umlagerungsion
2406	0485	Umrechnungsfaktor
2407	2494	Umschlagsbereich
2408	2492	Umwandlung, Transformation
2409	1102	unelastische Streuung
2410	1101	unelastischer Stoß
2411	2539	ungesteuerter Hochspannungs-funkengenerator

2412	2538	ungesteuerter Wechselstrombogen
2413	1523	Universalfunkenquellen, Vielzweckfunkenquellen
2414	1581	unspezifische Störungen
2415	0193	Untergrund
2416	2546	Untergrund
2417	0236	Untergrund der Blindprobe
2418	0194	Untergrundkorrektur
2419	0195	Untergrundmassenspektrum
2420	0196	Untergrundstrahlung
2421	2540	unterkritisch gedämpfte oszillierende Entladung
2422	2119	Untersetzer
2423	2120	Untersetzerschaltung
2424	1854	Urtiterlösung
2425	2548	Vakuum-Becherelektrode
2426	2549	Vakuumphotozelle
2427	2551	Valenzelektronen
2428	2558	Varianz
2429	0384	Variationskoeffizient
2430	0269	Verbreiterung
2431	2160	Verbreiterung durch Eigenabsorption, Verbreiterung durch Selbstabsorption
2432	0414	Verbrennung
2433	0823	Verdampfung
2434	2553	Verdampfung
2435	2567	Verdampfung
2436	0824	Verdampfungsgleichgewicht
2437	2569	Verdampfungshilfsstoffe
2438	2568	Verdampfungsrate
2439	0612	Verdampfungsunterdrücker, Verdampfungsbremse
2440	0670	Verdrängungschromatographie
2441	0643	Verdünnungsmittel, Verdünner
2442	0644	Verdünnungsreihe, Verdünnungstest
2443	0975	Verengung im Gasdurchlaß
2444	0416	Vergleichslösung
2445	2017	Vergleichsstrahl, Referenzstrahl
2446	0108	vernichten, annihilieren
2447	0253	verstärkte Glimmentladung in einem Magnetfeld
2448	0252	verstärkte Hohlkathode
2449	0807	Verstärkung
2450	0077	Verstärkungsreaktion
2451	2409	Versuchselektrode, Testelektrode, Arbeitselektrode
2452	2595	Versuchselektrode, Testelektrode, Arbeitselektrode
2453	0678	Verteilung
2454	1678	Verteilungschromatographie
2455	1681	Verteilungsfunktion
2456	0683	Verteilungsgesetze
2457	0681	Verteilungsgleichgewichte
2458	0682	Verteilungsgleichgewichtskonstante
2459	0679	Verteilungskoeffizient
2460	1679	Verteilungskoeffizient
2461	0680	Verteilungskonstante
2462	1680	Verteilungskonstante
2463	0684	Verteilungsverhältnis
2464	0453	Verunreinigung durch Einschluß
2465	0454	Verunreinigung einer Fällung, Kontamination eines Niederschlags
2466	1522	Vervielfachungsmethoden, Vermehrungsmethoden
2467	0577	verzögerte Fluoreszenz
2468	0822	verzögerte Fluoreszenz von E-Typ
2469	1873	verzögerte Fluoreszenz von P-Typ
2470	0578	verzögerte Neutronen
2471	0259	Verzweigungsanteil
2472	0260	Verzweigungsverhältnis
2473	1907	Vierteltransitionszeit-Potential
2474	1908	Viertelwellenlängen-Resonator
2475	2563	visueller Endpunkt
2476	2565	Voigt-Funktion
2477	2566	Voigt-Profil
2478	0953	Vollenergiepeak, Gesamtenergiepeak, Maximalenergiepeak
2479	0954	Vollenergiepeak-Wirkungsgrad, Gesamtenergiepeak-Wirkungsgrad
2480	0417	vollständige Verdampfung
2481	2575	Voltammetrie
2482	1316	Voltammetrie mit linearem Spannungsanstieg
2483	2338	Voltammetrie mit stationärer Elektrode
2484	2573	voltammetrische Analyse
2485	2574	voltammetrische Konstante
2486	2577	Voltamperogramm
2487	2341	Volumen der stationären Phase
2488	2578	Volumenkapazität
2489	0217	Volumenkapazität der Trennschicht
2490	2581	Volumenströmungsgeschwindigkeit
2491	2579	Volumenverteilungskoeffizient
2492	1822	Vorbogenperiode
2493	1837	Vorfiltereffekt
2494	1044	Vorfunken mit großer Wiederholungsrate
2495	1840	Vorfunkenperiode
2496	0225	vorgespannter linearer Impulsverstärker
2497	1835	Vorläufer-Ion, Mutterion
2498	1858	Vorläufer-Ion
2499	1836	Vorläufer-Isotop, Mutterisotop
2500	1833	Vorsäulenreaktor
2501	1834	Vorstufe Mutternuklid, Vorgänger, Vorläufer
2502	2197	Vorzeichen-Regeln
2503	0317	Vorzeichen-Regeln für Kathodenprozesse
2504	0203	Waage
2505	2591	Wägetitration
2506	2589	Wägeverfahren
2507	1999	wahre Absorption
2508	0381	Wandern des Lichtbogens
2509	2582	wandstabilisierter Lichtbogen
2510	2417	Wärmeleitfähigkeit
2511	2418	Wärmeleitfähigkeitsdetektor
2512	2428	Wärmequelle
2513	2427	Wärmestrahlung
2514	1020	Wärmestromdichte-Differentialscanningkalorimetrie
2515	2470	Wartezeitkurve
2516	2138	waschen
2517	2139	Waschlösung
2518	1067	Wasserstoffelektrode
2519	0068	Wechselspannung
2520	1702	Wechselspannung
2521	0069	Wechselspannungsamplitude

2522 0070 Wechselspannungschrono-potentiometrie
2523 0071 Wechselspannungspolarographie
2524 1701 Wechselstrom
2525 0066 Wechselstrom-Chronopotentiometrie
2526 0067 Wechselstrom-Polarographie
2527 0064 Wechselstromamplitude
2528 0065 Wechselstrombogen
2529 1138 Wechselwirkungszeit, Einwirkungzeit
2530 1083 weißglühende Körper
2531 1084 weißglühende Strahlung
2532 0478 wellenformgesteuerter Funken
2533 0013 Wellenlänge der Absorptionskante
2534 2585 Wellenlängendispersion
2535 2584 Wellenlängenkonfiguration
2536 2586 Wellenlängenmodulation
2537 2587 Wellenzahl
2538 2592 Westcott-Wirkungsquerschnitt
2539 2057 Widerstand
2540 1700 Wiederfindungsrate
2541 2009 Wiederfindungstest
2542 2008 Wiedergewinnungsfaktor
2543 2035 Wiederzündungsspannung
2544 0101 Winkel Dispersion, angulare Dispersion
2545 0102 Winkelzerstäuber
2546 0724 Wirkungsgrad
2547 1060 Wirt-Gast-Chemie
2548 1061 Wirtskörper
2549 2520 Wolframglühfadenlampe
2550 2521 Wolframhalogenlampe
2551 2597 Xenonlampe
2552 0512 Zählgenauigkeit, Zählpräzision
2553 0510 Zählgeometrie
2554 0513 Zählrate
2555 0514 Zählraten
2556 0508 Zählrohr
2557 0505 Zählung, Anzahl
2558 0509 Zählungswirkungsgrad
2559 0511 Zählverluste
2560 2616 Zeeman-Effekt
2561 2615 Zeeman-Untergrundkorrektur
2562 2059 Zehn Prozent-Tal-Definition der Auflösung
2563 2471 zeitaufgelöste Spektroskopie
2564 2472 zeitauflösende Techniken
2565 0470 zeitgesteuerter Funken
2566 2468 Zeitkonstante
2567 2404 zeitliche Auflösung
2568 2406 zeitliche Stabilisierung
2569 0325 Zementation
2570 0568 Zerfall
2571 0570 Zerfallskonstante
2572 0665 Zerfallskonstante
2573 0571 Zerfallskurve
2574 0666 Zerfallsrate
2575 0569 Zerfallsreihe
2576 0572 Zerfallsschema
2577 0258 Zerfallsverzweigung
2578 1620 Zersetzung
2579 0164 Zerstäuber
2580 1536 Zerstäuber
2581 2309 Zerstäuber, Sprüher
2582 0471 Zerstäuber mit kontrollierter Flüssigkeitszufuhr
2583 0705 Zerstäuber mit Tropfenerzeugung
2584 1537 Zerstäuber-Flamme-Systeme
2585 2307 Zerstäuberkammer
2586 1534 Zerstäubung
2587 1535 Zerstäubungswirkungsgrad
2588 1576 zerstörungsfreie Aktivierungsanalyse
2589 0329 zertifiziertes Referenzmaterial
2590 0376 Zirkularpolarisation
2591 2619 Zone
2592 1987 zufällige Koinzidenz
2593 2559 Zufallsvariable
2594 1076 Zündfunken
2595 1075 Zündimpuls
2596 1074 Zündstromkreis
2597 1077 Zündstromkreis
2598 1073 Zündung
2599 0900 zurückschlagen
2600 0040 Zusätze, Additive
2601 2375 zusätzliche Elektrode
2602 2374 zusätzliche Entladung
2603 0613 zweiatomig
2604 2530 zweidimensionale Chromatographie
2605 0227 Zweierstöße
2606 2529 zweifarbiger Indikator
2607 0230 zweikernige Komplexe, weikernkomplexe
2608 0699 Zwei(spektral)strahlspektrometer
2609 0692 Zweistrahlspektrometer
2610 0693 Zweistrahlsystem
2611 0700 Zwei(synchron)strahlspektrometer
2612 2528 Zwillingszerstäuber
2613 1150 Zwischenplasmagas
2614 1151 Zwischenrohr
2615 1169 Zwischensystemübergang, Intersystem crossing
2616 0554 Zyklotron, Teilchenbeschleuniger
2617 1186 Zyklotronresonanzmassen-spektrometer, Ionenzyklotron-resonanzmassenspektrometer
2618 0556 Zylinderauffänger
2619 0557 Zylinderform

ADDENDA

A01 A02 Atomabsorptionsspektroskopie
A02 A03 Atomemissionsspektroskopie
A03 A04 Atomfluoreszenzspektroskopie
A04 A06 Flammenspektroskopie
A05 A07 innere Bezugselektrode
A06 A08 Molekülllumineszenzespektroskopie
A07 A01 optisch-akustische Verschlüsse
A08 A09 optische Emissionsspektroskopie
A09 A10 Röntgen-Fluoreszenzspektroskopie
A10 A05 Schwärzung

Spanish

0001	0006	absorbancia
0002	1153	absorbancia interna
0003	0009	absorbedor
0004	0010	absorción
0005	0146	absorción atómica
0006	0017	absorción de partículas
0007	0016	absorción de radiación gamma
0008	1123	absorción integral
0009	1999	absorción real
0010	2210	absorción singlete–singlete
0011	2515	absorción triplete–triplete
0012	0021	absortividad
0013	1489	absortividad molar
0014	1239	abundancia isotópica
0015	1531	abundancia isotópica natural
0016	1528	abundancia natural
0017	1257	acción láser
0018	1305	acelerador lineal de electrones
0019	0027	acidimetría
0020	0072	ácido aminopolicarboxílico
0021	2048	acoplado a distancia
0022	0348	actinómetro químico
0023	0032	activación
0024	1746	activación fotónica
0025	1555	activación neutrónica
0026	0338	activación por partículas cargadas
0027	0037	actividad
0028	2260	actividad específica
0029	1486	actividad molal
0030	1490	actividad molar
0031	2036	actividad relativa
0032	1402	adaptación de matrices
0033	2522	adaptador de sintonización
0034	0041	aditividad
0035	0040	aditivos
0036	1578	admitancia no faradaica
0037	0044	adsorbente
0038	0043	adsorbido, adsorbato
0039	0045	adsorción
0040	0380	adsorción de gases residuales
0041	0049	aerosol
0042	0709	aerosol seco
0043	1389	agente enmascarante
0044	0418	agente formador de complejos
0045	2045	agentes liberadores
0046	0051	aglomeración
0047	0053	agregación
0048	0052	agregado
0049	0576	agua desionizada
0050	0581	agua desmineralizada
0051	1208	aguja de ionización
0052	1690	ajuste de picos
0053	0057	alcalimetría
0054	0058	alcalinizar
0055	0459	alimentación continua
0056	1043	alta reflectancia
0057	1027	altura de observación
0058	1612	altura de observación
0059	2583	altura de onda
0060	1691	altura de pico
0061	2214	altura de rendija
0062	0812	altura de rendija de entrada
0063	2346	altura de salto
0064	1026	altura equivalente a un plato teórico
0065	1025	altura equivalente a un plato teórico efectivo
0066	1760	amontonamiento
0067	0075	amperometría
0068	0618	amperometría diferencial
0069	0966	amplificación gaseosa
0070	1874	amplificador de impulsos
0071	1314	amplificador lineal de impulsos
0072	0225	amplificador lineal de impulsos polarizado
0073	0064	amplitud de corriente alterna
0074	1160	amplitud de oscilación interna
0075	0069	amplitud de voltaje alterno
0076	1247	análisis cinético
0077	1259	análisis con láser
0078	1268	análisis con microsonda láser
0079	1896	análisis cualitativo
0080	1898	análisis cuantitativo
0081	2122	análisis de barrido de láser
0082	2365	análisis de dilución isotópica subestequiométrica
0083	0725	análisis de efluvios gaseosos
0084	1713	análisis de fase
0085	0909	análisis de fluorescencia
0086	0797	análisis de fluorescencia de rayos X de dispersión de energía
0087	1719	análisis de fosforescencia
0088	0806	análisis de fosforescencia exaltada (aumentada)
0089	0825	análisis de gases desprendidos
0090	1434	análisis de mesotrazas
0091	1683	análisis de picos
0092	0361	análisis de quimioluminiscencia
0093	2599	análisis de rayos X
0094	2366	análisis de subtrazas
0095	2382	análisis de superficies
0096	2481	análisis de torsión en trenza

0097	2488	análisis de trazas
0098	2536	análisis de ultratrazas
0099	0598	análisis derivatográfico
0100	0776	análisis elemental
0101	0456	análisis en flujo continuo
0102	2282	análisis espectroquímico
0103	1676	análisis granulométrico
0104	1240	análisis isotópico
0105	1344	análisis local
0106	1265	análisis local con láser
0107	0033	análisis por activación
0108	1119	análisis por activación instrumental
0109	1121	análisis por activación instrumental de neutrones
0110	1556	análisis por activación neutrónica
0111	1576	análisis por activación no destructiva
0112	1588	análisis por activación nuclear
0113	1236	análisis por dilución isotópica
0114	0905	análisis por inyección en flujo (FIA)
0115	2360	análisis por redisolución
0116	2415	análisis térmico
0117	0595	análisis térmico de derivadas
0118	0787	análisis térmico de emanaciones
0119	0629	análisis térmico diferencial
0120	1899	análisis térmico diferencial cuantitativo
0121	0630	análisis térmico diferencial en medio isotérmico
0122	2201	análisis térmico diferencial simultáneo
0123	2450	análisis termogranular
0124	2440	análisis termogravimétrico
0125	0632	análisis termogravimétrico diferencial
0126	0716	análisis termogravimétrico dinámico
0127	1232	análisis termogravimétrico isotérmico
0128	2445	análisis termomecánico
0129	2457	análisis termovaporimétrico
0130	2573	análisis voltamétrico
0131	2478	análisis volumétrico
0132	0080	analito
0133	1882	analizador de altura de impulsos
0134	2204	analizador de altura de impulsos monocanal
0135	1519	analizador de altura de impulsos multicanal
0136	1875	analizador de amplitud de impulsos
0137	1763	analizador de campo magnético de π radianes
0138	1764	analizador de campo magnético de $\pi/2$ radianes
0139	1765	analizador de campo magnético de $\pi/3$ radianes
0140	1914	analizador de campos electrostáticos radiales
0141	1895	analizador de masas cuadripolar
0142	1518	analizador multicanal
0143	2593	anchura
0144	0205	anchura de banda
0145	2268	anchura de banda espectral
0146	2594	anchura de haz de luz
0147	1322	anchura de línea
0148	1465	anchura de línea mínima
0149	1696	anchura de pico
0150	1697	anchura de pico en la semialtura
0151	2215	anchura de rendija
0152	0813	anchura de rendija de entrada
0153	1009	anchura de semi-intensidad de una línea de absorción
0154	1010	anchura de semi-intensidad de una línea de emisión
0155	0955	anchura en la semialtura (FWHM)
0156	0256	ángulo de Bragg
0157	0427	ángulo de Compton
0158	0100	ángulo de divergencia
0159	0242	ángulo de ranura
0160	2219	ángulo sólido
0161	2317	anillos de estabilización
0162	0419	anión complejo
0163	1496	anión molecular
0164	0109	aniquilación
0165	1667	aniquilación de partículas
0166	0108	aniquilar
0167	0107	anisotropía
0168	0111	ánodo anular
0169	0121	antipartícula
0170	1780	antorcha de plasma
0171	2137	apantallado
0172	1620	apertura
0173	2283	aplicaciones espectroquímicas
0174	0131	arco
0175	0296	arco capilar
0176	0439	arco condensado
0177	1030	arco de alta corriente
0178	1041	arco de alta presión
0179	1357	arco de baja corriente
0180	0316	arco de capa catódica
0181	0939	arco de combustión
0182	0065	arco de corriente alterna
0183	1072	arco de corriente alterna cebado
0184	0465	arco de corriente alterna controlada
0185	2538	arco de corriente alterna incontrolada
0186	2010	arco de corriente alterna rectificado
0187	0647	arco de corriente continua
0188	1425	arco de corriente media
0189	0306	arco de destilación en portador
0190	2424	arco de encendido térmico
0191	0446	arco de temperatura constante
0192	1438	arco de vapor de metal
0193	0691	arco doble
0194	0733	arco eléctrico
0195	0994	arco esferoidal
0196	2582	arco estabilizado en las paredes
0197	0135	arco voltaico, lamparo de arco
0198	2318	arcos estabilizados
0199	0977	arcos estabilizados con gas
0200	1379	arcos estabilizados magnéticamente
0201	1164	arcos interrumpidos
0202	0275	arcos regulados
0203	2154	arcos sembrados
0204	1684	área de pico
0205	0140	aspiración
0206	0142	aspirador
0207	0139	aspirar
0208	0165	atenuación

0209	0987	atenuación geométrica
0210	0466	atmósfera controlada
0211	0132	atmósfera de arco
0212	0656	atmósfera de descarga
0213	1103	atmósfera de descarga inerte
0214	0161	atomización
0215	1260	atomización y excitación por láser
0216	0164	atomizador
0217	0302	atomizador de cápsula de carbón
0218	0999	atomizador de cápsula de grafito
0219	0303	atomizador de filamento de carbón
0220	1435	atomizador de filamento metálico
0221	1261	atomizadores de láser
0222	1062	átomo de alta energía
0223	0082	átomos analito
0224	2510	atrapamiento
0225	1419	atrapamiento mecánico
0226	2159	autoabsorción
0227	2166	autoapantallamiento
0228	2163	autoelectrodo
0229	2165	autoinversión
0230	0174	autoionización
0231	0176	automatización
0232	0177	automatización
0233	0175	automatizar
0234	0178	automatizar
0235	0180	autooxidación
0236	0179	autorradiografía
0237	1360	baja pérdida
0238	0203	balanza
0239	0085	balanza analítica
0240	1448	balanza microquímica
0241	0463	banda de control
0242	0782	banda de elución
0243	1497	bandas moleculares
0244	0549	bandas moleculares del cianógeno
0245	0206	barn, barnio
0246	1313	barrido lineal de potencial
0247	0207	base
0248	1685	base de pico
0249	1857	base de prisma
0250	0212	basicidad
0251	0215	becquerel, becquerelio
0252	0223	biamperometría
0253	0232	bipotenciometría
0254	0235	blanco
0255	2399	blanco
0256	2233	blanco de disolvente
0257	1091	blanco de indicador
0258	0245	bloque
0259	1098	bobina de inducción
0260	2408	bobina de Tesla
0261	0251	bolómetro
0262	0148	bombardeo atómico
0263	0873	bombardeo con átomos rápidos (FAB)
0264	1668	bombardeo con partículas
0265	1183	bombardeo iónico
0266	0627	bombeo diferencial
0267	1632	bombeo óptico
0268	0315	brillo catódico, luminiscencia catódico
0269	2111	bucle de muestra
0270	0569	cadena de desintegración
0271	1946	cadena radioactiva
0272	0402	cadenas de choques
0273	0086	calibrado analítico
0274	1734	calibrado de emulsiones fotográficas
0275	0288	calorimetría
0276	0628	calorimetría de barrido diferencial
0277	1816	calorimetría de barrido diferencial a potencia compensada
0278	1020	calorimetría de barrido diferencial con flujo calórico (de calor)
0279	0713	calorimetría dinámica diferencial
0280	2244	cámara de chispa
0281	1203	cámara de ionización
0282	2307	cámara de nebulización
0283	2116	cámara de saturación
0284	0378	cámara limpia
0285	1063	cámara radioactiva
0286	0105	cambiador aniónico
0287	0324	cambiador catiónico
0288	1191	cambiador iónico
0289	0226	cambiador iónico bifuncional
0290	1512	cambiador iónico monofuncional
0291	1796	cambiador iónico polifuncional
0292	2013	cambiadores iónicos rédox
0293	1381	campo magnético
0294	1056	campo magnético homogéneo
0295	0524	campos eléctrico y magnético cruzados
0296	1579	campos magnéticos inhomogéneos
0297	1900	cantidad
0298	0997	capa de gradiente
0299	1996	capa de reacción
0300	1014	capa de semivalor
0301	1375	capa lumínica
0302	0293	capacidad
0303	1514	capacidad de monocapa
0304	2072	capacidad de respuesta
0305	0264	capacidad de saturación
0306	1821	capacidad específica práctica
0307	2413	capacidad específica teórica
0308	0274	capacidad tamponadora
0309	2578	capacidad volumétrica
0310	0217	capacidad volumétrica del lecho
0311	2110	cápsula de muestro
0312	0299	captura
0313	0960	captura de radiación gamma
0314	0753	captura electrónica
0315	1557	captura neutrónica
0316	1938	captura radiativa
0317	2571	características de voltaje–corriente
0318	0527	características del cristal
0319	2269	características espectrales
0320	1342	carga
0321	1777	carga de plasma
0322	2242	carga espacial
0323	0957	cascada de rayos gamma
0324	0311	cataforesis
0325	1498	catión molecular
0326	1064	cátodo hueco caliente
0327	1039	cátodo hueco de alta frecuencia
0328	1619	cátodo hueco de extremo abierto
0329	0393	cátodo hueco frío
0330	0252	cátodo hueco reforzado
0331	0907	caudal
0332	1117	caudal instantáneo

0333	1309	caudal lineal
0334	0185	caudal promedio
0335	2581	caudal volumétrico
0336	2398	cavidad cónica
0337	1886	cavidad de bombeo
0338	2068	cavidad resonante
0339	1246	célula de Kerr
0340	1785	célula de Pockels
0341	1623	célula funcional de pH
0342	0325	cementación
0343	2134	centelleador
0344	2130	centelleo
0345	1076	chispa de encendido
0346	1001	chispa de grafito
0347	2542	chispa unidireccional
0348	2550	chispas en vacío
0349	0400	choque
0350	0731	choque elástico
0351	1101	choque inelástico
0352	0227	choques binarios
0353	1358	choques de baja energía
0354	1139	choques interatómicos
0355	1776	chorro de plasma
0356	2555	chorro de vapor
0357	0950	ciclo de combustible
0358	0711	ciclo de trabajo
0359	0554	ciclotrón
0360	2368	cima
0361	0343	circuito de carga
0362	0389	circuito de coincidencia
0363	0657	circuito de descarga
0364	1377	circuito de elementos concentrados
0365	1074	circuito de encendido
0366	1662	circuito de parálisis
0367	2120	circuito desmultiplicador
0368	1077	circuito encendedor
0369	2397	circuito oscilante
0370	0449	clasificación por constituyentes
0371	2108	clasificación por pesos de muestra
0372	0382	coagulación
0373	0060	coeficiente α
0374	0011	coeficiente de absorción
0375	1310	coeficiente de absorción fotoeléctrica lineal
0376	0038	coeficiente de actividad
0377	1192	coeficiente de actividad iónica
0378	1410	coeficiente de actividad iónica medio
0379	1408	coeficiente de actividad medio
0380	1487	coeficiente de actividad molal
0381	1491	coeficiente de actividad molar
0382	0166	coeficiente de atenuación
0383	1657	coeficiente de atenuación de pares
0384	1726	coeficiente de atenuación fotoeléctrica
0385	1302	coeficiente de atenuación lineal
0386	1390	coeficiente de atenuación másico
0387	1155	coeficiente de conversión interna
0388	1688	coeficiente de difracción de pico
0389	0638	coeficiente de difusión
0390	1315	coeficiente de dispersión lineal
0391	0679	coeficiente de distribución
0392	2579	coeficiente de distribución de volúmenes
0393	1055	coeficiente de distribución homogéneo
0394	1352	coeficiente de distribución logarítmico
0395	0728	coeficiente de Einstein para la emisión espontánea
0396	0860	coeficiente de extracción
0397	1204	coeficiente de ionización
0398	1126	coeficiente de reflexión integral
0399	1679	coeficiente de reparto
0400	2158	coeficiente de selectividad
0401	1810	coeficiente de selectividad potenciométrica
0402	0384	coeficiente de variación
0403	1987	coincidencia aleatoria
0404	1866	coincidencia inmediata
0405	2516	coincidencia real
0406	0392	coiones
0407	0396	colector
0408	0556	colector cilíndrico
0409	0871	colector de jaula de Faraday
0410	0397	colimación
0411	0398	colimador
0412	0408	columna
0413	0133	columna de arco
0414	1655	columna de relleno
0415	1801	columna positiva
0416	2416	columna térmica
0417	1622	columna tubular abierta
0418	0949	combustible
0419	1593	combustible nuclear
0420	0414	combustión
0421	1845	combustión primaria
0422	2141	combustión secundaria
0423	0415	comparador
0424	0106	complejo aniónico
0425	0423	complejometría
0426	1515	complejos binarios mononucleares
0427	0230	complejos binucleares
0428	1472	complejos de ligando mixto
0429	1473	complejos de metal mixto
0430	1287	complejos de puente de ligando
0431	1469	complejos mixtos
0432	1797	complejos polinucleares
0433	1516	complejos ternarios mononucleares
0434	0424	complexona
0435	0425	componente
0436	1663	compuestos paramagnéticos
0437	0431	concentración
0438	0125	concentración aparente
0439	0334	concentración característica
0440	0039	concentración de actividad
0441	0277	concentración en el seno de la disolución
0442	2384	concentración superficial
0443	0438	concomitantes
0444	1399	condensador adaptador
0445	1036	condensador cortocircuitante de alta frecuencia
0446	0845	condición exoenergética
0447	0426	condiciones de compromiso
0448	1636	condiciones óptimas de chispa
0449	1626	conductancia óptica
0450	0442	conductimetría
0451	1035	conductimetría de alta frecuencia

0452	0649	conductimetría de corriente continua
0453	0440	conductividad
0454	0734	conductividad eléctrica
0455	2417	conductividad térmica
0456	1661	configuración de espejos paralelo–paralelo
0457	2584	configuración de longitudes de onda
0458	1467	configuración especular
0459	1786	configuración punto a plano
0460	1787	configuración punto a punto
0461	2462	configuración trielectródica
0462	2113	conmutador de colorante saturable
0463	1892	conmutador Q
0464	0444	constante
0465	0370	constante cronoculombimétrica
0466	0372	constante cronopotenciométrica
0467	0028	constante de acidez
0468	0558	constante de amortiguación
0469	0640	constante de corriente de difusión
0470	0570	constante de desintegración
0471	0665	constante de desintegración
0472	1665	constante de desintegración parcial
0473	0672	constante de disociación
0474	0680	constante de distribución
0475	0817	constante de equilibrio
0476	0434	constante de equilibrio de concentraciones
0477	0682	constante de equilibrio de distribución
0478	2315	constante de estabilidad
0479	0537	constante de estabilidad acumulada
0480	0861	constante de extracción
0481	0870	constante de Faraday
0482	0934	constante de formación
0483	0535	constante de formación acumulada
0484	1650	constante de formación global
0485	2349	constante de formación por etapas
0486	1870	constante de protonación
0487	0536	constante de protonación acumulada
0488	1680	constante de reparto
0489	2069	constante de respuesta
0490	2468	constante de tiempo
0491	0346	constante de tiempo de carga
0492	2493	constante de transformación
0493	1398	constante de velocidad del electrolisis controlada por el transporte de masa
0494	0614	constante dieléctrica
0495	2574	constante voltamétrica
0496	0432	constantes de concentración
0497	1649	constantes de distribución globales
0498	2316	constantes de estabilidad de los complejos metal-iónicos
0499	2434	constantes termodinámicas
0500	0448	constituyente
0501	1466	constituyente minoritario
0502	2489	constituyente traza
0503	1386	constituyentes mayoritarios
0504	0452	constricción de plasma de arco
0505	1989	contador
0506	2131	contador de centelleo
0507	1337	contador de centelleo de líquido
0508	1929	contador de radiaciones
0509	1901	contadores cuánticos
0510	0454	contaminación de un precipitado
0511	0453	contaminación por atrapamiento
0512	2385	contaminación superficial
0513	0450	contenido de constituyentes
0514	0462	continuo
0515	2270	continuo espectral
0516	0506	contraelectrodo
0517	0507	contraión
0518	2466	control por tiristores
0519	0481	convección
0520	0213	convenio de Bates–Guggenheim
0521	1078	convenio de signos de la ecuación de Ilkovič
0522	0317	convenio de signos para los procesos catódicos
0523	2197	convenios de signo
0524	1154	conversión interna
0525	0078	convertidor analógico–digital (ADC)
0526	1081	convertidor de impedancias
0527	0641	convertidor digital–analógico (DAC)
0528	0490	coprecipitación
0529	2615	corrección de fondo de Zeeman
0530	1092	corrección de indicador
0531	0501	corrección de pesadas directas
0532	0237	corrección del blanco
0533	0194	corrección del fondo
0534	2062	corrección del tiempo de resolución
0535	0565	corrección del tiempo muerto
0536	0548	corrector de curvas
0537	0541	corriente
0538	0310	corriente catalítica
0539	1292	corriente catalítica límite
0540	1248	corriente cinética
0541	1295	corriente cinética límite
0542	0445	corriente constante
0543	0646	corriente continua
0544	0047	corriente de adsorción
0545	1291	corriente de adsorción límite
0546	0294	corriente de capacidad
0547	0344	corriente de carga
0548	2369	corriente de cima
0549	0658	corriente de descarga
0550	1695	corriente de descarga máxima
0551	0868	corriente de desmodulación faradaica
0552	0639	corriente de difusión
0553	1294	corriente de difusión límite
0554	0696	corriente de doble capa
0555	1462	corriente de migración
0556	1296	corriente de migración límite
0557	2313	corriente de onda cuadrada
0558	1687	corriente de pico
0559	0869	corriente de rectificación faradaica
0560	2034	corriente de regeneración
0561	0740	corriente eléctrica
0562	0867	corriente faradaica
0563	1550	corriente faradaica neta
0564	1104	corriente inicial
0565	1116	corriente instantánea
0566	1185	corriente iónica
0567	2483	corriente iónica total

0568	1293	corriente límite
0569	1569	corriente no aditiva
0570	0559	corriente oscura
0571	1701	corriente periódica
0572	2051	corriente residual
0573	2218	cortinilla
0574	0518	cráter
0575	1470	cristal mixto
0576	0368	cromatografía
0577	2403	cromatografía a temperatura programada
0578	2530	cromatografía bidimensional
0579	0046	cromatografía de adsorción
0580	0670	cromatografía de desplazamiento
0581	2099	cromatografía de desplazamiento salino
0582	0783	cromatografía de elución
0583	2084	cromatografía de fase invertida
0584	1703	cromatografía de filtración
0585	0985	cromatografía de filtración por gel
0586	0906	cromatografía de flujo programado
0587	0967	cromatografía de gases
0588	1188	cromatografía de intercambio iónico
0589	1323	cromatografía de líquidos
0590	1678	cromatografía de reparto
0591	0620	cromatografía diferencial
0592	2461	cromatografía en (de) capa fina
0593	0409	cromatografía en (de) columna
0594	0879	cromatografía en (de) filamento
0595	1659	cromatografía en (de) papel
0596	1621	cromatografía en (de) tubo abierto
0597	0946	cromatografía frontal
0598	0972	cromatografía gas–líquido
0599	0976	cromatografía gas–sólido
0600	1184	cromatografía iónica
0601	1325	cromatografía líquido–gel
0602	1329	cromatografía líquido–líquido
0603	1339	cromatografía líquido–sólido
0604	0367	cromatografiar
0605	0366	cromatógrafo
0606	0365	cromatograma
0607	0619	cromatograma diferencial
0608	1124	cromatogramas integrales
0609	0369	cronoamperometría
0610	0482	cronoamperometría de convección
0611	0697	cronoamperometría de doble salto de potencial
0612	0115	cronoamperometría de redisolución anódica con barrido lineal de potencial
0613	0371	cronoculombimetría
0614	0483	cronoculombimetría de convección
0615	0698	cronoculombimetría de doble salto de potencial
0616	1808	cronoculombimetría de saltos de potencial
0617	0375	cronopotenciometría
0618	1863	cronopotenciometría a corriente programada
0619	0550	cronopotenciometría cíclica
0620	0551	cronopotenciometría cíclica de saltos de corriente
0621	0066	cronopotenciometría de corriente alterna
0622	0588	cronopotenciometría de derivadas
0623	0543	cronopotenciometría de interrupción de corriente
0624	0545	cronopotenciometría de inversión de corriente
0625	0070	cronopotenciometría de voltaje alterno
0626	1169	cruce intersistémico
0627	0912	cuanto de fluorescencia
0628	0958	cuanto de rayos gamma
0629	0914	cuanto total de fluorescencia
0630	0579	cubeta (capsula) de Delves
0631	0233	cuerpo negro
0632	1083	cuerpos incandescentes
0633	0504	culombimetría
0634	0474	culombimetría a potencial controlado
0635	0467	culombimetría de corriente controlada
0636	0706	culombimetría de electrodo de gotas
0637	0116	culombimetría de redisolución anódica a potencial controlado
0638	0540	curie, curio
0639	0088	curva analítica
0640	1022	curva de calentamiento
0641	0287	curva de calibrado
0642	0791	curva de calibrado de emulsiones
0643	0547	curva de corriente–tiempo
0644	1007	curva de crecimiento de la actividad
0645	0571	curva de desintegración
0646	1205	curva de eficacia de ionización
0647	0784	curva de elución
0648	1065	curva de Hurter y Driffield
0649	2280	curva de respuesta espectral
0650	2470	curva de tiempo de espera
0651	2475	curva de valoración
0652	1317	curva de valoración lineal
0653	1353	curva de valoración logarítmica
0654	2441	curva termogravimétrica
0655	0488	curvas de enfriamiento
0656	0092	curvas de evaluación analítica
0657	1145	curvas de interferencia
0658	0489	curvas de velocidad de enfriamiento
0659	1176	curvas de velocidad de enfriamiento inversas
0660	0560	datos
0661	1303	decremento lineal
0662	1351	decremento logarítmico
0663	1802	densidad de corriente de una columna positiva
0664	1919	densidad de energía radiante
0665	2278	densidad de energía radiante espectral
0666	0920	densidad de flujo
0667	2531	densidad de flujo a 2200 metros por segundo
0668	0798	densidad de flujo de energía
0669	1671	densidad de flujo de partículas
0670	1751	densidad de flujo fotónico
0671	1560	densidad de flujo neutrónico
0672	1558	densidad neutrónica
0673	1627	densidad óptica
0674	2040	densidad relativa

0675	0583	densitómetro
0676	0921	depresión de flujo
0677	2128	depuración
0678	2127	depurador
0679	0702	deriva
0680	0587	derivada
0681	0599	derivatografía
0682	0655	descarga
0683	0522	descarga amortiguada críticamente
0684	1651	descarga amortiguada hipercríticamente
0685	0181	descarga auxiliar
0686	1051	descarga de cátodo hueco
0687	2246	descarga de chispa
0688	2164	descarga de chispas autoencendidas
0689	0853	descarga de chispas encendidas externamente
0690	0470	descarga de duración controlada
0691	0478	descarga de forma de onda controlada
0692	2308	descarga de pulverización
0693	0136	descarga de tipo arco
0694	1427	descarga de voltaje medio
0695	2212	descarga deslizante
0696	0735	descarga eléctrica
0697	1363	descarga eléctrica a baja presión
0698	2190	descarga eléctrica apantallada
0699	0270	descarga en corona, descarga radiante
0700	2193	descarga en cortocircuito
0701	0995	descarga lumíniscente
0702	0254	descarga lumíniscente amplificada
0703	0001	descarga lumíniscente anormal
0704	1382	descarga lumíniscente de campo magnético
0705	0253	descarga lumíniscente de campo magnético amplificada
0706	1584	descarga lumíniscente normal
0707	2074	descarga lumíniscente restringida
0708	1641	descarga oscilante
0709	2540	descarga oscilante amortiguada subcríticamente
0710	0648	descarga oscilante polarizada de corriente continua
0711	1044	descarga previa a alta velocidad de repetición
0712	1846	descarga primaria
0713	0451	descarga restringida
0714	2142	descarga secundaria
0715	2388	descarga superficial
0716	2374	descarga suplementaria
0717	2541	descarga unidireccional
0718	1045	descargas de alta repetición
0719	1047	descargas de alto voltaje
0720	1365	descargas de baja tensión
0721	0739	descargas eléctricas
0722	0661	descargas en vacío
0723	0968	descargas gaseosas
0724	2255	descargas por semiciclo
0725	2252	descargas pseudodisruptivas (de tipo chispa)
0726	1442	descomposición metaestable
0727	2419	descomposición térmica
0728	0580	desenmascaramiento
0729	0404	desexcitación por choque
0730	1939	desexcitación radiativa
0731	0568	desintegración
0732	0061	desintegración alfa
0733	0219	desintegración beta
0734	0850	desintegración exponencial
0735	1590	desintegración nuclear
0736	1949	desintegración radioactiva
0737	0258	desintegración ramificada
0738	2119	desmultiplicador
0739	0247	desplazamiento al azul
0740	2016	desplazamiento al rojo
0741	0214	desplazamiento batocrómico
0742	1321	desplazamiento de líneas
0743	2274	desplazamiento de líneas espectrales
0744	0688	desplazamiento Doppler
0745	1068	desplazamiento hipercrómico
0746	1069	desplazamiento hipocrómico
0747	1070	desplazamiento hipsocrómico
0748	0407	desplazamiento por choque
0749	0357	desplazamiento químico
0750	2098	desplazamiento salino
0751	0584	despolarizador
0752	0602	dessolvatación
0753	0676	destilación
0754	0936	destilación fraccionada
0755	0611	desviación
0756	1380	desviación magnética
0757	2322	desviación típica
0758	2323	desviación típica de recuento
0759	2043	desviación típica relativa
0760	0612	desvolatilizador
0761	0603	detección
0762	0073	detección amperométrica del punto final
0763	0221	detección biamperométrica del punto final
0764	0726	detección de efluvios gaseosos
0765	0826	detección de gases desprendidos
0766	0606	detector
0767	2135	detector centelleador
0768	0754	detector de captura electrónica
0769	2132	detector de centelleo
0770	1338	detector de centelleo de líquido
0771	0326	detector de Cerenkov
0772	2418	detector de conductividad térmica
0773	0645	detector de diodos en fila
0774	2224	detector de estado sólido
0775	0932	detector de hojas
0776	1669	detector de partículas
0777	1930	detector de radiaciones
0778	1603	detector de rastros nucleares
0779	2169	detector de semiconductor
0780	2383	detector de semiconductor de barrera superficial
0781	0636	detector de semiconductor de unión difusa
0782	1762	detector de semiconductor PIN
0783	0621	detector diferencial
0784	1125	detector integral
0785	0974	detector proporcional de gases
0786	1891	detectores piroeléctricos
0787	1021	determinación de la curva de calentamiento

0788	2590	determinación de variaciones de peso
0789	1226	determinación isobárica de variaciones de masa
0790	1545	determinación nefelométrica del punto final
0791	0520	diámetro de cráter
0792	0123	diámetro de diafragma
0793	0928	diámetro de mancha focal
0794	2533	diámetro de mancha focal definitivo
0795	0613	diatómico
0796	0616	dielectrometría
0797	0617	diferencial
0798	0529	difracción cristalina
0799	0755	difracción de electrones
0800	2601	difracción de rayos X
0801	1115	difracción de rayos X in situ
0802	2030	difracción electrónica de alta energía de reflexión (RHEED)
0803	2504	difracción electrónica de alta energía de transmisión (THEED)
0804	1359	difracción electrónica de baja energía (LEED)
0805	2155	difracción electrónica en áreas seleccionadas (SAED)
0806	0637	difusión
0807	0642	dilatometría
0808	0589	dilatometría de derivadas
0809	0622	dilatometría diferencial
0810	1235	dilución isotópica
0811	0643	diluyente
0812	2286	diluyente espectroquímico
0813	0664	discriminador
0814	0671	disociación
0815	0675	disolución
0816	2230	disolución
0817	0240	disolución de blanco
0818	0416	disolución de comparación
0819	2139	disolución de lavado
0820	2107	disolución de muestra
0821	2329	disolución de patrón
0822	1854	disolución de patrón primario
0823	2151	disolución de patrón secundario
0824	2361	disolución de reextracción (recuperación)
0825	2328	disolución de referencia normalizada
0826	0881	disolución de relleno
0827	2325	disolución de valorante normalizada
0828	1158	disolución interna de relleno
0829	1159	disolución interna de relleno de un electrodo de vidrio
0830	1586	disolución normal
0831	2324	disolución normalizada
0832	0267	disolución puente
0833	0268	disolución puente de un electrodo de referencia de doble unión
0834	2114	disolución saturada
0835	2372	disolución sobresaturada
0836	2221	disolución sólida
0837	2025	disoluciones de referencia
0838	2232	disolvente
0839	0076	disolvente anfiprótido
0840	1277	disparos de láser
0841	0667	dispersión
0842	2124	dispersión
0843	2125	dispersión
0844	0101	dispersión angular
0845	0430	dispersión de Compton
0846	0796	dispersión de la energía
0847	2585	dispersión de la longitud de onda
0848	2126	dispersión de la radiación
0849	1460	dispersión de Mie
0850	1985	dispersión de Raman
0851	1993	dispersión de Rayleigh
0852	0669	dispersión de un material
0853	0239	dispersión del blanco
0854	0732	dispersión elástica
0855	0668	dispersión en el análisis por inyección en flujo
0856	1102	dispersión inelástica
0857	1304	dispersión lineal
0858	2004	dispersión lineal inversa
0859	1132	dispositivo de calibrado de la intensidad
0860	1355	dispositivo de tubo largo
0861	1079	dispositivos de imagen
0862	2237	distancia de migración del disolvente
0863	0926	distancia focal
0864	0678	distribución
0865	0533	distribución acumulada
0866	1876	distribución de amplitud de impulsos
0867	1392	distribución de masas
0868	1677	distribución granulométrica
0869	1330	distribución líquido–líquido
0870	0882	dosímetro fotográfico en placa
0871	0350	dosímetro químico
0872	0689	dosis
0873	0007	dosis absorbida
0874	0690	dosis equivalente
0875	0720	dosis equivalente efectiva
0876	1880	duración de impulso
0877	0257	ecuación de Bragg
0878	0567	ecuación de Debye–Hückel
0879	0685	ecuación de Doerner y Hoskins
0880	1354	ecuación de Lomakin–Scheibe
0881	1548	ecuación de Nernst
0882	1481	ecuación de Nernst modificada
0883	2112	ecuación de Sand
0884	1945	edad radioactiva
0885	2038	efectividad biológica relativa
0886	0718	efecto
0887	0103	efecto aniónico
0888	0167	efecto Auger
0889	0333	efecto canal
0890	0322	efecto catiónico
0891	0327	efecto Cerenkov
0892	0428	efecto Compton
0893	2394	efecto de cola
0894	0947	efecto de cola frontal
0895	0280	efecto de combustión
0896	1761	efecto de constricción (autoconstricción)
0897	1024	efecto de los átomos pesados
0898	2392	efecto de Szilard–Chalmers
0899	1894	efecto electroóptico cuadrático
0900	1306	efecto electroóptico lineal

0901	1727	efecto fotoeléctrico
0902	1517	efecto Mössbauer
0903	1638	efecto orgánico
0904	1383	efecto Penning magnético
0905	2616	efecto Zeeman
0906	2161	efectos de la autoabsorción
0907	1401	efectos de la matriz
0908	1842	efectos de la presión
0909	2032	efectos de la refracción
0910	2402	efectos de la temperatura
0911	2234	efectos del disolvente
0912	0852	efectos externos de los átomos pesados
0913	1142	efectos interelementales
0914	1805	efectos post-filtro
0915	1837	efectos pre-filtro
0916	0724	eficacia
0917	0004	eficacia absoluta de fotopico
0918	0003	eficacia absoluta del pico de máxima energía
0919	1902	eficacia cuántica de la fluorescencia
0920	1903	eficacia cuántica de la luminiscencia
0921	2484	eficacia cuántica total de la fluorescencia
0922	0516	eficacia de acoplamiento
0923	0162	eficacia de atomización
0924	0604	eficacia de detección
0925	0910	eficacia de la fluorescencia
0926	2241	eficacia de la fuente
0927	1817	eficacia de la potencia de fluorescencia
0928	1535	eficacia de nebulización
0929	0509	eficacia de recuento
0930	0608	eficacia del detector
0931	0954	eficacia del pico de maxima energía
0932	1173	eficacia intrínseca
0933	1175	eficacia intrínseca de fotopico
0934	1172	eficacia intrínseca del detector
0935	1174	eficacia intrínseca del pico de máxima energía
0936	1345	eficacia local de la atomización
0937	1372	eficacia lumínica
0938	1950	efluvio radioactivo
0939	0774	electroaerosol
0940	0090	electrodo analítico
0941	0182	electrodo auxiliar
0942	0189	electrodo axial
0943	0413	electrodo combinado
0944	0532	electrodo cristalino selectivo de iones
0945	0539	electrodo de copa
0946	2548	electrodo de copa al vacío
0947	1431	electrodo de cubeta de mercurio
0948	2409	electrodo de ensayo
0949	1067	electrodo de gas de hidrógeno
0950	0707	electrodo de gotas de mercurio
0951	2085	electrodo de matriz rígida
0952	1029	electrodo de membrana heterogénea
0953	1057	electrodo de membrana homogénea
0954	2102	electrodo de muestra
0955	2460	electrodo de película fina
0956	1706	electrodo de pH
0957	1475	electrodo de portador móvil
0958	2018	electrodo de referencia
0959	1161	electrodo de referencia interno de un electrodo de vidrio
0960	0814	electrodo de sustrato enzimático
0961	2595	electrodo de trabajo
0962	0992	electrodo de vidrio
0963	1054	electrodo hueco
0964	1093	electrodo indicador
0965	1574	electrodo no cristalino
0966	0703	electrodo perforado
0967	1783	electrodo plano
0968	0307	electrodo portador
0969	1848	electrodo selectivo de iones primarios
0970	2196	electrodo tamizador
0971	0297	electrodos capilares
0972	1800	electrodos de cubeta porosa
0973	1222	electrodos específicos de iones
0974	1059	electrodos horizontales
0975	1799	electrodos porosos
0976	1220	electrodos selectivos de iones
0977	2375	electrodos suplementarios
0978	0747	electroerosión
0979	0770	electroforesis
0980	0771	electroforetograma
0981	0748	electrografía
0982	0749	electrogravimetría
0983	0475	electrogravimetría a potencial controlado
0984	2303	electrogravimetría espontánea
0985	1156	electrogravimetría interna
0986	1157	electrolisis interna
0987	0750	electrolito
0988	0208	electrolito de base
0989	1094	electrolito indiferente
0990	2377	electrolito soporte
0991	2560	electrómetro de láminas vibratorias (langüeta vibrante)
0992	0168	electrón de Auger
0993	0429	electrón de Compton
0994	0484	electrón de conversión
0995	0008	electrones absorbidos
0996	2551	electrones de valencia
0997	0635	electrones difractados
0998	0940	electrones libres
0999	0198	electrones retrodispersos
1000	2144	electrones secundarios
1001	0772	electroseparación
1002	0476	electroseparación a potencial controlado
1003	0773	electroseparación a potencial controlado
1004	0775	elemento
1005	0951	elemento combustible
1006	0079	elemento de análisis
1007	1766	elemento de imagen
1008	2019	elemento de referencia
1009	2211	elemento monoionizado
1010	0996	elución en gradiente
1011	2348	elución escalanado (por etapas)
1012	2156	elución selectiva
1013	0779	eluido
1014	0781	eluir
1015	0780	eluyente
1016	0149	emisión atómica

1017	1917	emisividad radiante
1018	1073	encendido
1019	2250	encendido de chispa
1020	1660	encendido en paralelo
1021	2189	encendido en serie
1022	0793	endoterma
1023	0760	energía cinética electrónica
1024	0014	energía de absorción
1025	0127	energía de aparición
1026	1887	energía de bombeo
1027	0229	energía de enlace
1028	0830	energía de excitación
1029	0730	energía de expulsión
1030	1881	energía de impulso
1031	1206	energía de ionización
1032	1931	energía de la radiación
1033	0171	energía de pico de Auger
1034	2064	energía de resonancia
1035	1271	energía de salida del láser
1036	0756	energía electrónica
1037	0844	energía excitatriz
1038	0911	energía fluorescente
1039	1492	energía molar de ionización
1040	1918	energía radiante
1041	2027	energía reflejada
1042	2465	energía umbral
1043	0719	energía umbral efectiva del cadmio
1044	1947	enfriamiento radioactivo
1045	0228	enlace
1046	1388	enmascaramiento
1047	0808	enriquecimiento
1048	0269	ensanchamiento
1049	0686	ensanchamiento de Doppler
1050	1356	ensanchamiento de Lorentz
1051	2332	ensanchamiento de Stark
1052	1529	ensanchamiento natural
1053	2160	ensanchamiento por autoabsorción
1054	0401	ensanchamiento por choque
1055	2063	ensanchamiento por resonancia
1056	0644	ensayo de dilución
1057	2009	ensayo de recuperación
1058	0054	envejecimiento
1059	1754	envejecimiento físico
1060	0349	envejecimiento químico
1061	2414	envejecimiento térmico
1062	1283	equilibrado en capas
1063	2251	equilibrio de descargas disruptivas
1064	0824	equilibrio de evaporación
1065	1755	equilibrio físico
1066	0351	equilibrio químico
1067	1951	equilibrio radioactivo
1068	2153	equilibrio secular
1069	2420	equilibrio térmico
1070	1350	equilibrio térmico local
1071	2435	equilibrio termodinámico
1072	0681	equilibrios de distribución
1073	0421	equilibrios de formación de complejos
1074	1871	equilibrios de protonación
1075	2231	equilibrios en disolución
1076	1331	equilibrios líquido–líquido
1077	0819	equivalente
1078	1464	equivalente en miligramos de la lectura mínima
1079	0352	equivalente químico
1080	1264	erosión con láser
1081	0820	error
1082	2476	error de valoración
1083	0993	error del electrodo de vidrio
1084	1699	error porcentual
1085	2041	error relativo
1086	0801	escala de energía
1087	1624	escala funcional de pH
1088	1710	escala funcional de pH
1089	0381	escalada del arco
1090	2532	esfera de Ulbricht
1091	1128	esfera integradora
1092	0464	espacia (trampa) de control
1093	2405	espaciado temporal
1094	0112	espacio oscuro anódico
1095	0312	espacio oscuro catódico
1096	0143	espacio oscuro de Aston
1097	0523	espacio oscuro de Crookes
1098	0872	espacio oscuro de Faraday
1099	1048	espacio oscuro de Hittorf
1100	2298	espectro
1101	0496	espectro corregido de polarización de emisión de luminiscencia
1102	0497	espectro corregido de polarización de excitación de luminiscencia
1103	0493	espectro de emisión corregido
1104	1747	espectro de emisión fotónica
1105	0494	espectro de excitación corregido
1106	0841	espectro de excitación de fluorescencia
1107	0842	espectro de excitación de fosforescencia
1108	1720	espectro de excitación–emisión de fosforescencia
1109	2389	espectro de fluorescencia excitada sincrónicamente
1110	2390	espectro de fosforescencia excitada sincrónicamente
1111	2299	espectro de la radiación
1112	1397	espectro de masas
1113	1541	espectro de masas de iones negativos
1114	0195	espectro de masas del fondo
1115	1937	espectro de radiaciones
1116	0963	espectro de rayos gamma
1117	2614	espectro de rayos X
1118	0594	espectrofotometría de derivadas
1119	2291	espectrografía
1120	2289	espectrógrafo
1121	2288	espectrograma
1122	2294	espectrometría
1123	0147	espectrometría de absorción atómica
1124	1219	espectrometría de dispersión iónica (ISS)
1125	0150	espectrometría de emisión atómica
1126	2602	espectrometría de emisión de rayos X
1127	1672	espectrometría de emisión de rayos X inducida por partículas
1128	1628	espectrometría de emisión óptica
1129	0152	espectrometría de fluorescencia atómica
1130	0898	espectrometría de llama
1131	1370	espectrometría de luminiscencia

1132	1396	espectrometría de masas
1133	2147	espectrometría de masas de iones secundarios
1134	0555	espectrometría de masas de resonancia de ciclotrón
1135	2396	espectrometría de masas en tándem
1136	1505	espectrometría molecular
1137	1494	espectrometría molecular de absorción
1138	1499	espectrometría molecular de emisión
1139	1502	espectrometría molecular de luminiscencia
1140	2292	espectrómetro
1141	2133	espectrómetro de centelleo
1142	0530	espectrómetro de difracción cristalina
1143	0692	espectrómetro de doble haz
1144	0699	espectrómetro de doble haz espectral
1145	0700	espectrómetro de doble haz sincrónico
1146	0913	espectrómetro de fluorescencia
1147	1721	espectrómetro de fosforescencia
1148	2203	espectrómetro de haz simple
1149	1369	espectrómetro de luminiscencia
1150	1395	espectrómetro de masas
1151	0695	espectrómetro de masas bifocal
1152	1301	espectrómetro de masas de aceleración lineal
1153	0714	espectrómetro de masas de campo dinámico
1154	2337	espectrómetro de masas de campo estático
1155	1186	espectrómetro de masas de resonancia del ion ciclotrón
1156	2469	espectrómetro de masas de tiempo de vuelo
1157	0935	espectrómetro de masas de transformada de Fourier
1158	1865	espectrómetro de masas de trocoide alargada
1159	2207	espectrómetro de masas monofocal
1160	1936	espectrómetro de radiaciones
1161	0962	espectrómetro de rayos gamma
1162	0120	espectrómetro de rayos gamma anti-Compton
1163	2067	espectrómetro de resonancia
1164	0157	espectros atómicos
1165	0204	espectros de bandas
1166	2544	espectros de bandas irresolubles
1167	2545	espectros de bandas no resueltas
1168	0790	espectros de emisión
1169	0839	espectros de excitación
1170	0829	espectros de excitación–emisión
1171	2194	espectros de Shpol'skii
1172	0766	espectros electrónicos
1173	1730	espectros fotoelectrónicos
1174	1196	espectros iónicos
1175	2297	espectroscopía
1176	0159	espectroscopía atómica
1177	0172	espectroscopía Auger
1178	2603	espectroscopía de emisión de rayos X
1179	1673	espectroscopía de emisión de rayos X inducida por partículas (PIXES)
1180	1748	espectroscopía de emisión fotónica
1181	1733	espectroscopía de fotoemisión
1182	1740	espectroscopía de fotoionización
1183	1371	espectroscopía de luminiscencia
1184	0757	espectroscopía de pérdida de energía electrónica
1185	1046	espectroscopía de pérdida de energía de alta resolución (HRELS)
1186	2028	espectroscopía de pérdida de energía electrónica de reflexión (REELS)
1187	2501	espectroscopía de pérdida de energía electrónica de transmisión (TEELS)
1188	2471	espectroscopía de resolución temporal
1189	0199	espectroscopía de retrodispersión (BSS)
1190	0169	espectroscopía electrónica de Auger
1191	0767	espectroscopía electrónica para el análisis químico (ESCA)
1192	1731	espectroscopía fotoelectrónica (PES)
1193	2612	espectroscopía fotoelectrónica de rayos X (XPS)
1194	2537	espectroscopía fotoelectrónica ultravioleta (UPS)
1195	1506	espectroscopía molecular
1196	1495	espectroscopía molecular de absorción
1197	1500	espectroscopía molecular de emisión
1198	1986	espectroscopía Raman
1199	2296	espectroscopio
1200	0930	espejo de enfoque
1201	1269	espejo de láser
1202	2174	espejo semitransparente
1203	2458	espesor de la capa de reacción
1204	1015	espesor de semivalor
1205	0721	espesor infinito efectivo
1206	2248	espinterómetro
1207	0572	esquema de desintegración
1208	2259	estabilización espacial
1209	2406	estabilización temporal
1210	2344	estado estacionario
1211	1006	estado fundamental
1212	1445	estado metaestable
1213	1781	estados de plasmón
1214	0843	estados excitados
1215	1228	estados isómeros
1216	1272	estela de láser
1217	2557	estela de vapor
1218	1924	estela radiante
1219	2353	estequiometría
1220	0975	estrangulador de gases
1221	2358	estriaciones
1222	0823	evaporación
1223	1666	evaporación parcial
1224	2421	evaporación térmica
1225	0417	evaporación total
1226	0022	exactitud
1227	2281	exactitud espectroquímica

1228	0828	excitación
1229	1414	excitación de fluorescencia medida
1230	1416	excitación de fosforescencia medida
1231	0736	excitación eléctrica
1232	0743	excitación inelectródica
1233	2245	excitación por descarga cruzada
1234	1940	excitación radiativa
1235	0846	exoterma
1236	2118	expansión de escala
1237	0851	exposición
1238	1920	exposición a la radiación
1239	1911	extinción
1240	1368	extinción de fluorescencia
1241	0859	extracción
1242	2235	extracción con disolventes
1243	1324	extracción con líquidos
1244	1332	extracción líquido–líquido
1245	0858	extractante
1246	0855	extracto
1247	0856	extraer
1248	0857	extraibilidad
1249	2310	eyección atómica
1250	0015	factor de absorción
1251	2162	factor de autoabsorción
1252	1897	factor de calidad
1253	0295	factor de capacidad
1254	0485	factor de conversión
1255	2479	factor de conversión volumétrico
1256	1843	factor de corrección de gradiente de presión
1257	0809	factor de enriquecimiento
1258	0866	factor de Fano
1259	2008	factor de recuperación
1260	2029	factor de reflexión
1261	2076	factor de retraso (R_f)
1262	2184	factor de separación
1263	2503	factor de transmisión
1264	1308	factor dependiente de la transferencia lineal de energía
1265	0988	factor geométrico
1266	2339	fase estacionaria
1267	1340	fase estacionaria líquida
1268	1334	fase líquida
1269	1476	fase móvil
1270	1948	fechado radioactivo
1271	0455	fenómenos de contaminación en la precipitación
1272	1629	fibras ópticas
1273	0984	filtración por geles
1274	0884	filtrar
1275	0883	filtro
1276	1630	filtros ópticos
1277	0885	fisil
1278	0887	fisión
1279	2305	fisión espontánea
1280	1591	fisión nuclear
1281	2422	fisión térmica
1282	0888	fisionable
1283	1758	fisisorción
1284	0904	floculación
1285	0919	flujo
1286	0925	flujo a tráves del monocromador
1287	1670	flujo de partículas
1288	1921	flujo energético
1289	1750	flujo fotónico
1290	1253	flujo laminar
1291	1568	flujo lineal nominal
1292	1374	flujo lumínico
1293	1559	flujo neutrónico
1294	2527	flujo turbulento
1295	0908	fluorescencia
1296	0122	fluorescencia anti-Stokes
1297	0151	fluorescencia atómica
1298	0653	fluorescencia de línea directa
1299	2605	fluorescencia de rayos X
1300	2354	fluorescencia de Stokes
1301	2350	fluorescencia escalonada
1302	0577	fluorescencia retrasada
1303	0822	fluorescencia retrasada de tipo E
1304	1873	fluorescencia retrasada de tipo P
1305	2145	fluorescencia secundaria
1306	1912	fluorimetría de extinción
1307	1714	fluorimetría de fase
1308	0901	fluorimetría instantánea
1309	0193	fondo
1310	2546	fondo
1311	0236	fondo de blanco
1312	0896	fondo de llama
1313	2267	fondo espectral
1314	0557	forma cilíndrica
1315	0521	forma de cráter
1316	2273	forma de las líneas espectrales
1317	1320	forma de línea
1318	0969	forma gaseosa
1319	0420	formación de complejos
1320	0933	formalidad
1321	1718	fosforescencia
1322	1722	fosforimetría
1323	1723	fosforoscopios
1324	1724	fotocorriente
1325	1725	fotodetector
1326	0979	fotodetectores conmutados
1327	2198	fotodiodos de silicio
1328	1729	fotoelectrón
1329	1739	fotoionización
1330	1743	fotometría
1331	1909	fotometría cuasi-monocromática
1332	1737	fotometría fotográfica
1333	1028	fotometría heterocromática
1334	1745	fotón
1335	1082	fotón incidente
1336	2613	fotones de rayos X
1337	1753	fotopico
1338	2549	fototubo de vacío
1339	0163	fracción atomizada
1340	2340	fracción de fase estacionaria
1341	0601	fracción dessolvatada
1342	1165	fracción intersticial
1343	1346	fracción local atomizada
1344	1347	fracción local dessolvatada
1345	1348	fracción local volatilizada
1346	0285	fracción quemada
1347	0259	fracción ramificada
1348	0889	fragmentos de fisión
1349	0943	frecuencia
1350	2247	frecuencia de descarga
1351	0945	frecuencia de modulación
1352	1483	frecuencia de modulación
1353	0893	frecuencia fija

1354	0903	frecuencia fluctuante
1355	0727	frecuencias propias
1356	2236	frente del disolvente
1357	2240	fuente
1358	1769	fuente de descarga lumínica de cátodo plano
1359	0838	fuente de excitación
1360	1617	fuente de excitación monoetápica
1361	1278	fuente de láser
1362	1221	fuente iónica
1363	1852	fuente primaria
1364	2011	fuente rectificada
1365	2428	fuente térmica
1366	0387	fuentes coherentes
1367	1053	fuentes de cátodo hueco
1368	1523	fuentes de chispa polivalentes (para usos multiples)
1369	2319	fuentes de corriente estabilizadas
1370	1879	fuentes de impulsos de láser
1371	1935	fuentes de radiaciones
1372	1572	fuentes ópticas incoherentes
1373	1643	fuerza del oscilador
1374	0752	fuerza electromotriz
1375	1198	fuerza iónica
1376	1017	función de acidez de Hammett
1377	0792	función de calibrado de emulsiones
1378	1681	función de reparto
1379	2565	función de Voigt
1380	1319	función del perfil de línea
1381	0942	funcionamiento autóexcitado
1382	0461	funcionamiento en onda continua (CWO)
1383	0093	funciones analíticas
1384	0029	funciones de acidez
1385	0087	funciones de calibrado analítico
1386	0956	fusión
1387	1594	fusión nuclear
1388	0659	gas de descarga
1389	1910	gas de extinción
1390	1107	gas de inyección
1391	1773	gas de plasma
1392	1150	gas de plasma intermedio
1393	0880	gas de relleno
1394	2596	gas de trabajo
1395	0309	gas portador
1396	0050	gas portador de aerosol
1397	1856	gas principal
1398	1566	gases nobles
1399	2052	gases residuales
1400	2606	generación de rayos X
1401	2607	generación de rayos X por los electrones
1402	2609	generación de rayos X por los iones positivos
1403	2608	generación de rayos X por los fotones
1404	0472	generador de descargas de alto voltaje controlado
1405	2539	generador de descargas de alto voltaje incontrolado
1406	1428	generador de descargas de voltaje medio
1407	2249	generador de descargas disruptivas
1408	0704	generador de gotas
1409	1404	geometría de Mattauch–Herzog
1410	1565	geometría de Nier–Johnson
1411	0510	geometría de recuento
1412	2588	gradiente de potencial débil
1413	0573	grado de disociación
1414	0574	grado de ionización
1415	2261	grado de quemado
1416	0575	grado de valoración
1417	1005	gray
1418	1217	grupos ionogénicos
1419	0383	guía de ondas coaxial
1420	1133	haz de intensidad modulada
1421	2101	haz de muestra
1422	2017	haz de referencia
1423	1071	histéresis
1424	1002	horno de tubo de grafito
1425	0304	horno de varilla de carbón
1426	1000	horno de varilla de grafito
1427	1774	impedancia de un plasma
1428	1530	impedancias naturales
1429	1075	impulso de encendido
1430	1274	impulso de láser
1431	1088	impulso de radiación incoherente
1432	2208	impulso simple
1433	2301	impulsos breves, picos, saltos
1434	1933	impulsos de radiación
1435	1225	impulsos de radiación irregulares
1436	1279	impulsos parásitos de láser
1437	2467	inclinación de la placa fotográfica
1438	1087	inclusión
1439	1090	indicador
1440	0023	indicador ácido–base
1441	2136	indicador apantallado
1442	2529	indicador bicolor
1443	0048	indicador de adsorción
1444	0862	indicador de extracción
1445	1018	indicador de Hammett
1446	1653	indicador de oxido–reducción
1447	1707	indicador de pH
1448	1826	indicador de precipitación
1449	1436	indicador metalocrómico
1450	1437	indicador metalofluorescente
1451	1471	indicador mixto
1452	1615	indicador monocolor
1453	0362	indicador quimioluminiscente
1454	1954	indicador radioactivo
1455	2012	indicador rédox
1456	2564	indicador visual
1457	0435	índice de concentración
1458	2033	índice de refracción
1459	2078	índice de retención
1460	0276	índice de tampón
1461	1100	inductor
1462	1775	iniciación de un plasma
1463	1106	iniciar
1464	1080	inmunoensayo
1465	1122	instrumentación
1466	2290	instrumentos espectrográficos
1467	0922	integración de flujo
1468	1129	intensidad
1469	0495	intensidad corregida
1470	1135	intensidad de la radiación
1471	1136	intensidad de línea espectral
1472	1692	intensidad de pico
1473	2610	intensidad de rayos X
1474	2022	intensidad de referencia

1475	2271	intensidad espectral
1476	1415	intensidad medida
1477	1922	intensidad radiante
1478	1137	intensidad referida al pico de base
1479	0807	intensificación
1480	0024	interacción ácido–base
1481	2228	interacción soluto–disolvente
1482	0104	intercambio aniónico
1483	0323	intercambio catiónico
1484	0340	intercambio de cargas
1485	0356	intercambio de isótopos químicos
1486	1187	intercambio iónico
1487	1237	intercambio isotópico
1488	1143	interfase
1489	0746	interfase electrodo–disolución
1490	1144	interferencia
1491	1281	interferencia de difusión lateral
1492	0897	interferencia de geometría de la llama
1493	1756	interferencia física
1494	1526	interferencia mutua
1495	1582	interferencia no espectral
1496	0354	interferencia química
1497	0673	interferencias de disociación
1498	2257	interferencias de distribución espacial
1499	0831	interferencias de excitación
1500	1207	interferencias de ionización
1501	2508	interferencias de transporte
1502	2229	interferencias de volatilización de soluto
1503	2556	interferencias en fase vapor
1504	2262	interferencias específicas
1505	2272	interferencias espectrales
1506	1581	interferencias inespecíficas
1507	1146	interferente
1508	2600	interpretación de datos de rayos X
1509	1708	interpretación del pH
1510	0364	interruptor periódico
1511	1988	intervalo
1512	1413	intervalo de masas medio
1513	2105	intervalo de muestra
1514	1995	intervalo de reacción
1515	2400	intervalo de Tast
1516	2494	intervalo (zona) de viraje
1517	0864	intervalo extrapolado
1518	1412	intervalo lineal medio
1519	1798	inversión de la población
1520	1336	inyección de muestras líquidas
1521	1110	inyector
1522	2104	inyector de muestras
1523	0286	inyector en derivación
1524	1182	ion
1525	2002	ion de transposición (reorganización)
1526	0937	ion fragmentario
1527	1242	ion isotópico
1528	1501	ion molecular
1529	2001	ion molecular transpuesto (de reorganización)
1530	1540	ion negativo
1531	1835	ion precursor
1532	1847	ion primario
1533	1664	ion progenitor
1534	1858	ion progenitor
1535	2146	ion secundario
1536	0083	iones analito
1537	0894	iones fijos
1538	1803	iones positivos
1539	1200	ionización
1540	0878	ionización de campo
1541	2254	ionización de fuente de chispas
1542	2263	ionización específica
1543	0759	ionización por impacto de electrones
1544	1202	ionización por pulverización
1545	1263	ionización por rayos láser
1546	0355	ionización química
1547	2053	ionización residual
1548	2423	ionización térmica
1549	1224	irradiación
1550	1223	irradiancia
1551	1596	isobaras nucleares
1552	1227	isoformación
1553	1597	isómeros nucleares
1554	1231	isoterma
1555	1189	isoterma de intercambio iónico
1556	2239	isoterma de sorción
1557	1233	isotonos
1558	1234	isótopo
1559	1836	isótopo precursor
1560	1955	isótopo radioactivo
1561	1238	isótopos de los átomos
1562	1255	lámpara
1563	1439	lámpara de arco de vapor de metal
1564	1888	lámpara de bombeo
1565	1052	lámpara de cátodo hueco
1566	1134	lámpara de cátodo hueco de intensidad modulada
1567	0487	lámpara de cátodo hueco enfriado
1568	2140	lámpara de cátodo hueco sellada
1569	1878	lámpara de descarga de impulsos
1570	0742	lámpara de descarga inelectródica
1571	2520	lámpara de filamento de tungsteno
1572	2521	lámpara de tungsteno–halógeno
1573	2205	lámpara de un elemento
1574	2597	lámpara de xenón
1575	1042	lámpara de xenón de alta presión
1576	0902	lámpara estroboscópica
1577	1361	lámparas de arco de baja presión
1578	1362	lámparas de descarga de baja presión
1579	0983	lámparas de Geissler
1580	1256	láser
1581	0712	láseres de colorante
1582	2225	láseres de estado sólido
1583	0971	láseres gaseosos
1584	1328	láseres líquidos
1585	2138	lavado de extractos
1586	1998	lectura
1587	1567	lectura en vacío (sin carga)
1588	1120	lectura instrumental
1589	1997	lectura mínima
1590	0929	lente de enfoque
1591	1282	ley de acción de masas
1592	1405	ley de Maxwell–Boltzmann
1593	1768	ley de Planck
1594	2096	ley de Saha–Eggert
1595	2351	ley de Stern–Volmer
1596	0683	leyes de la distribución

1597	2044	liberadores
1598	0308	libre de portador
1599	1286	ligando
1600	1299	límite de cuantificación
1601	0605	límite de detección
1602	1297	límite de detección
1603	1298	límite de determinación
1604	0710	límite de longitudes de onda cortas de Duane–Hunt
1605	0918	limpiado
1606	1300	línea
1607	0153	línea atómica
1608	2334	línea de arranque
1609	0209	línea de base
1610	2023	línea de referencia
1611	1162	línea de referencia interna
1612	2065	línea de resonancia
1613	2335	línea de salida
1614	0137	línea del arco
1615	1058	línea homóloga
1616	1193	línea iónica
1617	1908	líneas en cuarto de onda
1618	0158	líneas espectrales atómicas
1619	1197	líneas espectrales iónicas
1620	0989	líneas homólogas de Gerlach
1621	1147	líneas interferentes
1622	0895	llama
1623	2191	llama apantallada
1624	0952	llama enriquecida en combustible
1625	1252	llama laminar
1626	2182	llama separada
1627	2526	llama turbulenta
1628	0353	llamas químicas
1629	1952	lluvia radioactiva
1630	1284	longitud de base de prisma efectiva
1631	0243	longitud de onda de ranura
1632	0013	longitud de onda del límite de absorción
1633	0018	longitud de paso de absorción
1634	1195	longitud de trayectoria iónica
1635	1366	luminiscencia
1636	0114	luminiscencia anódica
1637	0827	luminiscencia de excímeros
1638	1539	luminiscencia negativa
1639	2181	luminiscencia sensibilizada
1640	2380	luminiscencia supratérmica
1641	2356	luz parásita
1642	0991	macroimpulso
1643	2086	maduración
1644	1646	maduración de Ostwald
1645	1384	magnetrón
1646	1757	magnitud física
1647	1376	magnitud lumínica
1648	1926	magnitud radiante
1649	1934	magnitudes de la radiación
1650	2279	magnitudes de la radiación espectral
1651	0986	magnitudes geométricas
1652	1633	magnitudes ópticas
1653	2306	mancha
1654	0927	mancha focal
1655	1250	marcador
1656	1387	marcador
1657	1956	marcador radioactivo
1658	1251	marcar
1659	0154	masa atómica
1660	2037	masa atómica relativa
1661	0335	masa característica
1662	1610	masa nuclídica
1663	0031	material activable
1664	2129	material centelleante
1665	1964	material de desecho radioactivo
1666	2024	material de referencia
1667	0329	material de referencia certificado
1668	2327	material de referencia normalizado
1669	0876	material fértil
1670	1061	material hospedante
1671	1580	material inhomogéneo
1672	1957	materiales radioactivos
1673	1400	matriz
1674	2056	matriz de resina
1675	1693	máximo de pico
1676	1421	mecanismo
1677	0833	mecanismo de excitación
1678	2046	mecanismo de liberación
1679	1422	mecanización
1680	1423	mecanizar
1681	0279	mechero Bunsen
1682	0652	mechero de inyección directa
1683	2083	mechero de inyección directa invertida
1684	1429	mechero de Méker
1685	1839	mechero de mezcla previa
1686	2464	mechero de tres rendijas
1687	1524	mecheros de multirendijas
1688	2216	mecheros de rendijas
1689	1407	media
1690	1418	medida
1691	1709	medida de pH
1692	0238	medida del blanco
1693	1819	medida práctica de pH
1694	1735	medidas de intensidad fotográficas
1695	2611	medidas de rayos X
1696	1118	medidas instantáneas
1697	1424	medio
1698	0035	medio activo
1699	1194	medio iónico
1700	1430	membrana
1701	1190	membrana de intercambio iónico
1702	0973	membrana permeable a los gases
1703	0377	mesa de trabajo limpia
1704	1782	meseta
1705	2117	meseta de saturación
1706	1432	meso
1707	1433	mesomuestra
1708	1440	metaestable
1709	1441	metaestable
1710	2321	método de las adiciones de patrón
1711	1474	método de las disoluciones mezcladas
1712	2183	método de las disoluciones separadas
1713	2331	método de las sustracciones de patrón
1714	1503	métodos moleculares
1715	1522	métodos multiplicativos
1716	1698	mezcla gaseosa de Penning
1717	1446	microanálisis
1718	0764	microanálisis con sonda electrónica (EPMA)

1719	1218	microanálisis con sonda iónica (IPMA)
1720	0765	microanálisis de rayos X con sonda electrónica (EXPMA)
1721	1114	microanálisis in situ
1722	1275	microanálisis Raman de láser (LRMA)
1723	1447	microbalanza
1724	1449	microcomponente
1725	1451	microculombimetría
1726	1267	microespectrometría de masas con láser (LAMMS)
1727	1266	microespectroscopía de emisión de láser (LAMES)
1728	1452	microfotómetro
1729	1454	micromuestra
1730	1457	microonda
1731	1450	microordenador
1732	1453	microprocesador
1733	2502	microscopía de transmisión electrónica (TEM)
1734	2123	microscopía de transmisión electrónica de barrido (STEM)
1735	0761	microscopía electrónica
1736	0091	microscopía electrónica analítica
1737	2121	microscopía electrónica de barrido (SEM)
1738	1215	microscopía iónica
1739	1456	microtraza
1740	1461	migración
1741	1463	miliculombimetría
1742	1477	moderación
1743	1478	moderador
1744	1482	modificador
1745	1403	modificadores de matriz
1746	0834	modo de excitación
1747	2509	modo electromagnético transversal
1748	1479	modos
1749	2586	modulación de la longitud de onda
1750	1289	modulación de la luz
1751	1484	modulación de radiación lineal polarizada
1752	1634	modulación de señales ópticas
1753	0517	modulo de acoplamiento
1754	1488	molalidad
1755	1493	molaridad
1756	1507	molécula
1757	0923	monitor de flujo
1758	1509	monoalternancia
1759	1513	monocapa
1760	1511	monocromador
1761	0788	monocromador de emisión
1762	0835	monocromador de excitación
1763	0248	montaje de tipo Boersma
1764	0289	montaje de tipo Calvet
1765	2517	montaje de tubos
1766	2266	montaje portamuestras
1767	2100	muestra
1768	2265	muestra
1769	0278	muestra bruta
1770	2106	muestra de cantidad limitada
1771	2222	muestra de disolución sólida
1772	1814	muestra en polvo
1773	1335	muestra líquida
1774	2220	muestra sólida
1775	0097	muestras analíticas
1776	1520	multicapa
1777	0944	multiplicación de la frecuencia
1778	0762	multiplicación electrónica (de electrones)
1779	1561	multiplicación neutrónica (de neutrones)
1780	2143	multiplicador de electrones secundarios
1781	1527	nanotraza
1782	2109	navecilla de muestra
1783	1534	nebulización
1784	1536	nebulizador
1785	1003	nebulizador alimentado por gravedad
1786	0102	nebulizador angular
1787	0437	nebulizador concéntrico
1788	0332	nebulizador de cámara
1789	0471	nebulizador de flujo controlado
1790	0705	nebulizador de gotas
1791	2367	nebulizador de succión
1792	2528	nebulizador doble
1793	1784	nebulizador neumático
1794	2535	nebulizador ultrasónico
1795	2031	nebulizadores de reflujo
1796	1547	nefelometría
1797	1543	negatrón
1798	1553	neutro(a)
1799	1554	neutrón
1800	2066	neutrón de resonancia
1801	1867	neutrón instantáneo
1802	1149	neutrón intermedio
1803	0890	neutrones de fisión
1804	0815	neutrones epicádmicos
1805	0816	neutrones epitérmicos
1806	0394	neutrones fríos
1807	2217	neutrones lentos
1808	0874	neutrones rápidos
1809	0578	neutrones retardados
1810	2425	neutrones térmicos
1811	1468	niebla
1812	1285	nivel de valoración
1813	0799	niveles de energía
1814	0832	niveles de excitación
1815	2543	no polarizado(a)
1816	1585	normalidad
1817	2554	nube de vapor
1818	1273	nube de vapor producida por láser
1819	1606	nucleación
1820	1607	nucleón
1821	1959	núcleos radioactivos
1822	1609	núclido
1823	0886	núclido fisil
1824	1960	núclido radioactivo
1825	0877	núclidos fértiles
1826	0156	número atómico
1827	1288	número de ligandos
1828	1562	número de neutrones
1829	2587	número de onda
1830	2411	número de platos teóricos
1831	0722	número de platos teóricos efectivo
1832	1872	número de protones
1833	1394	número másico
1834	1608	número nucleónico
1835	0188	número promedio de ligandos

1836	1611	observación
1837	2326	observador normal
1838	2090	obturadores de disco giratorio
1839	2091	obturadores de espejo giratorio
1840	0769	obturadores electro-ópticos
1841	1420	obturadores mecánicos
1842	1614	oclusión
1843	1583	ojo normal
1844	0460	onda continua (CW)
1845	1893	operaciones de conmutación Q
1846	1637	orden de un espectro
1847	1639	orificio
1848	1642	oscilador
1849	0528	oscilador de cristal
1850	2519	oscilador de líneas sintonizadas
1851	1644	oscilografía
1852	1645	oscilopolarografía
1853	1652	oxidante
1854	1864	pantalla de proyección
1855	2192	pantallas
1856	0062	parámetro alfa
1857	1318	parámetro de ensanchamiento de línea
1858	0738	parámetro de fuente eléctrica
1859	0737	parámetro eléctrico
1860	1736	parámetro fotográfico
1861	1367	parámetros de luminiscencia
1862	2079	parámetros de retención
1863	1778	parámetros del plasma
1864	0063	partícula alfa
1865	0220	partícula beta
1866	0777	partícula elemental
1867	1212	partícula ionizante
1868	1598	partícula nuclear
1869	0339	partículas cargadas
1870	1923	partículas radiantes
1871	2026	patrón de pH de valor de referencia
1872	0854	patrón externo
1873	1127	patrón integral
1874	1163	patrón interno
1875	1853	patrón primario
1876	2150	patrón secundario
1877	0098	patrones analíticos
1878	1711	patrones de pH
1879	1625	patrones de pH funcionales
1880	1849	patrones de pH primarios
1881	1851	patrones de referencia primarios
1882	2459	película fina
1883	2070	pendiente de respuesta
1884	0511	pérdida de recuento
1885	0019	perfil de absorción
1886	0789	perfil de emisión
1887	0585	perfil de profundidad
1888	2566	perfil de Voigt
1889	1889	periodo de bombeo
1890	2362	periodo de modulación
1891	1822	periodo de pre-arco
1892	1840	periodo de pre-descarga
1893	2275	periodo espectral
1894	1704	permselectividad
1895	0924	perturbación del flujo
1896	0261	perturbaciones de descarga, inestibilidad de raptura
1897	0160	peso atómico
1898	2343	peso estadístico
1899	0865	peso factorial
1900	1705	pH
1901	0126	pH aparente
1902	1682	pico
1903	2482	pico de absorción total
1904	0211	pico de base
1905	0694	pico de doble fuga
1906	0785	pico de elución
1907	0953	pico de energía total
1908	2604	pico de fuga de rayos X
1909	2206	pico de fuga simple
1910	0055	pico del aire
1911	0794	pico endotérmico
1912	0847	pico exotérmico
1913	1728	pico fotoeléctrico
1914	1443	pico iónico metaestable
1915	1444	pico metaestable
1916	1759	picotraza
1917	2452	pila térmica, termopila
1918	1767	PIXES
1919	0443	placa confinadora
1920	2378	placa soporte
1921	1772	plasma
1922	0290	plasma acoplado capacitativamente
1923	0542	plasma de arco portador de corriente
1924	1364	plasma de baja presión inducido por microondas
1925	0745	plasma de cavidad resonante electródica
1926	0744	plasma de cavidad resonante inelectródica
1927	0291	plasma de microondas capacitivas
1928	1616	plasma de microondas monoelectródico
1929	1458	plasma inducido por microondas
1930	0145	plasma inducido por microondas a presión atmosférica
1931	0544	plasma libre de corriente
1932	1631	plasma ópticamente fino
1933	2463	plasma trielectródico
1934	1099	plasmas acoplados inductivamente
1935	1459	plasmas de microondas
1936	1974	plasmas de radiofrecuencia
1937	1575	plasmas no portadores de corriente
1938	2491	plasmas transferidos
1939	2092	plataforma giratoria
1940	2256	plataformas de descarga
1941	1818	poder de detección
1942	2355	poder de frenado
1943	2060	poder de resolución
1944	1788	polaridad
1945	1789	polarización
1946	0376	polarización circular
1947	0778	polarización elíptica
1948	1770	polarización en un plano
1949	1311	polarización lineal de luminiscencia
1950	1790	polarizada
1951	1771	polarizada en un plano
1952	1791	polarizador
1953	1312	polarizador lineal
1954	1792	polarografía
1955	0477	polarografía a potencial controlado
1956	0552	polarografía cíclica de ondas triangulares

1957 0546 polarografía de barrido de corriente
1958 1525 polarografía de barrido múltiple
1959 1089 polarografía de carga incremental
1960 0067 polarografía de corriente alterna
1961 1031 polarografía de corriente alterna de armónicos superiores
1962 1032 polarografía de corriente alterna de armónicos superiores con rectificación sensible a la fase
1963 0650 polarografía de corriente continua
1964 0591 polarografía de derivadas
1965 0660 polarografía de descarga
1966 0582 polarografía de desmodulación
1967 0701 polarografía de doble tono
1968 1245 polarografía de Kalousek
1969 1485 polarografía de modulación
1970 2320 polarografía de onda escalonada
1971 2314 polarografía de ondas cuadradas
1972 2512 polarografía de ondas triangulares
1973 1975 polarografía de radiofrecuencia
1974 0318 polarografía de rayos catódicos
1975 0341 polarografía de saltos de carga
1976 2401 polarografía de Tast
1977 0071 polarografía de voltaje alterno
1978 0623 polarografía diferencial
1979 1883 polarografía impulsional
1980 0593 polarografía impulsional de derivadas
1981 0626 polarografía impulsional diferencial
1982 2209 polarografía oscilográfica de barrido simple
1983 2170 polarografía semi-integral
1984 1794 poliatómico(a)
1985 1795 policromador
1986 2014 polímeros rédox
1987 1573 polvo no conductor
1988 1049 portador de retención
1989 2285 portador espectroquímico
1990 1241 portador isotópico
1991 0305 portador(a)
1992 2103 portamuestras
1993 2021 portarreferencias
1994 1804 positrón
1995 1806 post-precipitación
1996 1890 potencia de bombeo
1997 2300 potencia de fuga
1998 1884 potencia de impulso
1999 1270 potencia del láser
2000 1085 potencia incidente
2001 1426 potencia media
2002 1925 potencia radiante por ángulo sólido
2003 1807 potencial
2004 0130 potencial aplicado
2005 0336 potencial característico
2006 0128 potencial de aparición
2007 0144 potencial de asimetría
2008 2370 potencial de cima
2009 1907 potencial de cuarto de tiempo de transición
2010 0674 potencial de disociación
2011 0836 potencial de excitación
2012 1209 potencial de ionización
2013 1016 potencial de semionda
2014 1012 potencial de semipico
2015 1327 potencial de unión líquida
2016 2055 potencial de unión líquida residual
2017 1813 potenciometría
2018 0469 potenciometría de corriente controlada
2019 1829 potenciometría de punto cero de precisión
2020 0625 potenciometría diferencial
2021 1825 precipitación en disolución homogénea
2022 1823 precipitado
2023 1824 precipitar
2024 1828 precisión
2025 0512 precisión de recuento
2026 1830 precisión de una balanza
2027 1832 precisión de una lectura
2028 1831 precisión de una pesada
2029 1834 precursor
2030 1838 pre-ionización
2031 1841 presión
2032 0763 presión electrónica
2033 0186 presión gaseosa promedio
2034 2486 primer coeficiente de Townsend
2035 1210 probabilidad de ionización
2036 2495 probabilidad de transición
2037 0020 probabilidad de transición de absorción
2038 2496 probabilidad de transición de absorción
2039 2304 probabilidad de transición de emisión espontánea
2040 2497 probabilidad de transición de emisión espontánea
2041 2498 probabilidad de transición de emisión estimulada
2042 0729 probabilidad de transmisión de Einstein
2043 0457 procedimiento continuo
2044 0255 procedimiento de horquillaje
2045 2589 procedimiento de pesada
2046 0246 procedimientos de insuflado
2047 0662 procedimientos discontinuos
2048 0562 proceso de desactivación
2049 2185 proceso de separación
2050 0406 procesos de choque
2051 0758 procesos de transferencia de energía electrónica
2052 1941 procesos radiativos
2053 2426 procesos térmicos
2054 1658 producción de pares
2055 2227 producto de solubilidad
2056 0561 productos de filiación
2057 0891 productos de fisión
2058 0519 profundidad de cráter
2059 0821 profundidad de fuga
2060 1859 programa
2061 1861 programa
2062 1860 programar
2063 1862 programar
2064 0184 promedio
2065 2287 propiedades espectroquímicas
2066 1869 protón
2067 1130 puente de intensidad
2068 0319 pulverización catódica
2069 2309 pulverizador
2070 2618 punto cero de la escala

2071	2617	punto cero de un electrodo de vidrio
2072	2333	punto de arranque
2073	0863	punto de arranque extrapolado
2074	0281	punto de combustión
2075	0249	punto de ebullición
2076	0818	punto de equivalencia
2077	2336	punto de partida
2078	2073	punto de reposo
2079	0795	punto final
2080	1033	punto final conductimétrico de alta frecuencia
2081	0373	punto final cronopotenciométrico
2082	0590	punto final de derivadas
2083	0810	punto final entalpimétrico
2084	2352	punto final estequiométrico
2085	0916	punto final fluorimétrico
2086	1741	punto final fotométrico
2087	1809	punto final potenciométrico
2088	1978	punto final radiométrico
2089	2410	punto final teórico
2090	2447	punto final termométrico
2091	2523	punto final turbidimétrico
2092	2563	punto final visual
2093	1230	punto isopotencial
2094	0563	punto muerto final
2095	1981	pureza radionuclídica
2096	1968	pureza radioquímica
2097	1967	purificación radioanalítica
2098	0284	quemado
2099	1060	química anfitrión–huésped
2100	1928	química de las radiaciones
2101	1589	química nuclear
2102	1966	química radioanalítica
2103	0359	quimio-ionización
2104	0360	quimioluminiscencia
2105	0363	quimiosorción
2106	1913	rad
2107	1927	radiación
2108	0386	radiación coherente
2109	0458	radiación continua
2110	1577	radiación continua no discreta
2111	0502	radiación cósmica
2112	0110	radiación de aniquilación
2113	0328	radiación de Cerenkov
2114	0196	radiación de fondo
2115	0266	radiación de frenado
2116	1112	radiación de frenado interna
2117	1674	radiación de partículas
2118	0234	radiación del cuerpo negro
2119	2357	radiación dispersa (parásita)
2120	0751	radiación electromagnética
2121	0385	radiación electromagnética coherente
2122	1571	radiación electromagnética incoherente
2123	2342	radiación estacionaria
2124	1752	radiación fotónica
2125	0959	radiación gamma
2126	0301	radiación gamma de captura
2127	1084	radiación incandescente
2128	1086	radiación incidente
2129	1213	radiación ionizante
2130	1504	radiación molecular
2131	1510	radiación monocromática
2132	1532	radiación natural
2133	1563	radiación neutrónica
2134	1850	radiación primaria
2135	1984	radiación Raman
2136	2149	radiación secundaria
2137	2427	radiación térmica
2138	2598	radiación X
2139	0337	radiación X característica
2140	1915	radiancia
2141	2276	radiancia espectral
2142	2277	radiancia espectral del cuerpo negro
2143	1965	radioactividad
2144	0138	radioactividad artificial
2145	1096	radioactividad inducida
2146	1533	radioactividad natural
2147	1944	radioactivo(a)
2148	1973	radiocoloide
2149	1972	radiocromatógrafo
2150	1976	radioisótopo
2151	1977	radiólisis
2152	1980	radionúclido
2153	1971	radioquímica
2154	0095	radioquímica analítica
2155	1982	radiorreactivo
2156	1602	rastro nuclear
2157	0961	rayo gamma
2158	0119	rayo gamma anti-Compton
2159	1262	rayo láser
2160	0077	reacción de amplificación
2161	1592	reacción de fisión nuclear
2162	1595	reacción de fusión nuclear
2163	1216	reacción ion–molécula
2164	1599	reacción nuclear
2165	1640	reacción oscilante
2166	2387	reacción superficial
2167	0848	reacciones exotérmica
2168	1097	reacciones inducidas
2169	2432	reacciones termoquímicas
2170	1833	reactor en pre-columna
2171	1600	reactor nuclear
2172	0395	recogida
2173	1942	recombinaciones radiativas
2174	1040	rectificación faradaica de alto nivel
2175	2386	recubrimiento superficial
2176	0505	recuento
2177	0002	recuento absoluto
2178	0390	recuento por coincidencias
2179	2039	recuento relativo
2180	2007	recuperación
2181	1700	recuperación porcentual
2182	0531	red cristalina
2183	2223	red cristalina de estado sólido
2184	2359	redisolución
2185	0320	redisolución catódica
2186	2095	re-dispersión de Rutherford (RBS)
2187	0192	re-extracción
2188	0190	re-extracto
2189	0191	re-extraer
2190	1983	refinado
2191	0837	región de excitación
2192	0981	región de Geiger–Müller
2193	1170	región interzonal
2194	2006	registro
2195	2302	regla de la conservación del espín

2196	1618	relación activo/inactivo
2197	0436	relación de concentraciones
2198	0684	relación de distribución
2199	0433	relación de distribución de concentraciones
2200	1393	relación de distribución de masas
2201	1715	relación de fases
2202	1131	relación de puentes de intensidad
2203	0260	relación de ramificación
2204	0012	relación de salto del borde de absorción
2205	2505	relación de transmisión
2206	1717	relación de volúmenes de fase
2207	2580	relación de volúmenes de hinchamiento de los cambiadores iónicos
2208	1391	relación masa/carga
2209	1844	relación presión–voltaje–corriente
2210	2572	relación voltaje–tiempo de un arco
2211	1656	relleno
2212	0998	relleno en gradiente
2213	2047	rem
2214	0990	remoción de gases residuales
2215	2213	rendija
2216	0534	rendimiento acumulado de la fisión
2217	1904	rendimiento cuántico
2218	1905	rendimiento cuántico de fluorescencia
2219	1906	rendimiento cuántico de luminiscencia
2220	0498	rendimiento cuántico de luminiscencia corregido
2221	1417	rendimiento cuántico de luminiscencia medido
2222	0173	rendimiento de Auger
2223	0410	rendimiento de columna
2224	1749	rendimiento de emisión fotónica
2225	2312	rendimiento de evaporación catódica
2226	0892	rendimiento de fisión
2227	0651	rendimiento de fisión directa
2228	0330	rendimiento de fisión en cadena
2229	0915	rendimiento de fluorescencia
2230	2148	rendimiento de iones secundarios
2231	0170	rendimiento electrónico de Auger
2232	0805	rendimiento energético de luminiscencia
2233	1732	rendimiento fotoelectrónico
2234	1373	rendimiento lumínico
2235	0358	rendimiento químico
2236	1970	rendimiento radioquímico
2237	2049	reprecipitación
2238	2057	resistencia
2239	0345	resistencia de carga
2240	1343	resistencia de carga
2241	0965	resistencia de intervalo
2242	2058	resolución
2243	2059	resolución al 10% del valle
2244	0800	resolución de energía
2245	1694	resolución de picos
2246	0586	resolución de profundidad
2247	0609	resolución de un detector
2248	2293	resolución de un espectrómetro
2249	0399	resolución del colimador
2250	2258	resolución espacial

2251	1820	resolución práctica
2252	2404	resolución temporal
2253	1276	resonador de láser
2254	1549	respuesta nernstiana
2255	2075	resultado
2256	0096	resultado analítico
2257	2077	retención
2258	2042	retención relativa
2259	2097	retención salina
2260	2005	retroceso
2261	0900	retroceso de la llama
2262	0197	retrodispersión
2263	0610	revelar
2264	2089	roentgen
2265	0059	ruptura en α
2266	0218	ruptura en β
2267	2345	salto
2268	2115	saturación
2269	0331	saturación de cámara
2270	0525	sección eficaz
2271	0034	sección eficaz de activación
2272	0300	sección eficaz de captura
2273	0403	sección eficaz de choque
2274	0526	sección eficaz de ionización
2275	2592	sección eficaz de Westcott
2276	1378	sección eficaz macroscópica
2277	1455	sección eficaz microscópica
2278	0723	sección eficaz térmica efectiva
2279	2347	sector graduado
2280	2187	secuestro
2281	1877	selector de amplitud de impulsos
2282	0405	semianchura de choque
2283	0687	semianchura Doppler de una línea espectral
2284	2168	semiconductor
2285	2173	semiconmutado a Q
2286	1013	semiespesor
2287	2171	semimicro
2288	2172	semimicromuestra
2289	1011	semivida
2290	1953	semivida radioactiva
2291	2176	sensibilidad
2292	2179	sensibilidad de un sistema de introducción
2293	2180	sensibilidad de una fuente iónica
2294	2178	sensibilidad de una sonda directa
2295	2177	sensibilidad para una carga dada
2296	1243	separación isotópica
2297	1969	separación radioquímica
2298	2188	serie
2299	1961	serie radioactiva
2300	0224	sesgo
2301	2195	sievert
2302	1885	sistema bombeado
2303	1111	sistema de admisión (entrada)
2304	0693	sistema de doble haz
2305	2518	sistema de medida de amplificador sintonizado
2306	1521	sistema de paso múltiple
2307	0875	sistema de re-alimentación
2308	1004	sistema de tamizado de polvos alimentado por gravedad
2309	0030	sistema de tamizado por ondas de choque acústicas
2310	0388	sistema de unidades coherente

2311	0715	sistema dinámico
2312	1601	sistema nuclear
2313	0099	sistemas analíticos
2314	2200	sistemas analíticos simultáneos
2315	1537	sistemas de nebulización en llama
2316	1635	sistemas ópticos
2317	2373	sobresaturación
2318	1019	solapamiento de armónicos
2319	0036	sólido activo
2320	1480	sólido activo modificado
2321	0677	soluto en distribución
2322	0654	sonda directa
2323	2376	soporte
2324	2226	soporte sólido
2325	2238	sorción
2326	2363	sublimación
2327	2364	submicroanálisis
2328	0538	suma acumulada
2329	2381	superficie
2330	0849	superficie experimental
2331	0948	superficie frontal
2332	0379	superficie limpia
2333	2167	superficie semiconductora
2334	2371	superposición
2335	2379	supresores
2336	1258	sustancia activa al láser
2337	0244	sustancia descolorable
2338	0741	sustancia electroactiva
2339	1148	sustancia interferente
2340	1855	sustancia primaria
2341	2330	sustancias patrón
2342	1962	sustancias radioactivas
2343	1675	tamaño de partícula
2344	0271	tampón
2345	1199	tampón de ajuste de la fuerza iónica
2346	2284	tampón espectroquímico
2347	0272	tamponar
2348	1201	tampones de ionización
2349	1249	técnica de Kossel
2350	0084	técnica de la adición analítica
2351	0081	técnica de la adición de analito
2352	0089	técnica de la curva analítica
2353	0210	técnica de la línea de base en espectroscopía atómica
2354	2199	técnica de simulación
2355	2020	técnica del elemento de referencia
2356	0273	técnicas de adición de tampón
2357	1815	técnicas de polvos
2358	2243	técnicas de resolución espacial
2359	2472	técnicas de resolución temporal
2360	2202	técnicas simultáneas
2361	0515	técnicas simultáneas acopladas
2362	0663	técnicas simultáneas discontinuas
2363	2430	técnicas termoanalíticas
2364	0005	temperatura absoluta
2365	0129	temperatura de aparición
2366	0411	temperatura de columna
2367	0250	temperatura de ebullición
2368	0840	temperatura de excitación
2369	1108	temperatura de inyección
2370	1211	temperatura de ionización
2371	0899	temperatura de la llama
2372	1916	temperatura de radiancia
2373	2080	temperatura de retención
2374	2186	temperatura de separación
2375	0978	temperatura del gas
2376	1779	temperatura del plasma
2377	0768	temperatura electrónica
2378	1105	temperatura inicial
2379	1349	temperatura local de un gas
2380	1564	temperatura neutrónica
2381	1587	temperatura norma
2382	2094	temperatura rotacional
2383	2436	temperatura termodinámica
2384	2562	temperatura vibracional
2385	2407	tensametría
2386	0068	tensión alterna, voltaje alterno
2387	2429	termoacustimetría
2388	2431	termobalanza
2389	2433	termodilatometría
2390	2437	termoelectrometría
2391	2456	termoespectrometría
2392	2451	termofotometría
2393	2439	termografía
2394	2438	termograma
2395	0631	termograma diferencial
2396	2442	termogravimetría
2397	0596	termogravimetría de derivadas
2398	0633	termogravimetría diferencial
2399	2443	termoluminiscencia
2400	2444	termomagnetometría
2401	2446	termomecanometría
2402	0717	termomecanometría dinámica
2403	2449	termomicroscopía
2404	2453	termoptometría
2405	2454	termorrefractometría
2406	2455	termosonimetría
2407	0708	tiempo de goteo
2408	1138	tiempo de interacción
2409	2003	tiempo de recepción
2410	2050	tiempo de residencia
2411	2061	tiempo de resolución
2412	0391	tiempo de resolución de coincidencias
2413	2071	tiempo de respuesta
2414	2081	tiempo de retención
2415	0499	tiempo de retención corregido
2416	1551	tiempo de retención neto
2417	0262	tiempo de ruptura (de descarga)
2418	2087	tiempo de subida
2419	2499	tiempo de transición
2420	2500	tiempo de tránsito
2421	2511	tiempo de viaje
2422	0564	tiempo muerto
2423	0607	tiempo muerto del detector
2424	2000	tiempo real
2425	0491	tiempos de amortiguamentio de fluorescencia corregidos
2426	0492	tiempos de amortiguamentio de fosforescencia corregidos
2427	2477	título
2428	0964	trampa
2429	0094	trampa analítica
2430	0113	trampa anódica
2431	0134	trampa de arco
2432	0183	trampa de chispa auxiliar
2433	0479	trampa de chispa de control
2434	2391	trampa de giro sincrónico
2435	2395	trampa doble

2436 2514 trampa estacionaria activable
2437 0342 transferencia de carga
2438 0803 transferencia de energía
2439 1508 transferencia de momentos
2440 1307 transferencia lineal de energía
2441 2492 transformación
2442 0202 transformación de Baker–Sampson –Seidel
2443 1604 transformación nuclear
2444 1037 transformador elevador de alta frecuencia
2445 2093 transición de energía rotacional
2446 2561 transición de energía vibracional
2447 0804 transición energética
2448 1140 transición intercromofórica no radiativa
2449 1152 transición intermolecular no radiativa
2450 1171 transición intracromofórica no radiativa
2451 1229 transición isómera
2452 0938 transición libre–enlazado
2453 0941 transición libre–libre
2454 1932 transición no radiativa
2455 1605 transición nuclear
2456 1943 transiciones radiativas
2457 2506 transmitancia
2458 1738 transmitancia fotográfica
2459 2507 transporte
2460 1406 transposición de McLafferty
2461 2487 traza
2462 2490 trazador
2463 1244 trazador isotópico
2464 1963 trazador radioactivo
2465 0298 tubo capilar
2466 0508 tubo contador
2467 0980 tubo contador de Geiger–Müller
2468 1868 tubo contador proporcional
2469 1109 tubo de inyección
2470 1744 tubo fotomultiplicador
2471 1151 tubo intermedio
2472 2393 tubos en T
2473 2525 turbidimetría
2474 2534 ultramicroanálisis
2475 0802 umbral de energía
2476 0982 umbral de Geiger–Müller
2477 1280 umbral de láser
2478 0155 unidades de masa atómica
2479 1326 unión líquida
2480 2054 unión líquida residual
2481 1333 unión líquido–líquido
2482 1254 vaina laminar
2483 1994 valor de R_B
2484 2552 valor de una división
2485 1008 valor G
2486 2474 valoración
2487 0026 valoración acidimétrica
2488 0025 valoración ácido–base
2489 0056 valoración alcalimétrica
2490 0074 valoración amperométrica
2491 0222 valoración biamperométrica
2492 0231 valoración bipotenciométrica
2493 0422 valoración complejométrica
2494 0441 valoración conductimétrica
2495 1034 valoración conductimétrica de alta frecuencia
2496 0374 valoración cronopotenciométrica
2497 0503 valoración culombimétrica
2498 0473 valoración culombimétrica a potencial controlado
2499 1038 valoración de alta frecuencia
2500 0480 valoración de control
2501 1716 valoración de fases
2502 1654 valoración de oxido–reducción
2503 1712 valoración de pH
2504 1827 valoración de precipitación
2505 0241 valoración del blanco
2506 0615 valoración dielectrométrica
2507 1570 valoración en medio no acuoso
2508 0811 valoración entalpimétrica
2509 2295 valoración espectrofotométrica
2510 0917 valoración fluorimétrica
2511 1742 valoración fotométrica
2512 2591 valoración gravimétrica
2513 1812 valoración gravimétrica potenciométrica
2514 1095 valoración indirecta
2515 1546 valoración nefelométrica
2516 1793 valoración polarométrica
2517 0201 valoración por retroceso
2518 1811 valoración potenciométrica
2519 0468 valoración potenciométrica de corriente controlada
2520 0592 valoración potenciométrica de derivadas
2521 1177 valoración potenciométrica de derivadas inversas
2522 2152 valoración potenciométrica de segunda derivada
2523 0624 valoración potenciométrica diferencial
2524 0347 valoración quelatométrica
2525 1979 valoración radiométrica
2526 2448 valoración termométrica
2527 2524 valoración turbidimétrica
2528 2015 valoraciones rédox
2529 1178 valoraciones yodimétricas
2530 1180 valoraciones yodométricas
2531 2473 valorante
2532 0200 valorar por retroceso
2533 1538 válvula de aguja
2534 2553 vaporización
2535 0118 vaporización anódica
2536 0321 vaporización catódica
2537 2559 variable aleatoria
2538 2558 varianza
2539 0931 velo
2540 0141 velocidad de aspiración
2541 1990 velocidad de aspiración de líquido
2542 1023 velocidad de calentamiento
2543 0282 velocidad de combustión
2544 2547 velocidad de consumo
2545 1991 velocidad de consumo de líquido
2546 0666 velocidad de desintegración
2547 2311 velocidad de eyección
2548 1648 velocidad de flujo de salida
2549 1992 velocidad de nucleación
2550 0513 velocidad de recuento
2551 2088 velocidad de subida

2552	2568	velocidad de volatilización
2553	1166	velocidad intersticial
2554	1167	velocidad intersticial a la presión de salida
2555	1409	velocidad intersticial media de un gas portador
2556	0514	velocidades de recuento
2557	2253	velocidades de repetición de chispa
2558	0124	vértice, ápice
2559	1411	vida media
2560	1958	vida media radioactiva
2561	0187	vida promedio
2562	1544	vidrio de neodimio
2563	2567	volatilización
2564	2157	volatilización selectiva
2565	2569	volatilizadores
2566	2570	voltaje
2567	0447	voltaje constante
2568	0314	voltaje de caída catódica
2569	0283	voltaje de combustión
2570	1385	voltaje de red
2571	2035	voltaje de re-encendido
2572	0263	voltaje de ruptura
2573	0292	voltaje de un condensador
2574	1214	voltaje ionizante
2575	1702	voltaje periódico
2576	1686	voltaje pico de un condensador
2577	2575	voltametría
2578	0553	voltametría cíclica de ondas triangulares
2579	1316	voltametría de barrido lineal
2580	0486	voltametría de barrido lineal con integral de convolución
2581	0597	voltametría de derivadas
2582	2338	voltametría de electrodo estacionario
2583	2513	voltametría de ondas triangulares
2584	0117	voltametría de redisolución anódica
2585	0634	voltametría diferencial
2586	1066	voltametría hidrodinámica
2587	2576	voltamograma
2588	2577	voltamperograma
2589	0412	volumen de columna
2590	0786	volumen de elución
2591	1689	volumen de elución de pico
2592	2341	volumen de fase estacionaria
2593	1341	volumen de líquido
2594	2082	volumen de retención
2595	0042	volumen de retención ajustado
2596	0500	volumen de retención corregido
2597	1050	volumen de retención de eluyente no corregido
2598	0970	volumen de retención de gas portador no corregido
2599	2264	volumen de retención específico
2600	1552	volumen de retención neto
2601	2412	volumen de retención teórico
2602	2485	volumen de retención total
2603	0265	volumen de saturación
2604	0216	volumen del lecho
2605	1168	volumen intersticial
2606	1290	volumen limitado
2607	0566	volumen muerto
2608	0600	volumen nominal
2609	2175	volumen sensible de un detector
2610	2480	volumetría
2611	1179	yodimetría
2612	1181	yodometría
2613	2619	zona
2614	0313	zona de caída catódica
2615	1613	zona de observación de un plasma
2616	1647	zona externa
2617	1141	zona intercónica
2618	1113	zona interna
2619	1542	zonas negativas

ADDENDA

A01	A07	electrodo de referencia interno
A02	A05	ennegrecimiento
A03	A02	espectroscopía de absorción atómica
A04	A03	espectroscopía de emisión atómica
A05	A09	espectroscopía de emisión optica
A06	A04	espectroscopía de fluorescencia atomica
A07	A10	espectroscopía de fluorescencia de rayos X
A08	A06	espectroscopía de llama
A09	A08	espectroscopía de luminescencia molecular
A10	A01	obturadores acustico-opticas

Russian

0001	абсолютная температура	0005
0002	абсолютная эффективность полной энергии пика	0003
0003	абсолютная эффективность фотопика	0004
0004	абсолютный отсчет	0002
0005	абсорбер	0009
0006	автоионизация	0174
0007	автоматизация	0176,0177
0008	автоматизировать	0175,0178
0009	автоокисление	0180
0010	авторадиограф	0179
0011	агломерация	0051
0012	агрегат	0052
0013	агрегация	0053
0014	аддитивность	0041
0015	адсорбат, адсорбированное вещество	0043
0016	адсорбент	0044
0017	адсорбционная хроматография	0046
0018	адсорбционный индикатор	0048
0019	адсорбционный ток	0047
0020	адсорбция	0045
0021	аксиальный (осевой) электрод	0189
0022	активационный анализ	0033
0023	активация	0032
0024	активация заряженными частицами	0338
0025	активация нейтронами	1555
0026	активная концентрация	0039
0027	активная среда	0035
0028	активное вещество лазера	1258
0029	активное модифицированное твердое тело	1480
0030	активное твердое тело	0036
0031	активность	0037
0032	акустическая термометрия	2429
0033	алкалиметрическое титрование	0056
0034	алкалиметрия	0057
0035	альфа-коэффициент	0060
0036	альфа-параметр	0062
0037	альфа-распад	0061
0038	альфа-расщепление	0059
0039	альфа-частица	0063
0040	аминокарбоновая кислота	0072
0041	амперометрическое детектирование конечной точки	0073
0042	амперометрическое титрование	0074
0043	амперометрия	0075
0044	амплитуда внутренних колебаний (осцилляций)	1160
0045	амплитуда переменного напряжения	0069
0046	амплитуда переменного тока	0064
0047	амфипротный растворитель	0076
0048	анализ выделенных газов	0825
0049	анализ методом добавок	0084
0050	анализ методом изотопного разбавления	1236
0051	анализ методом субстехиометрического изотопного разбавления	2365
0052	анализ методом усиленной флуоресценции	0806
0053	анализ отходящих газов	0725
0054	анализ поверхности	2382
0055	анализ сканирующим лазером	2122
0056	анализ частиц по размерам	1676
0057	анализатор амплитуды импульсов	1875
0058	анализатор высоты импульсов	1882
0059	анализатор с магнитным полем под углом π–радиан	1763
0060	анализатор с магнитным полем под углом $\pi/2$-радиан	1764
0061	анализатор с магнитным полем под углом $\pi/3$-радиан	1765
0062	анализатор с радиальным электростатическим полем	1914

0063	аналит	0080
0064	аналитическая градуировка	0086
0065	аналитическая градуировочная функция	0087
0066	аналитическая кривая	0088
0067	аналитическая радиохимия	0095
0068	аналитическая электронная микроскопия (АЭМ)	0091
0069	аналитические весы	0085
0070	аналитические оценочные кривые	0092
0071	аналитические пробы	0097
0072	аналитические системы	0099
0073	аналитический промежуток (зазор)	0094
0074	аналитические функции	0093
0075	аналитический электрод	0090
0076	аналого-цифровой преобразователь (АЦП)	0078
0077	анизотропия	0107
0078	анионообменник	0105
0079	анионный комплекс	0106
0080	анионный обмен	0104
0081	анионный эффект	0103
0082	аннигилировать	0108
0083	аннигиляционная радиация	0110
0084	аннигиляция	0109
0085	аннигиляция частиц	1667
0086	анодная инверсионная вольтамперометрия (ИВА)	0117
0087	анодная инверсионная кулонометрия с контролируемым потенциалом	0116
0088	анодная инверсионная хроноамперметрия с линейной разверткой потенциала	0115
0089	анодное парообразование	0118
0090	анодное свечение	0114
0091	анодный промежуток (зазор)	0113
0092	аномальный тлеющий разряд	0001
0093	антикомптоновский гамма-луч	0119
0094	антикомптоновский гамма-спектрометр	0120
0095	антистоксова флуоресценция	0122
0096	античастица	0121
0097	апекс	0124
0098	аппаратурное обеспечение	1122
0099	аспиратор	0142
0100	аспирировать (всасывать)	0139
0101	аспирация	0140
0102	атмосфера дуги	0132
0103	атмосфера разряда	0656
0104	атомизатор	0164
0105	атомизатор в виде графитовой чашки	0999
0106	атомизатор в виде металлической нити	1435
0107	атомизатор с угольной нитью	0303
0108	атомизатор с угольной чашечкой	0302
0109	атомизация	0161
0110	атомизированная доля (фракция)	0163
0111	атомизированная локальная фракция	1346
0112	атомная абсорбция	0146
0113	атомная единица массы (а.е.м.)	0155
0114	атомная линия	0153
0115	атомная масса	0154
0116	атомная спектроскопия	0159
0117	атомная флуоресценция	0151
0118	атомная эмиссия	0149
0119	атомно-абсорбционная спектрометрия	0147
0120	атомно-флуоресцентная спектрометрия	0152
0121	атомно-эмиссионная спектрометрия	0150
0122	атомные изотопы	1238
0123	атомные спектральные линии	0158
0124	атомный вес	0160
0125	атомный номер	0156
0126	атомный спектр	0157
0127	атомы определяемого элемента	0082
0128	аттестованный материал сравнения	0329
0129	ацидиметрическое титрование	0026
0130	ацидиметрия	0027
0131	аэрозоль	0049
0132	барн	0206
0133	батохромный сдвиг	0214
0134	безизлучательный переход	1932
0135	безэлектродная разрядная лампа	0742
0136	безэлектродная резонансная полостная плазма	0744
0137	безэлектродное возбуждение	0743
0138	беккерель	0215
0139	бестоковая плазма	1544,1575
0140	бета-распад	0219

0141 бета-расщепление 0218
0142 бета-частица 0220
0143 биамперометрическое детектирование конечной точки 0221
0144 биамперометрическое титрование 0222
0145 биамперометрия 0223
0146 бинарные столкновения 0227
0147 бипотенциометрическое титрование 0231
0148 бипотенциометрия 0232
0149 бифункциональный ионообменник 0226
0150 биядерные комплексы 0230
0151 благородные газы 1566
0152 блокирование 0245
0153 болометр 0251
0154 бомбардировка атомами 0148
0155 бомбардировка быстрыми атомами (ББА) 0873
0156 бомбардировка частицами 1668
0157 бумажная хроматография 1659
0158 буфер 0271
0159 буфер для регулирования ионной силы 1199
0160 буферированные дуги 0275
0161 буферировать 0272
0162 буферная емкость 0274
0163 буферный индекс 0276
0164 быстрые нейтроны 0874
0165 биологический эквивалент рентгена (б.э.р.) 2047
0166 вакуум разряда 0661
0167 вакуумные искры 2550
0168 вакуумный фотоэлемент 2549
0169 валентные электроны 2551
0170 введенная мощность (энергия) 1085
0171 введенное излучение 1086
0172 вероятность абсорбционного перехода 0020,2496
0173 вероятность ионозации 1210
0174 вероятность перехода 2495
0175 вероятность перехода для индуцированного излучения 2498
0176 вероятность перехода для спонтанной эмиссии 2497
0177 вероятность спонтанного излучательного перехода 2304
0178 вероятность эйнштейновского перехода 0729
0179 вершина (пик) 2368
0180 весовая классификация проб 2108
0181 весовое титрование 2591
0182 вес факториальный 0865
0183 весы 0203
0184 вещество, способное к обесцвечиванию 0244
0185 взаимное наложение гармоник 1019
0186 взаимные мешающие влияния (помехи) 1526
0187 взаимодействие растворенного вещества с растворителем 2228
0188 вибрационная температура 2562
0189 вибрационный язычковый электрометр 2560
0190 визуальное детектирование конечной точки 2563
0191 визуальный индикатор 2564
0192 включение (инклюзия) 1087
0193 влияние (эффект) 0718
0194 влияние давления 1842
0195 влияние органических веществ 1638
0196 влияние растворителя 2234
0197 внешнее влияние тяжелых атомов 0852
0198 внешний стандарт 0854
0199 внешняя зона 1647
0200 внутреннее поглощение 1153
0201 внутреннее торможение потока 1112
0202 внутренний раствор заполнения 1158
0203 внутренний раствор заполнения стеклянного электрода 1159
0204 внутренний стандарт 1163
0205 внутренний электрод сравнения стеклянного электрода 1161
0206 внутренний электролиз 1157
0207 внутренняя зона 1113
0208 внутренняя конверсия 1154
0209 внутренняя линия сравнения 1162
0210 внутренняя электрогравиметрия 1156
0211 водородный газовый электрод 1067
0212 возбуждающая энергия 0844
0213 возбуждение 0828
0214 возбужденное состояние 0843
0215 возвратный носитель 1049
0216 воздушный пик 0055
0217 возмущение потока 0924
0218 возникающая температура 0127

0219	возникающая энергия	0128
0220	возникающий потенциал	0129
0221	возраст по отсчету радиоактивности	1945
0222	волатилизаторы	2569
0223	волновое число	2587
0224	волновые (характеристические) сопротивления	1530
0225	вольтамперная кривая	2576
0226	вольтамперные характеристики	2571
0227	вольтамперограмма	2577
0228	вольтамперометрическая константа	2574
0229	вольтамперометрический анализ	2573
0230	вольтамперометрия	2575
0231	вольтамперометрия с линейной разверткой потенциала	1316
0232	вольтамперометрия с линейной разверткой потенциала и интегральной сверткой информации	0486
0233	вольтамперометрия со стационарным электродом	2338
0234	вольтамперометрия с треугольной разверткой потенциала	2513
0235	вольфрамо-галогеновая лампа	2521
0236	воспроизводящие нуклиды	0877
0237	воспроизводящий материал	0876
0238	вращательная температура	2094
0239	вращающаяся платформа	2092
0240	вращающиеся дисковий затворы	2090
0241	вращающиеся зеркальные затворы	2091
0242	временная стабилизация	2406
0243	временное перемещение в пространстве	2405
0244	временное разрешение	2404
0245	время взаимодействия	1138
0246	время достижения	2003
0247	время отклика	2071
0248	время перемещения	2511
0249	время подъема	2087
0250	время пребывания (в приборе)	2050
0251	время пробоя	0262
0252	времяпролетный масс-спектрометр	2469
0253	время разрешения	2061
0254	время средней жизни радиоактивности	1958
0255	время удерживания	2081
0256	всаливание	2097
0257	всасывающий распылитель	2367
0258	вскрытие проб	1620
0259	всплескная мощность	2300
0260	вспомогательный искровой зазор	0183
0261	вспомогательный разряд	0181
0262	вспомогательный электрод	0182
0263	вторичная флуоресценция	2145
0264	вторичное горение	2141
0265	вторичное излучение	2149
0266	вторичные электроны	2144
0267	вторичный ион	2146
0268	вторичный разряд	2142
0269	вторичный стандарт	2150
0270	вторичный стандартный раствор	2151
0271	вторичноэлектронный умножитель	2143
0272	вуаль (на фотопластинке)	0931
0273	выборочная поверхностная дифракция электронов (ВПДЭ)	2155
0274	выгорать	0284
0275	выделенное пламя	2182
0276	вымывать (элюировать)	0781
0277	выпаривание	0823
0278	выпрямитель	2011
0279	высаливание	2098
0280	высаливающая хроматография	2099
0281	высвобождающие агенты	2045
0282	высвобождающие вещества (высвободители)	2044
0283	высокая отражательная способность	1043
0284	высоковольтные искры	1047
0285	высоковольтный искровой генератор с контролируемым напряжением	0472
0286	высокоинтенсивный полый катод	1039
0287	высокоскоростное повторение импульсов предварительного обыскривания	1044
0288	высокочастотная кондуктометрия	1035
0289	высокочастотное титрование	1038
0290	высокочастотное кондуктометрическое титрование	1034
0291	высокочастотный ступенчатый преобразователь	1037
0292	высокочастотный шунтирующий конденсатор	1036
0293	высота волны	2583
0294	высота входной щели	0812
0295	высота наблюдения	1027,1612

0296 высота пика 1691
0297 высота ступени 2346
0298 высота щели 2214
0299 выход вторичных ионов 2148
0300 выход оже-электронов 0170
0301 выход процессов деления 0892
0302 выход прямого деления 0651
0303 выход распыления 2312
0304 выход флуоресценции 0915
0305 выход фотонной эмиссии 1749
0306 выход фотоэлектронов 1732
0307 выход цепного деления 0330
0308 выход энергии лазера 1271
0309 выход энергии люминесценции 0805
0310 выходящая энергия лазера 1270
0311 газ заполнения 0880
0312 газ-носитель 0309
0313 газ-носитель аэрозоля 0050
0314 газ тушения 1910
0315 газовая смесь Пеннинга 1698
0316 газовая хроматография 0966
0317 газовое усиление 0965
0318 газовые лазеры 0970
0319 газовые разряды 0967
0320 газовый пропорциональный детектор 0973
0321 газовый резистор 0974
0322 газожидкостная хроматография 0971
0323 газообразное состояние 0968
0324 газоограничительная насадка (дроссель) 0975
0325 газопоглощение 0990
0326 газопроницаемая мембрана 0972
0327 газостабилизированные дуги 0977
0328 газо-твердофазная хроматография 0976
0329 газ разряда 0659
0330 гамма-излучение 0959
0331 гамма-каскад 0957
0332 гамма-квант 0958
0333 гамма-пучок 0961
0334 гамма-спектр 0963
0335 гамма-спектрометр 0962
0336 гашение (тушение) 1911
0337 гель-проникающая хроматография 0985
0338 гель-фильтрация 0984
0339 генератор 1642
0340 генератор искры 2249
0341 генератор искры среднего напряжения 1428
0342 генератор капель 0704
0343 генератор неконтролируемой высоковольтной искры 2539
0344 генератор с перестраиваемой длиной контура 2519
0345 генерирование рентгеновских лучей 2606
0346 генерирование рентгеновских лучей положительными ионами 2609
0347 генерирование рентгеновских лучей фотонами 2608
0348 генерирование рентгеновских лучей электронами 2607
0349 геометрические параметры 0986
0350 геометрический фактор 0988
0351 геометрическое ослабление 0987
0352 геометрия Маттауха-Герцога 1404
0353 геометрия Нира-Джонсона 1565
0354 геометрия счетного устройства 0510
0355 гетерохромная фотометрия 1028
0356 гидродинамическая вольтамперометрия 1066
0357 гиперхромный сдвиг 1068
0358 гипохромный сдвиг 1069
0359 гипсохромный сдвиг 1070
0360 гистерезис 1071
0361 глобульная дуга 0994
0362 глубина кратера 0519
0363 глубина проникновения 0821
0364 гомогенный мембранный электрод 1057
0365 гомологическая линия 1058
0366 гомологические линии Герлаха 0989
0367 горелка Бунзена 0279
0368 горелка Меккера 1429
0369 горелка с обращенной прямой инжекцией 2083
0370 горелка с предварительным смешением газов 1839
0371 горелка с прямой инжекцией 0652
0372 горение 0414
0373 горизонтальные электроды 1059

0374	горючий компонент	0951
0375	горячая ячейка	1063
0376	горячий атом	1062
0377	горячий полый катод	1064
0378	гравитационная просеивающая порошки система	1004
0379	гравитационный распылитель	1003
0380	градиентное заполнение	0998
0381	градиентное элюирование	0996
0382	градиентный слой	0997
0383	градуировка фотографической эмульсии	1734
0384	градуировочная кривая	0287
0385	грай	1005
0386	граница (поверхность) раздела электрод/раствор	0746
0387	графитовая стержневая печь	1000
0388	графитовая трубчатая печь	1002
0389	давление	1841
0390	данные	0560
0391	датирование по радиоактивности	1948
0392	двойная дуга	0691
0393	двойной выходной пик	0694
0394	двойной распылитель	2528
0395	двумерная хроматография	2530
0396	двухатомный	0613
0397	двухлучевая система	0693
0398	двухлучевой спектрометр	0692,0699
0399	двухцветный индикатор	2529
0400	двухчастотная полярография	0701
0401	действительное поглощение	1999
0402	деление	0887
0403	деление ядер	1591
0404	делящийся (расщепляющийся)	0888
0405	демаскирование	0580
0406	деминерализированная вода	0581
0407	демодуляционная полярография	0582
0408	денситометр	0583
0409	деполяризатор	0584
0410	держатель пробы	2103
0411	дериватографический анализ	0598
0412	дериватография	0599
0413	десольватированная фракция	0601
0414	десольватация	0602
0415	детектирование	0603
0416	детектирование выделенных газов	0826
0417	детектирование отходящих газов	0726
0418	детектор	0606
0419	детектор из полупроводника с поверхностным барьером	2383
0420	детектор из фольги	0932
0421	детектор по теплопроводности	2418
0422	детектор частиц	1669
0423	детектор Черенкова	0326
0424	детектор электронного захвата	0754
0425	детектор ядерных треков	1603
0426	диаметр действующей апертуры	0123
0427	диаметр кратера	0520
0428	диаметр сфокусированного пятна	0928
0429	диапазон (интервал, размах)	1988
0430	дилатометрия	0642
0431	динамическая разностная калориметрия	0713
0432	динамическая система	0715
0433	динамическая термомеханометрия	0717
0434	динамический полевой масс-спектрометр	0714
0435	динамический термогравиметрический анализ	0716
0436	диодно-матричный детектор	0645
0437	деионизированная вода	0576
0438	дискретные методы	0662
0439	дискретные одновременно применяемые методы	0663
0440	дискриминатор	0664
0441	дисперсия	0667
0442	дисперсия (стат.)	2558
0443	дисперсия в анализе с инжекцией пробы в поток	0668
0444	дисперсия вещества	0669
0445	дисперсия по длинам волн	2585
0446	диспесия по энергиям	0796
0447	диссоциация	0671
0448	дистанционно-связанный	2048
0449	дистилляция	0676
0450	дифракционирующие электроны	0635
0451	дифракция на кристалле	0529

0452	дифракция отраженных электронов высоких энергий (ДОЭВЭ)	2030
0453	дифракция прошедших электронов высоких энергий (ДПЭВЭ)	2504
0454	дифракция рентгеновских лучей	2601
0455	дифракция электронов	0755
0456	дифракция электронов низких энергий (ДЭНЭ)	1359
0457	дифференциальная импульсная полярография	0626
0458	диффузионный ток	0639
0459	диффузия	0637
0460	диэлектрическая постоянная	0614
0461	диэлектрометрическое титрование	0615
0462	диэлектрометрия	0616
0463	длина волны высвечивания	0243
0464	длина волны границы поглощения	0013
0465	длина пробега иона	1195
0466	длина эффективного основания призмы	1284
0467	длительность импульса	1880
0468	добавки	0040
0469	довозбуждение путем столкновений	0404
0470	доза	0689
0471	доза поглощения	0007
0472	докритический демпферированный осциллирующий разряд	2540
0473	дополнительные электроды	2375
0474	дополнительный разряд	2374
0475	допплеровская полуширина спектральной линии	0687
0476	допплеровское смещение	0688
0477	допплеровское уширение	0686
0478	дочерние продукты	0561
0479	дрейф	0702
0480	дуга	0131
0481	дуга выпрямленного переменного тока	2010
0482	дуга высокого давления	1041
0483	дуга высокого тока	1030
0484	дуга катодного слоя	0316
0485	дуга низкого тока	1357
0486	дуга переменного тока	0065
0487	дуга переменного тока с принудительным поджигом	1072
0488	дуга постоянного тока	0647
0489	дуга с выпариванием носителя	0306
0490	дуга с ионизационным буфером	2154
0491	дуга с постоянной температурой	0446
0492	дуга среднего тока	1425
0493	дуга, стабилизированная магнитным полем	1379
0494	дуга, стабилизированная стенкой	2582
0495	дуга с термическим поджигом	2424
0496	дуговая лампа	0135
0497	дуговая лампа низкого давления	1361
0498	дуговая линия	0137
0499	дуговой промежуток	0134
0500	дугообразный разряд	0136
0501	единичный выходной пик	2206
0502	единичный импульс	2208
0503	емкостная микроволновая плазма	0291
0504	емкостно-связанная плазма	0290
0505	емкостный ток	0294
0506	емкостный фактор	0295
0507	емкость до проскока	0264
0508	емкость конденсатора	2578
0509	емкость монослоя	1514
0510	емкость объема слоя	0217
0511	естественная радиация	1532
0512	естественная радиоактивность	1533
0513	естественная распространенность изотопа	1531
0514	естественное уширение	1529
0515	Е-тип задержки флуоресценции	0822
0516	жидкая проба	1335
0517	жидкая фаза	1334
0518	жидкостная гель-хроматография	1325
0519	жидкостная неподвижная фаза	1340
0520	жидкостная хроматография	1323
0521	жидкостная экстракция	1324
0522	жидкостное соединение	1326
0523	жидкостной сцитилляционный детектор	1338
0524	жидкостной сцитилляционный счетчик	1337
0525	жидкостные лазеры	1328
0526	жидкость-жидкостная хроматография	1329
0527	жидкость-жидкостная экстракция	1332
0528	жидкость-жидкостное равновесие	1331
0529	жидкость-жидкостное распределение	1330

0530	жидкость-жидкостное соединение	1333
0531	жидкость-твердофазная хроматография	1339
0532	G-значение	1008
0533	зависимость тока дуги от времени	2572
0534	загрязнение осадка	0454
0535	загрязнение путем захвата	0453
0536	задержанный объем газа	0969
0537	зажигание искрой	2250
0538	закон действия масс	1282
0539	закон Максвелла-Больцмана	1405
0540	закон Планка	1768
0541	закон Саха-Эггерта	2096
0542	закон Штерна-Фольмера	2351
0543	законы распределения	0683
0544	замедление	1477
0545	замедлидлитель	1478
0546	заместительная хроматография	0670
0547	запаздывающая люминесценция	0577
0548	запаздывающие нейтроны	0578
0549	запаянная лампа с полым катодом	2140
0550	заполненная колонка	1655
0551	зарядное сопротивление	0345
0552	заряженные частицы	0339
0553	затвор (зазор, промежуток)	0964
0554	затухание по логарифмическому закону	1351
0555	захват	0299
0556	захват гамма-излучения	0960
0557	захват нейтронов	1557
0558	захваченное гамма-излучение	0301
0559	зеемановская коррекция фона	2615
0560	знак маркировки	1250
0561	значение R_B	1994
0562	значение частного	2552
0563	зона	2619
0564	зонд прямого действия	0654
0565	игольчатый клапан	1538
0566	излучаемые частицы	1923
0567	излучение	1224
0568	излучение комбинационного рассеяния	1984
0569	излучение при свечении	1084
0570	излучение частиц	1674
0571	излучение Черенкова	0328
0572	излучение черного тела	0234
0573	излученные величины	1934
0574	измерение	1418
0575	измерение поправки холостого опыта	0238
0576	измерение pH	1709
0577	измерение pH на практике	1819
0578	измерение фотографической плотности	1735
0579	измеренная возбужденная флуоресценция	1414
0580	измереннаявозбужденная фосфоресценция	1416
0581	измеренная интенсивность	1415
0582	измеренный квантовый выход люминесценции	1417
0583	измерительная система с перестраиваемыми усилителями	2518
0584	изобарическое определение изменения массы	1226
0585	изомерные состояния	1228
0586	изомерный переход	1229
0587	изообразование	1227
0588	изопотенциальная точка	1230
0589	изотерма	1231
0590	изотерма ионного обмена	1189
0591	изотерма сорбции	2239
0592	изотермический термогравиметрический анализ	1232
0593	изотоны	1233
0594	изотоп	1234
0595	изотопная метка (трасер)	1244
0596	изотопное разбавление	1235
0597	изотопный анализ	1240
0598	изотопный индикатор	2490
0599	изотопный ион	1242
0600	изотопный носитель	1241
0601	изотопный обмен	1237
0602	изотоп-предшественник	1836
0603	иммуноанализ	1080
0604	импеданс плазмы	1774
0605	импедансный преобразователь	1081
0606	импульс некогерентного излучения	1088
0607	импульс огромной силы	0991

0608	импульсная полярография	1883
0609	импульсная разрядная лампа	1878
0610	импульсный лазерный источник	1879
0611	импульсный усилитель	1874
0612	импульсный усилитель линейного смещения	0225
0613	импульсы излучения	1933
0614	инверсионный анализ	2360
0615	инверсия популяции	1798
0616	индекс преломления (рефракциония)	2033
0617	индекс удерживания	2078
0618	индикатор	1090
0619	индикатор Гаммета	1018
0620	индикаторный электрод	1093
0621	индикатор pH	1707
0622	индифферентный электролит	1094
0623	индуктивно-связанная (индукционная) плазма (ИСП)	1099
0624	индуктионная катушка	1098
0625	индуктор	1100
0626	индуцированная микроволновым излучением плазма	1458
0627	индуцированная микроволновым излучением плазма при атмосферном давлении	0145
0628	индуцированная микроволновым излучением плазма при пониженном давлении	1364
0629	индуцированная реакция	1097
0630	инертная атмосфера разряда	1103
0631	инжектируемый газ	1107
0632	инжектор	1110
0633	инжектор пробы	2104
0634	инжекторная трубка	1109
0635	инжекция жидкой пробы	1336
0636	инициирование плазмы	1775
0637	инициировать	1106
0638	инициирующая искра	1076
0639	интегральная абсорбция (интегральное поглощение)	1123
0640	интегральные хроматограммы	1124
0641	интегральный детектор	1125
0642	интегральный стандарт	1127
0643	интегрирование потоков	0922
0644	интегрирующая сфера	1128
0645	интенсивностный мост	1130
0646	интенсивность	1129
0647	интенсивность излучения	1135
0648	интенсивность, отнесенная к базовому пику	1137
0649	интенсивность пика	1692
0650	интенсивность рентгеновского излучения	2610
0651	интенсивность спектральной линии	1136
0652	интервал перехода	2494
0653	интервал размеров пробы	2105
0654	интервал стробирования	2362
0655	интерпретация значений pH	1708
0656	интерпретация рентгеновских данных	2600
0657	инструментальная индикация	1120
0658	инструментальный активационный анализ	1119
0659	инструментальный нейтронно-активационный анализ (ИНАА)	1121
0660	интрахромный безизлучательный переход	1171
0661	иодиметрическое титрование	1178
0662	иодиметрия	1179
0663	иодометрическое титрование	1180
0664	иодометрия	1181
0665	ион	1182
0666	ионизационная игла	1208
0667	ионизационная камера	1203
0668	ионизационное распыление	1202
0669	ионизационные буферы	1201
0670	ионизационные помехи	1207
0671	ионизация	1200
0672	ионизация искровым источником	2254
0673	ионизация лазерным лучем	1263
0674	ионизация электронным ударом	0759
0675	ионизированная частица	1212
0676	ионизирующее излучение	1213
0677	ионизирующее напряжение	1214
0678	ион-молекулярная реакция	1216
0679	ионная бомбардировка	1183
0680	ионная линия	1193
0681	ионная микроскопия	1215
0682	ионная сила	1198
0683	ионная среда	1194
0684	ионная хроматография	1184
0685	ионно-зондовый микроанализ (ИЗМА)	1218

0686	ионные спектральные линии	1197
0687	ионные спектры	1196
0688	ионный источник	1221
0689	ионный обмен	1187
0690	ионный ток	1185
0691	ионогенные группы	1217
0692	ионообменная мембрана	1190
0693	ионообменная хроматография	1188
0694	ионообменник	1191
0695	ион-прародитель	1858
0696	ион-предшественник	1835
0697	ионселективные электроды (ИСЭ)	1220
0698	ионспецифичные электроды	1222
0699	ионы определяемого компонента	0083
0700	искра между графитовыми электродами	1001
0701	искра с контролируемой формой сигнала возбуждения	0478
0702	искра среднего напряжения	1427
0703	искровая камера	2244
0704	искровое кросс-возбуждение	2245
0705	искровое пятно	0281
0706	искровое равновесие	2251
0707	искровой промежуток (зазор)	2248
0708	искровой разряд	2246
0709	искровой разряд с внешним поджигом	0853
0710	искровые штативы	2256
0711	искроподобный разряд	2252
0712	искры с высокой скоростью повторения	1045
0713	искусственная радиоактивность	0138
0714	испарение	2553
0715	испарившаяся локальная фракция	1348
0716	испытуемый электрод	2409
0717	истинное совпадение	2516
0718	источник	2240
0719	источник возбуждения	0838
0720	источник первичного излучения	1852
0721	источник тлеющего разряда с плоским катодом	1769
0722	источники излучения	1935
0723	источники с полым катодом	1053
0724	кажущаяся концентрация	0125
0725	кажущееся значение pH	0126
0726	калильные тела	1083
0727	калориметрия	0288
0728	камера насыщения	2116
0729	камерное насыщение	0331
0730	каналирование	0333
0731	капающий ртутный электрод	0707
0732	капельный распылитель	0705
0733	капиллярная дуга	0296
0734	капиллярная коллонка	1622
0735	капиллярная трубка	0298
0736	капиллярная хроматография	1621
0737	капиллярная хроматография (хроматография в открытой трубке)	0879
0738	капиллярные электроды	0297
0739	каталитический ток	0310
0740	катафорез	0311
0741	катионный обмен	0323
0742	катионный эффект	0322
0743	катионообменник	0324
0744	катодная инверсия	0320
0745	катодное испарение	0321
0746	катодное падение потенциала	0314
0747	катодное распыление	0319
0748	катодное свечение	0315
0749	катодно-лучевая полярография	0318
0750	катушка Тесла	2408
0751	качественный анализ	1896
0752	квадратно-волновая полярография	2314
0753	квадратно-волновой ток	2313
0754	квадратичный электрооптический эффект	1894
0755	квадраупольный масс-анализатор	1895
0756	квазимонохроматичная фотометрия	1909
0757	квант флуоресценции	0912
0758	квантовая эффективность люминесценции	1903
0759	квантовая эффективность флуоресценции	1902
0760	квантовый выход	1904
0761	квантовый выход люминесценции	1906
0762	квантовый выход флуоресценции	1905
0763	квантовый счетчик	1901

0764	кинетическая энергия электрона	0760
0765	кинетические методы анализа	1247
0766	кинетический ток	1248
0767	кислотно-основное взаимодействие	0024
0768	кислотно-основное титрование	0025
0769	кислотно-основный индикатор	0023
0770	кислотные функции	0029
0771	кистевой разряд	0270
0772	классификация составляющих	0449
0773	коагуляция	0382
0774	коаксиальный волновод	0383
0775	когерентная система источников	0388
0776	когерентное излучение	0386
0777	когерентное электромагнитное излучение	0385
0778	когерентные источники	0387
0779	ко-ионы	0392
0780	колебательная реакция	1640
0781	количественный анализ	1898
0782	количественный разностный термический анализ	1899
0783	количество	1900
0784	количество излученной энергии	1926
0785	количество искр за полуцикл	2255
0786	количество света	1376
0787	коллектор	0396
0788	коллектор типа чашки Фарадея	0871
0789	коллиматор	0398
0790	коллимация	0397
0791	колонка	0408
0792	колоночная хроматография	0409
0793	кольцевой анод	0111
0794	комбинационное (рамановское) рассеяние	1985
0795	комбинированный электрод	0413
0796	компаратор	0415
0797	комплексант	0418
0798	комплексный анион	0419
0799	комплексообразование	0420
0800	комплексометрическое титрование	0422
0801	комплексометрия	0423
0802	комплексон	0424
0803	комплексы с мостиковой связью лигандов	1287
0804	компонент	0425
0805	компромиссные условия	0426
0806	комптоновское рассеяние	0430
0807	комптоновский электрон	0429
0808	конвективная хроноамперометрия	0482
0809	конвективная хронокулонометрия	0483
0810	конвекция	0481
0811	конверсионный электрон	0484
0812	конденсатор выравнивающего действия	1399
0813	конденсированная дуга	0439
0814	кондуктометрическое титрование	0441
0815	кондуктометрия	0442
0816	конечная точка высокочастотного кондуктометрического титрования	1033
0817	конечная точка титрования	0795
0818	конечная точка хронопотенциометрического титрования	0373
0819	конечная точка энтальпиметрического титрования	0810
0820	константа (постоянная)	0444
0821	константа времени заряжения	0346
0822	константа дезинтеграции	0665
0823	константа диссоциации	0672
0824	константа диффузионного тока	0640
0825	константа кислотности	0028
0826	константа концентрационного равновесия	0434
0827	константа кумулятивного образования	0535
0828	константа образования	0934
0829	константа преобразования	2493
0830	константа протонизации	1870
0831	константа процесса затухания	0558
0832	константа равновесия	0817
0833	константа равновесного распределения	0682
0834	константа распада	0570
0835	константа распределения	0680
0836	константа распределения	1680
0837	константа скорости электролиза с контролируемым массо-переносом	1398
0838	константа ступенчатого образования	2349
0839	константа устойчивости	2315
0840	константа устойчивости комплексных ионов металлов	2316
0841	константа Фарадея	0870

0842	константа экстракции	0861
0843	конструкция держателя образца	2266
0844	континиум (непрерывное пространство)	0462
0845	контролируемая атмосфера	0466
0846	контролируемая во времени дуга	0470
0847	контролируемая дуга переменного тока	0465
0848	контролируемый искровой промежуток (зазор)	0479
0849	контрольная полоса	0463
0850	контрольное титрование	0480
0851	контрольный промежуток (зазор)	0464
0852	контур масштабирования	2120
0853	контур разряда	0657
0854	конфигурация Бурсма	0248
0855	конфигурация длины волны	2584
0856	конфигурация параллельных зеркал	1661
0857	конфигурация системы зеркал	1467
0858	конфигурация типа "точка к плоскости"	1786
0859	конфигурация типа "точка к точке"	1787
0860	концентрационное отношение	0436
0861	концентрационное распределительное отношение	0433
0862	концентрационный индекс	0435
0863	концентрационные константы	0432
0864	концентрация	0431
0865	концентрирование	0395
0866	концентрический распылитель	0437
0867	коротковолновый предел Дюэйна-Ханта	0710
0868	корректор кривых	0548
0869	косвенное титрование	1095
0870	космическое излучение	0502
0871	коэффициент активности	0038
0872	коэффициент вариации	0384
0873	коэффициент внутренней конверсии	1155
0874	коэффициент гомогенного распределения	1055
0875	коэффициент дифракции пика	1688
0876	коэффициент диффузии	0638
0877	коэффициент интегрального отражения	1126
0878	коэффициент ионизации	1204
0879	коэффициент ионной активности	1192
0880	коэффициент ослабления	0166
0881	коэффициент парного ослабления	1657
0882	коэффициент поглощения (абсорбции)	0011
0883	коэффициент потенциометрической селективности	1810
0884	коэффициент пропускания	2503
0885	коэффициент распределения	0679,1679
0886	коэффициент селективности	2158
0887	коэффициент фотоэлектрического ослабления	1726
0888	коэффициент эйнштейновской спонтанной эмиссии	0728
0889	коэффициент экстракции	0860
0890	красное смещение	2016
0891	кратер	0518
0892	кремниевые фотодиоды	2198
0893	кривая времени ожидания	2470
0894	кривая вымывания (элюирования)	0784
0895	кривая нарастания активности	1007
0896	кривая распада	0571
0897	кривая спектрального отклика	2280
0898	кривая титрования	2475
0899	кривая ток-время	0547
0900	кривая Хертера-Дриффильда	1065
0901	кривая эффективности ионизации	1205
0902	кривые мешающих влияний	1145
0903	кривые нагревания	1022
0904	кривые охлаждения	0488
0905	кривые скорости охлаждения	0489
0906	кристаллическая решетка	0531
0907	кристаллическая решетка твердого состояния	2223
0908	кристаллический ионселективный электрод	0532
0909	кристаллический осциллятор с контролируемой частотой	0528
0910	круговая поляризация	0376
0911	ксеноновая лампа	2597
0912	ксеноновая лампа высокого давления	1042
0913	кулонометрическое титрование	0503
0914	кулонометрическое титрование при контролируемом потенциале	0473
0915	кулонометрическое титрование при контролируемом токе	0467
0916	кулонометрия	0504
0917	кулонометрия при контролируемом потенциале	0474
0918	кулонометрия с капающим электродом	0706
0919	кумулятивная (общая) константа протонизации	0536

0920 кумулятивная (общая) константа устойчивости 0537
0921 кумулятивное (общее) распределение 0533
0922 кумулятивный (общий) выход деления 0534
0923 кюри 0540
0924 лазер 1256
0925 лазерная атомизация и возбуждение 1260
0926 лазерная микромасс-спектрометрия (ЛММС) 1267
0927 лазерная эмиссионная микроспектрометрия (ЛЭМС) 1266
0928 лазерная эрозия 1264
0929 лазерное воздействие 1257
0930 лазерное зеркало 1269
0931 лазерные атомизаторы 1261
0932 лазерные пички 1277
0933 лазерные укалывающие воздействия 1279
0934 лазерный анализ 1259
0935 лазерный импульс 1274
0936 лазерный источник 1238
0937 лазерный локальный анализ (ЛЛА) 1265
0938 лазернй луч 1262
0939 лазерный микрозондовый анализ (ЛМЗА) 1268
0940 лазерный плюмаж 1272
0941 лазерный резонатор 1276
0942 лазеры на красителях 0712
0943 ламинарная защитная оболочка 1254
0944 ламинарное пламя 1252
0945 ламинарный поток 1253
0946 лампа 1255
0947 лампа накачки 1888
0948 лампа с вольфрамовой нитью (лампа накаливания) 2520
0949 лампа с охлаждаемым полым катодом 0487
0950 лампа с полым катодом 1052
0951 лампа с полым катодом и модулированной интенсивностью 1134
0952 лампы Гайсслера 0983
0953 лиганд 1286
0954 линейная дисперсия 1304
0955 линейная поляризованная люминесценция 1311
0956 линейная развертка потенциала 1313
0957 линейная скорость потока 1309
0958 линейное затухание 1303
0959 линейный импульсный ускоритель 1314
0960 линейный коэффициент ослабления 1302
0961 линейный коэффициент рассеяния 1315
0962 линейный коэффициент фотоэлектрического поглощения 1310
0963 линейный поляризатор 1312
0964 линейный преобразователь энергии 1307
0965 линейный электронный ускоритель 1305
0966 линейный электростатический эффект 1306
0967 линия 1300
0968 линия сравнения 2023
0969 линия старта 2334
0970 логарифмическая кривая титрования 1353
0971 логарифмический коэффициент распределения 1352
0972 лодочка для введения пробы 2109
0973 локальная температура газа 1349
0974 локальная эффективность ионизации 1345
0975 локальное термическое равновесие (ЛТР) 1350
0976 локальный анализ 1344
0977 луч с модулированной интенсивностью 1133
0978 лучеиспускание (блеск) 1223
0979 лучистая интенсивность 1922
0980 люминесцентная спектрометрия 1370
0981 люминесцентная спектроскопия 1371
0982 люминесцентный спектрометр 1369
0983 люминесценция 1366
0984 магнетрон 1384
0985 магнитное отклонение 1380
0986 магнитное поле 1381
0987 магнитный эффект Пеннинга 1383
0988 макроскопическое поперечное сечение 1378
0989 максимум пика 1693
0990 маркер (метка) 1387
0991 маркировать (метить) 1251
0992 маскирование 1388
0993 маскирующий агент 1389
0994 масса нуклида 1610
0995 массовое распределение 1392
0996 массовое число 1394
0997 массовый коэффициент ослабления 1390

0998	масс-спектр	1398
0999	масс-спектр отрицательных ионов	1541
1000	масс-спектрометр	1395
1001	масс-спектрометр ионного циклотронного резонанса	1186
1002	масс-спектрометр с двойной фокусировкой	0695
1003	масс-спектрометр с единичной фокусировкой	2207
1004	масс-спектрометр с линейным ускорением	1301
1005	масс-спектрометр с преобразованием Фурье	0935
1006	масс-спектрометр со статическим полем	2337
1007	масс-спектрометр циклотронного резонанса	0555
1008	масс-спектрометрия	1396
1009	масс-спектрометрия вторичных ионов (МСВИ)	2147
1010	масштаб титрования	1285
1011	материал, способный к активации	0031
1012	материал сравнения	2024
1023	материал типа "хозяин"	1061
1014	матрица	1400
1015	матрица из смолы	2056
1016	матричные эффекты	1401
1017	мгновенная скорость потока	1117
1018	мгновенная флуориметрия	0901
1019	мгновенное совпадение	1866
1020	мгновенные измерения	1118
1021	мгновенный нейтрон	1867
1022	мгновенный ток	1116
1023	медленные нейтроны	2217
1024	межатомные столкновения	1139
1025	межзональная область	1170
1026	межконусная область	1141
1027	межмолекулярный безизлучательный переход	1152
1028	межсистемное пересечение	1169
1029	межфазная граница	1143
1030	межхромофорное безизлучательное превращение	1140
1031	межэлементные влияния	1142
1032	мезо-	1432
1033	мезопроба	1433
1034	мезоследовый анализ	1434
1035	мембрана	1430
1036	"мертвая" конечная точка	0563
1037	"мертвое" время	0564
1038	"мертвое" время детектора	0607
1039	"мертвый" объем	0566
1040	металлопаровая дуга	1438
1041	металлопаровая дуговая лампа	1439
1042	металлофлуоресцентный индикатор	1437
1043	металлохромный индикатор	1436
1044	метастабильное состояние	1445
1045	метастабильность	1441
1046	метастабильный	1440
1047	метастабильный распад	1442
1048	метод аналитической кривой	0089
1049	метод базовой линии в атомной спектроскопии	0210
1050	метод буферирующих добавок	0273
1051	метод вдувания	0246
1052	метод дальностной вилки	0255
1053	метод добавки определяемого компонента	0081
1054	метод Косселя	1249
1055	метод отдельных растворов	2183
1056	метод смешанных растворов	1474
1057	метод стандартных вычитаний	2331
1058	метод стандартных добавок	2321
1059	метод элемента сравнения	2020
1060	методы моделирования	2199
1061	методы с временным разрешением	2472
1062	методы с применением порошков	1815
1063	методы с пространственным разрешением	2243
1064	методы умножения	1522
1065	механизация	1422
1066	механизировать	1423
1067	механизм	1421
1068	механизм возбуждения	0833
1069	механизм высвобождения	2046
1070	механическое задерживающее устройство	1419
1071	механические затворы (шторки)	1420
1072	мешающее вещество	1148
1073	мешающее влияние (помеха)	1144
1074	мешающие линии	1147
1075	мешающий компонент	1146

1076	миграционный ток	1462
1077	миграция	1461
1078	микроанализ	1446
1079	микроанализ in situ	1114
1080	микроанализ методом лазерного комбинационного рассеяния (ЛКРМА)	1275
1081	микровесы	1447
1082	микроволна (СВЧ-волна)	1457
1083	микроволновая плазма (СВЧ-плазма)	1459
1084	микрокомпонент	1449
1085	микрокомпьютер	1450
1086	микрокулонометрия	1451
1087	микрорентгеновская дифракция in situ	1115
1088	микропроба	1454
1089	микропроцессор	1453
1090	микроскопическое поперечное сечение	1455
1091	микроскопия проникающих электронов (МПЭ)	2502
1092	микроследы	1456
1093	микрофотометр	1452
1094	микрохимические весы	1448
1095	миллиграмм-эквивалент отсчета	1464
1096	милликулонометрия	1463
1097	минимальная ширина линии	1465
1098	минимально возможный отсчет	1997
1099	мишень	2399
1100	многовариантные методы	1521
1101	многоканальный анализатор	1518
1102	многоканальный анализатор по высоте импульсов	1519
1103	многоцелевые искровые источники	1523
1104	многощелевые горелки	1524
1105	мода (режим)	1479
1106	мода (режим) возбуждения	0834
1107	модификатор	1482
1108	модификаторы матрицы	1403
1109	модифицированное уравнение Нернста	1481
1110	модуляционная полярография	1485
1111	модуляция линейно поляризованного излучения	1484
1112	модуляция оптического сигнала	1634
1113	модуляция по длине волны	2586
1114	модуляция света	1289
1115	молекула	1507
1116	молекулярная абсорбционная спектрометрия	1494
1117	молекулярная абсорбционная спектроскопия	1495
1118	молекулярная люминесцентная спектрометрия	1502
1119	молекулярная спектрометрия	1505
1120	молекулярная спектроскопия	1506
1121	молекулярная эмиссионная спектрометрия	1499
1122	молекулярная эмиссионная спектроскопия	1500
1123	молекулярное излучение	1504
1124	молекулярные методы	1503
1125	молекулярные полосы	1497
1126	молекулярные полосы цианогенов	0549
1127	молекулярный анион	1496
1128	молекулярный ион	1501
1129	молекулярный катион	1498
1130	моляльная активность	1486
1131	моляльность	1488
1132	моляльный коэффициент активности	1487
1133	молярная активность	1490
1134	молярная энергия ионизации	1492
1135	молярное поглощение	1489
1136	молярность	1493
1137	молярный коэффициент активности	1491
1138	моновариантность	1509
1139	монослой	1513
1140	монофункциональный ионообменник	1512
1141	монохроматическое излучение	1510
1142	монохроматор	1511
1143	монохроматор возбуждения	0835
1144	моноядерные бинарные комплексы	1515
1145	моноядерные тройные комплексы	1516
1146	мощность детектирования	1818
1147	мощность импульса	1884
1148	мощность накачки	1890
1149	мощность флуоресценции	0911
1150	мультислой	1520
1151	набивка	1656
1152	наблюдаемая зона пламени	1613
1153	наблюдение	1611

1154	наведенная радиоактивность	1096
1155	нагреваемая колонка	2416
1156	нагруженный резистор	1343
1157	нагрузка	1342
1158	надтепловая люминесценция	2380
1159	наложения	2371
1160	наноследы	1527
1161	напряжение	2570
1162	напряжение горения	0283
1163	напряжение конденсатора	0292
1164	напряжение повторного поджига	2035
1165	напряжение сети	1385
1166	напряжение пробоя	0263
1167	насадка	1639
1168	настроечный шлейф	2522
1169	насыщение	2115
1170	насыщенный раствор	2114
1171	начальная температура	1105
1172	начальный ток	1104
1173	неаддитивный ток	1569
1174	неводное титрование	1570
1175	негатрон	1543
1176	недеструктивный активационный анализ	1576
1177	недискретное излучение сплошного спектра	1577
1178	незатухающая волна (НВ)	0460
1179	нейтральный	1553
1180	нейтрон	1554
1181	нейтронная плотность	1558
1182	нейтронная температура	1564
1183	нейтронно-активационный анализ (НАА)	1556
1184	нейтронное излучение	1563
1185	нейтронный поток	1559
1186	нейтроны деления	0890
1187	некогерентное электромагнитное излучение	1571
1188	неконтролируемая дуга переменного тока	2538
1189	некогерентные оптические источники	1572
1190	некристаллический электрод	1574
1191	неодимовое стекло	1544
1192	неоднородные магнитные поля	1579
1193	неоднородный материал	1580
1194	неосновной компонент	1466
1195	неполяризованный	2543
1196	неподвижная фаза	2339
1197	непрерывная подача	0459
1198	непрерывное излучение	0458
1199	непрерывный метод	0457
1200	непрерывный проточный анализ	0456
1201	непрерывный спектр	2270
1202	непроводящий ток порошок	1573
1203	неразрешаемая спектральная полоса	2544
1204	неразрешенная спектральная полоса	2545
1205	нерегулярные импульсы излучения	1225
1206	нернстовский отклик	1549
1207	несинхронизованная операция	0942
1208	неспектральная помеха	1582
1209	неспецифические помехи	1581
1210	несущий заряд	0307
1211	нетто фарадеевский ток	1550
1212	неупругое столкновение	1102
1213	неупругое рассеяние	1101
1214	нефарадеевский адмиттанс	1578
1215	нефелометрическое детектированиеконечной точки	1545
1216	нефелометрическое титрование	1546
1217	нефелометрия	1547
1218	низкие потери	1360
1219	низковольтные искры (искры низкого напряжения)	1365
1220	низкоэнергетические столкновения	1358
1221	номинальный линейный поток	1568
1222	нормальность	1585
1223	нормальный раствор	1586
1224	нормальный глаз	1583
1225	нормальный тлеющий разряд	1584
1226	нормативная температура	1587
1227	носитель	0305
1228	нуклид	1609
1229	нуклон	1607
1230	нулевая точка стеклянного электрода	2617
1231	нулевая точка шкалы	2618

1232	обводной инжектор	0286
1233	области применения спектрального анализа	2283
1234	область возбуждения	0837
1235	область Гейгера-Мюллера	0981
1236	область катодного спада	0313
1237	область экстраполяции	0864
1238	обмен зарядами	0340
1239	обогащение	0808
1240	образец	2265
1241	образование пары	1658
1242	образование "хвостов"	2394
1243	обратно рассеянные электроны	0198
1244	обратное значение линейной дисперсии	2004
1245	обратное резерфордовское рассеяние (ОРР)	2095
1246	обратное титрование	0201
1247	обратный рассеиватель	0197
1248	обращеннофазовая хроматография	2084
1249	обращенные кинетические кривые охлаждения	1176
1250	объем до проскока	0265
1251	объем жидкости	1341
1252	объем колонки	0412
1253	объем неподвижной фазы	2341
1254	объем, отвечающий пику элюирования	1689
1255	объем слоя	0216
1256	объем удерживания	2082
1257	объем элюирования	0786
1258	объемная концентрация	0277
1259	объемная проба	0278
1260	объемная скорость потока	2581
1261	объемное соотношение фаз	1717
1262	объемный резонатор	2068
1263	объемный коэффициент распределения	2579
1264	общая константа образования	1650
1265	общая константа распределения	1649
1266	общее количество флуоресценции	0914
1267	обязательный цикл	0711
1268	ограниченный объем	1290
1269	ограниченный тлеющий разряд	2074
1270	ограничительная пластина	0443
1271	одновременно задействованные аналитические системы	2200
1272	одновременно применяемые методы	2202
1273	одновременный разностный термический анализ	2201
1274	одноканальный анализатор импульсов по высоте	2204
1275	однолучевой спектрометр	2203
1276	однонаправленная искра	2542
1277	однонаправленный разряд	2541
1278	однородное магнитное поле	1056
1279	одноступенчатый источник возбуждения	1617
1280	одноцветный индикатор	1615
1281	одноэлектродная микроволновая плазма	1616
1282	одноэлементная лампа	2205
1283	оже-выход	0173
1284	оже-спектроскопия	0172
1285	оже-электрон	0168
1286	оже-электронная спектроскопия (ОЭС)	0169
1287	оже-эффект	0167
1288	окислитель	1652
1289	окислительно-восстановительный индикатор	1653
1290	окислительно-восстановительное титрование	1654
1291	окклюзия	1614
1292	операция незатухающей волны	0461
1293	операция переключения добротности (Q-операция)	1893
1294	опорная плита	2378
1295	опорный держатель	2021
1296	определение изменения веса	2590
1297	определение по кривым нагревания	1021
1298	определяемый элемент	0079
1299	оптимальные условия получения искры	1636
1300	оптическая накачка	1632
1301	оптическая плотность	1627
1302	оптическая проводимость	1626
1303	оптическая эмиссионная спектрометрия	1628
1304	оптически тонкая плазма	1631
1305	оптические волокна	1629
1306	оптические параметры	1633
1307	оптические системы	1635
1308	оптические фильтры	1630
1309	осадительное титрование	1827

1310	осадительный индикатор	1826
1311	осадок	1824
1312	осаждать	1823
1313	осаждение из гомогенного раствора	1825
1314	осколки деления	0889
1315	ослабление	0165
1316	основание	0207
1317	основание пика	1685
1318	основание призмы	1857
1319	основная линия	0209
1320	основное состояние (уровень)	1006
1321	основной газ	1856
1322	основной пик	0211
1323	основной электролит	0208
1324	основность	0212
1325	основные компоненты	1386
1326	остаточная ионизация	2053
1327	остаточное жидкостное соединение	2054
1328	остаточные газы	2052
1329	остаточный ток	2051
1330	остаточный потенциал жидкостного соединения	2055
1331	осциллирующая искра	1641
1332	осциллографическая полярография с однократной разверткой	2209
1333	осциллография	1644
1334	осциллополярография	1645
1335	отдельно ионизованный элемент	2211
1336	отклонение (девиация)	0611
1337	отмывка струей жидкости	0918
1338	относительная активность	2036
1339	относительная атомная масса	2037
1340	относительная биологическая эффективность	2038
1341	относительная объемная набухаемость ионообменников	2580
1342	относительная плотность	2040
1343	относительная погрешность в процентах	1699
1344	относительная ошибка	2041
1345	относительное стандартное отклонение	2043
1346	относителное удерживание	2042
1347	относительный отсчет	2039
1348	отношение выхода побочного распада	0260
1349	отношение масса/заряд	1391
1350	отношение фаз	1715
1351	отношение шунтирования интенсивности	1131
1352	отображающее устройство	1079
1353	отражающая способность	2027
1354	отраженный фотон	1082
1355	отрицательное свечение тлеющего разряда	1539
1356	отрицательные зоны	1542
1357	отрицательный ион	1540
1358	отскок (отражение)	2005
1359	отсчет	1998
1360	отсчет без нагрузки	1567
1361	отсчет скоростей	0514
1362	отфильтровывать	0884
1363	оценка разрешения по 10%-му промежутку между пиками	2059
1364	очиститель	2127
1365	очистка	2128
1366	паразитное излучение	2357
1367	паразитный свет	2356
1368	параллельный поджиг	1660
1369	параллельный резонансный контур	2397
1370	парамагнитные соединения (парамагнетики)	1663
1371	параметр уширения линии	1318
1372	параметр электрического источника	0738
1373	параметры люминесценции	1367
1374	параметры плазмы	1778
1375	параметры спектрального излучения	2279
1376	параметры удерживания	2079
1377	паровая струя	2555
1378	паровое облако	2554
1379	паровое облако, полученное с помощью лазера	1273
1380	парофазные помехи	2556
1381	первичное вещество	1855
1382	первичное горение	1845
1383	первичное излучение	1850
1384	первичные pH-стандарты	1849
1385	первичные стандартные образцы	1851
1386	первичный ион	1847
1387	первичный ионоселективный электрод	1848

1388	первичный разряд	1846
1389	первичный стандарт	1853
1390	первичный стандартный раствор	1854
1391	первый коэффициент Таунсенда	2486
1392	перегруппировка МакЛафферти	1406
1393	перегруппировочный ион	2002
1394	перегруппировочный молекулярный ион	2001
1395	передаточное отношение	2505
1396	передача энергии	0804
1397	передача энергии вибрации	2561
1398	переключатель добротности	1892
1399	переключатель полудобротности	2173
1400	переключатель с насыщенным красителем	2113
1401	переключатель стационарного зазора	2514
1402	переменное напряжение	0068
1403	переменно-токовая полярография	0067
1404	переменно-токовая полярография на высших гармониках	1031
1405	переменно-токовая полярография на высших гармониках с фазочувствительным выпрямлением	1032
1406	переменно-токовая хронопотенциометрия	0066
1407	перемещенные плазмы	2491
1408	перенос заряда	0342
1409	перенос количества движения	1508
1410	перенос энергии	0803
1411	переосаждение	2049
1412	пересекающиеся электрическое и магнитное поля	0524
1413	пересчетное устройство	2119
1414	пересыщение	2373
1415	пересыщенный раствор	2372
1416	переход вращательной энергии	2093
1417	переходное время	2499
1418	петля для ввода пробы	2111
1419	период капания	0708
1420	период накачки	1889
1421	период полураспада радиоактивности	1953
1422	периодическое напряжение	1702
1423	периодический ток	1701
1424	перо паров	2557
1425	печь графитовая стержневого типа	0304
1426	пик	1682
1427	пик вымывания (элюирования)	0785
1428	пик метастабильного иона	1443
1429	пик метастабильного состояния	1444
1430	пик полного поглощения	2482
1431	пик проникающего рентгеновского излучения	2604
1432	пиковое напряжение конденсатора	1686
1433	пиковый ток искры	1695
1434	пикоследы	1759
1435	ПИН полупроводниковый детектор	1762
1436	пинч-эффект	1761
1437	пироэлектрические детекторы	1891
1438	пички	2301
1439	пирамидальный резонатор	2398
1440	плавление	0956
1441	плазма	1772
1442	плазменная загрузка	1777
1443	плазменная струя	1776
1444	плазменный факел	1780
1445	плазменные состояния	1781
1446	плазмообразующий газ	1773
1447	пламенная спектрометрия	0898
1448	пламя	0895
1449	пламя, богатое топливом	0952
1450	плато	1782
1451	плато насыщения	2117
1452	платрод (вращающаяся платформа-электрод)	1783
1453	плоскополяризованный	1771
1454	плотность нейтронного потока	1560
1455	плотность потока	0920
1456	плотность потока при 2200 м/с	2531
1457	плотность потока частиц	1671
1458	плотность потока энергии	0798
1459	плотность тока положительного столба	1802
1460	плотность фотонного потока	1751
1461	плотность энергии излучения	1919
1462	плотность энергии спектрального излучения	2278
1463	площадь пика	1684
1464	плюмаж излучения	1924

1465	пневматический распылитель	1784
1466	побочная фракция	0259
1467	побочный распад	0258
1468	поверхностная искра	2388
1469	поверхностная концентрация	2384
1470	поверхностная реакция	2387
1471	поверхностное загрязнение	2385
1472	поверхностное покрытие	2386
1473	поверхность	2381
1474	поглотительная способность	0021
1475	поглощаемость	0006
1476	поглощение (абсорбция)	0010
1477	поглощение гамма-излучения	0016
1478	поглощение частиц	0017
1479	пиковый анализ	1683
1480	поглощенные электроны	0008
1481	погрешность	0820
1482	погрешность стеклянного электрода	0993
1483	погрешность титрования	2476
1484	подавители	2379
1485	подавитель парообразования	0612
1486	подавление потока	0921
1487	подвижная фаза	1476
1488	поджиг	1073
1489	поджигающий импульс	1075
1490	подложка (субстрат)	2376
1491	подщелачивать	0058
1492	позитрон	1804
1493	показатель качества	1897
1494	полевая ионизация	0878
1495	полиатомный	1794
1496	полифункциональный ионообменник	1796
1497	полихроматор	1795
1498	полиядерные комплексы	1797
1499	полная ширина на половине максимума (ПШПМ)	0955
1500	полная эффективность энергетического пика	0954
1501	полное выпаривание (выпаривание досуха)	0417
1502	полный ионный ток	2483
1503	полный квантовый выход люминесценции	2484
1504	полный объем удерживания	2485
1505	половина значения толщины	1015
1506	половина слоя	1014
1507	половина толщины	1013
1508	положительные ионы	1803
1509	положительный столб	1801
1510	полоса вымывания (элюирования)	0782
1511	полосатый спектр	0204
1512	полосчатости	2358
1513	полуинтегральная полярография	2170
1514	полумикро	2171
1515	полумикропроба	2172
1516	полупериод	1011
1517	полупроводник	2168
1518	полупроводниковый детектор	2169
1519	полупроводниковый детектор с диффузионным соединением	0636
1520	полупроводящая поверхность	2167
1521	полупрозрачное зеркало	2174
1522	полуширина за счет столкновений	0405
1523	полуширина интенсивности линии поглощения	1009
1524	полый катод с открытой трубкой	1619
1525	полый электрод	1054
1526	поляризатор	1791
1527	поляризация	1789
1528	поляризация в плоскости	1770
1529	поляризованный	1790
1530	поляриметрическое титрование	1793
1531	полярность	1788
1532	полярография	1792
1533	полярография Калоусека	1245
1534	полярография с контролируемым потенциалом	0477
1535	полярография с многократной разверткой потенциала	1525
1536	полярография с нарастающим зарядом	1089
1537	полярография со сканированием тока	0546
1538	полярография со ступенчатым изменением заряда	0341
1539	полярография со ступенчатым потенциалом	2320
1540	полярография с переменным напряжением	0071
1541	полярография с треугольным напряжением	2512
1542	помеха геометрии пламени	0897

1543	помеха поперечной диффузии	1281
1544	помехи возбуждения	0831
1545	помехи диссоциации	0673
1546	помехи с пространственным распределением	2257
1547	помехи улетучивания растворенного вещества	2229
1548	поперечное сечение	0525
1549	поперечное сечение активации	0034
1550	поперечное сечение захвата	0300
1551	поперечное сечение ионизации	0526
1552	поперечное сечение столкновений	0403
1553	поперечное сечение Уэсткотта	2592
1554	поправка на время разрешения	2062
1555	поправка на индикатор	1092
1556	поправка на мертвое время	0565
1557	поправка на фон	0194
1558	поправка прямого взвешивания	0501
1559	поправка холостого опыта	0237
1560	поправочный фактор на перепад давления	1843
1561	пористые электроды чашечного типа ("пород")	1799,1800
1562	порог Гейгера-Мюллера	0982
1563	порог лазера	1280
1564	пороговая энергия	2465
1565	порошковая проба	1814
1566	порядок спектра	1637
1567	последовательное осаждение	1806
1568	последовательный поджиг	2189
1569	послефильтрационные эффекты	1805
1570	послойное равновесие	1283
1571	постоянная времени	2468
1572	постоянная отклика	2069
1573	постоянное напряжение	0447
1574	постоянно-токовая кондуктометрия	0649
1575	постоянно-токовая полярография	0650
1576	постоянный ток	0445,0646
1577	потенциал	1807
1578	потенциал ассиметрии	0144
1579	потенциал возбуждения	0836
1580	потенциал диссоциации	0674
1581	потенциал жидкостного соединения	1327
1582	потенциал ионизации	1209
1583	потенциал пика	2370
1584	потенциал полуволны	1016
1585	потенциал полупика	1012
1586	потенциал четверти переходного времени	1907
1587	потенциометрическая индикация конечной точки	1809
1588	потенциометрическое титрование	1811
1589	потенциометрическое титрование по второй производной	2152
1590	потенциометрическое титрование по массе	1812
1591	потенциометрическое титрование с контролируемым током	0468
1592	потенциометрическое титрование с обращенной производной	1177
1593	потенциометрия	1813
1594	потенциометрия с контролируемым током	0469
1595	потери при счете	0511
1596	поток	0919
1597	поток излучения	1921
1598	поток через монохроматор	0925
1599	поток частиц	1670
1600	правила знаков	2197
1601	правило Бэйтса-Гуггенбейма	0213
1602	правило записи катодного процесса	0317
1603	правило записи уравнения Ильковича	1078
1604	правильность	0022
1605	правильность спектрального анализа	2281
1606	практическая удельная емкость	1821
1607	практическое разрешение	1820
1608	предварительная ионизация	1838
1609	преддуговой период	1822
1610	предел обнаружения	0605,1297
1611	предел определения	1298
1612	предел получения количественных оценок	1299
1613	предельная нагрузка	0293
1614	предельный адсорбционный ток	1291
1615	предельный диаметр фокального пятна	2533
1616	предельный диффузионный ток	1294
1617	предельный каталитический ток	1292
1618	предельный кинетический ток	1295
1619	предельный миграционный ток	1296
1620	предельный ток	1293

1621	предискровый период	1840
1622	предколоночный реактор	1833
1623	предфильтрационные эффекты	1837
1624	предшественник	1834
1625	преобразование	2492
1626	преобразование Бейкера-Сэмпсона-Зайделя	0202
1627	прерыватель	0364
1628	прерывистые дуги	1164
1629	прецизионная потенциометрия с нулевой точкой	1829
1630	прецизионность	1828
1631	приложенный потенциал	0130
1632	принудительно-возбужденные эмиссионные спектры	0829
1633	принцип рассеяния Мио	1460
1634	проба	2100
1635	проба ограниченного размера	2106
1636	проба твердого раствора	2222
1637	проверка полноты извлечения	2009
1638	проводимость	0440
1639	проводить обратное титрование	0200
1640	программа	1859,1861
1641	программировать	1860,1862
1642	продукты деления	0891
1643	проекционный экран	1864
1644	произведение растворимости	2227
1645	производная	0587
1646	производная вольтамперометрия	0597
1647	производная дилатометрия	0589
1648	производная импульсная полярография	0593
1649	производная конечной точки титрования	0590
1650	производная полярография	0591
1651	производная спектрофотометрия	0594
1652	производная термогравиметрия	0596
1653	производная хронопотенциометрия	0588
1654	производное потенциометрическое титрование	0592
1655	производный термический анализ	0595
1656	прокачиваемая лампа	0902
1657	пролетное время	2500
1658	пролетный трохоидальный масс-спектрометр	1865
1659	промежуточная скорость	1166
1660	промежуточная скорость при выходном давлении	1167
1661	промежуточная трубка	1151
1662	промежуточная фракция	1165
1663	промежуточный нейтрон	1149
1664	промежуточный объем	1168
1665	промежуточный плазменный газ	1150
1666	промывка	2138
1667	промывной раствор	2139
1668	проникающая хроматография	1703
1669	пропорциональный счетчик	1868
1670	пропускание	2506
1671	просверленный электрод	0703
1672	просеивающая акустико-волновая ударная система	0030
1673	просеивающий электрод	2196
1674	проскок пламени горелки	0900
1675	пространственная стабилизация	2259
1676	пространственное расщепление	2258
1677	пространственно-угловая мощность излучения	1925
1678	пространственный заряд	2242
1679	пространственный угол	2219
1680	противоионы	0507
1681	противоточные распылители	2031
1682	противоэлектрод	0506
1683	протон	1869
1684	протонное число	1872
1685	проточно-инжекционный анализ (анализ с инжекцией в поток)	0905
1686	проточный монитор	0923
1687	протяженность поглощения	0018
1688	профиль поглощения (абсорбции)	0019
1689	профиль по глубине	0585
1690	профиль Фойта	2566
1691	процедура взвешивания	2589
1692	процентная мера правильности	1700
1693	процесс дезактивации	0562
1694	процесс разделения	2185
1695	процессы передачи энергии электронов	0758
1696	процессы столкновений	0406
1697	проэкстраполированная выборка	0863
1698	пульсация пробоя	0261

1699	пучок сравнения	2017
1700	пятно	2306
1701	pH	1705
1702	pH-метрическое титрование	1712
1703	pH-метрический электрод	1706
1704	P-тип затухания флуоресценции	1873
1705	рабочая поверхность	0849
1706	рабочая шкала pH	1624,1710
1707	рабочая pH-метрическая ячейка	1623
1708	рабочие стандартные растворы для pH-метрии	1625
1709	рабочий промежуток (зазор)	2596
1710	рабочий электрод	2595
1711	равновесие в растворе	2231
1712	равновесие испарения	0824
1713	равновесие за вековой период	2153
1714	равновесие реакции комплексообразования	0421
1715	равновесие реакции протонизации	1871
1716	равновесия распределения	0681
1717	рад	1913
1718	радиационная рекомбинация	1942
1719	радиационная химия	1928
1720	радиационное возбуждение	1940
1721	радиационное довозбуждение	1939
1722	радиационные переходы	1943
1723	радиационные процессы	1941
1724	радиационный детектор	1930
1725	радиационный захват	1938
1726	радиационный спектрометр	1936
1727	радиация (излучение)	1927
1728	радиоактивная метка	1963
1729	радиоактивная цепочка	1946
1730	радиоактивное охлаждения	1947
1731	радиоактивное равновесие	1951
1732	радиоактивность	1965
1733	радиоактивные вещества	1962
1734	радиоактивные материалы	1957
1735	радиоактивные отходы	1964
1736	радиоактивные ряды	1961
1737	радиоактивные ядра	1959
1738	радиоактивный	1944
1739	радиоактивный меченый атом	1956
1740	радиоактивный выходящий поток	1950
1741	радиоактивный изотоп	1955
1742	радиоактивный индикатор	1954
1743	радиоактивный нуклид	1960
1744	радиоактивный распад	1949
1745	радиоактивный реагент	1982
1746	радиоаналитическая очистка	1967
1747	радиоаналитическая химия	1966
1748	радиоизотоп	1976
1749	радиоколлоид	1973
1750	радиолиз	1977
1751	радиометрическое детектирование конечной точки	1978
1752	радиометрическое титрование	1979
1753	радионуклид	1980
1754	радионуклидная чистота	1981
1755	радиохимическая чистота	1968
1756	радиохимическое выделение	1969
1757	радиохимический выход	1970
1758	радиохимия	1971
1759	радиохроматография	1972
1760	радиочастотная плазма	1974
1761	радиочастотная полярография	1975
1762	разбавляющее вещество (разбавитель)	0643
1763	развитие дуги	0381
1764	разделение изотопов	1243
1765	размеры частиц	1675
1766	размножение нейтронов	1561
1767	размытие фронта	0947
1768	разнолигандные комплексы	1472
1769	разнометалльные комплексы	1473
1770	разностная амперометрия	0618
1771	разностная вольтамперометрия	0634
1772	разностная дилатометрия	0622
1773	разностная накачка	0627
1774	разностная сканирующая калориметрия с горячим потоком	1020
1775	разностная полярография	0623
1776	разностная потенциометрия	0625

1777	разностная сканирующая калориметрия	0628
1778	разностная сканирующая калориметрия с компенсацией энергии	1816
1779	разностная термогравиметрия	0633
1780	разностная термограмма	0631
1781	разностная хроматограмма	0619
1782	разностная хроматография	0620
1783	разностное потенциометрическое титрование	0624
1784	разностный	0617
1785	разностный детектор	0621
1786	разностный термический анализ	0629
1787	разностный термический анализ в изометрической среде	0630
1788	разностный термогравиметрический анализ	0632
1789	разрабатывать	0610
1790	разрешающая способность	2060
1791	разрешение	2058
1792	разрешение детектора	0609
1793	разрешение коллиматора	0399
1794	разрешение пиков	1694
1795	разрешение по глубине	0586
1796	разрешение по энергии	0800
1797	разрешение спектрометра	2293
1798	разряд	0655
1799	разряд в полом катоде	1051
1800	разряд в укороченной цепи	2193
1801	разряд критического ослабления	0522
1802	разрядная лампа низкого давления	1362
1803	разрядная полярография	0660
1804	разрядный ток	0658
1805	распад	0568
1806	распад по экспоненциальному закону	0850
1807	расположение типа Кальве	0289
1808	распределение	0678
1809	распределение импульсов по амплитуде	1876
1810	распределение частиц по размерам	1677
1811	распределенное вещество	0677
1812	распределительная хроматография	1678
1813	распределительное отношение	0684
1814	распространенность в природе	1528
1815	распространенность изотопа	1239
1816	распыление	1534,2310
1817	распыленный разряд	2308
1818	распылитель	1536,2309
1819	распылитель камерного типа	0332
1820	распылитель с контролируемым потоком	0471
1821	распылительная камера	2307
1822	рассеиватель	2124
1823	рассеяние	2125
1824	рассеяние изличения	2126
1825	расстояние миграции растворителя	2237
1826	раствор	2230
1827	раствор в инверсионном анализе (истощаемый раствор)	2361
1828	растворение	0675
1829	растворенная локальная фракция	1347
1830	раствор заполнения	0881
1831	растворитель	2232
1832	раствор пробы	2107
1833	раствор солевого мостика	0267
1834	раствор солевого мостика электрода сравнения с двойным соединением	0268
1835	раствор сравнения	0416
1836	раствор холостого опыта	0240
1837	растворы сравнения	2025
1838	расширение шкалы	2118
1839	расщепленное ядро	0886
1840	расщепленный	0885
1841	рафинат	1983
1842	реакционный интервал	1995
1843	реакционный слой	1996
1844	реакция ядерного деления	1592
1845	реакция ядерного синтеза	1595
1846	реальный масштаб времени	2000
1847	регенеративный ток	2034
1848	регистрировать	2006
1849	регулируемый удерживаемый объем	0042
1850	редокс-индикатор	2012
1851	редоксионнообменники	2013
1852	редоксполимеры	2014
1853	редокс-титроваие	2015
1854	режим поперечного электромагнитного поля	2509

1855	резистор	2057
1856	резонансная линия	2065
1857	резонансное уширение	2063
1858	резонансный нейтрон	2066
1859	резонансный спектрометр	2067
1860	резонатор накачки	1886
1861	результат	2075
1862	результат анализа	0096
1863	рентген	2089
1864	рентгеновская флуоресценция	2605
1865	рентгеновские измерения	2611
1866	рентгеновские фотоны	2613
1867	рентгеновский анализ	2599
1868	рентгеновский спектр	2614
1869	рентгеновское излучение	2598
1870	рентгеновский флуоресцентный анализ с дисперсией по энергиям	0797
1871	рентгеноэмиссионная спектрометрия	2602
1872	рентгеноэмиссионная спектрометрия, индуцированная частицами	1672
1873	рентгеноэмиссионная спектроскопия	2603
1874	рентгеноэмиссионная спектроскопия с применением заряженных частиц	1673
1875	рефракционные эффекты	2032
1876	реэкстрагировать	0191
1877	реэкстракт	0190
1878	реэкстракция	0192
1879	родительский ион	1664
1880	ртутный макроэлектрод	1431
1881	рэлеевское (релеевское) рассеяние	1993
1882	РЭСИЧ	1767
1883	ряды (последовательности)	2188
1884	самообращенный	2165
1885	самопоглощение	2159
1886	самоподжигающийся искровой разряд	2164
1887	самопроизвольная электрогравиметрия	2303
1888	самоэкранирование	2166
1889	самоэлектрод	2163
1890	сборно-разборная лампа (трубка)	2517
1891	сверхкритический затухающий разряд	1651
1892	световая действенность	1372
1893	световой поток	1374
1894	светочувствительный слой	1375
1895	свободногорящая дуга	0939
1896	свободно-свободный переход	0941
1897	свободно-связанный переход	0938
1898	свободные электроны	0940
1899	свободный носитель	0308
1900	связь	0228
1901	сглаживание пика	1690
1902	сгоревшая фракция	0285
1903	сдвиг (смещение) спектральной линии	2274
1904	секвестрация	2187
1905	селективная проницаемость	1704
1906	селективное вымывание (элюирование)	2156
1907	селективное испарение	2157
1908	селектор амплитуды импульсов	1877
1909	селекторные фотодетектории	0979
1910	сесибилизированная люминесценция	2181
1911	сиверт	2195
1912	сила осциллятора	1643
1913	сила торможения	2355
1914	синглет-синглетное поглощение	2210
1915	синее смещение	0247
1916	синхронно вращающийся промежуток (зазор)	2391
1917	система ввода	1111
1918	система накачки	1885
1919	система обратной связи	0875
1920	система распыления в пламя	1537
1921	сканирующая просвечивающая электронная микроскопия (СПЭМ)	2123
1922	сканирующая электронная микроскопия (СЭМ)	2121
1923	скачкообразное соотношение края поглощения	0012
1924	скользящая искра	2212
1925	скопление импульсов	1760
1926	скорость аспирации	0141
1927	скорость всасывания жидкости	1990
1928	скорость выходящего потока	1648
1929	скорость горения	0282
1930	скорость дезинтеграции	0666
1931	скорость нагревания	1023
1932	скорость повторения искры	2253

1933	скорость подъема	2088
1934	скорость потока	0907
1935	скорость потребления	2547
1936	скорость распыления	2311
1937	скорость расхода жидкости	1991
1938	скорость спинового перехода	2302
1939	скорость счета	0513
1940	скорость улетучивания	2568
1941	скорость ядрообразования	1992
1942	скорректированная интенсивность	0495
1943	скорректированное время удерживания	0499
1944	скорректированное время распада (исчезновения) флуоресценции	0491
1945	скорректированное время распада (исчезновения) фосфоресценции	0492
1946	скорректированный квантовый выход люминесценции	0498
1947	скорректированный люминесцентный эмиссионный поляризованный спектр	0496
1948	скорректированный объем удерживания	0500
1949	скорректированный поляризованный спектр возбуждения люминесценции	0497
1950	скорректированный спектр возбуждения	0494
1951	скорректированный эмиссионный спектр	0493
1952	слабый градиент потенциала	2588
1953	следовый анализ	2488
1954	следовый компонент	2489
1955	следы	2487
1956	случайная величина	2559
1957	случайное совпадение	1987
1958	смешанные комплексы	1469
1959	смешанный индикатор	1471
1960	смешанный кристалл	1470
1961	смещение (сдвиг)	0224
1962	смещение за счет столкновений	0407
1963	смещение линии	1321
1964	смещенный осциллирующий разряд постоянного тока	0648
1965	собственная эффективность	1173
1966	собственная эффективность полного энергетического пика	1174
1967	собственная эффективность фотопика	1175
1968	собственные частоты	0727
1969	совмещенные единицы	0517
1970	согласованность состава матрицы	1402
1971	согласованный отсчет	0390
1972	согласованное время разрешения	0391
1973	содержание составляющей	0450
1974	созревание	2086
1975	созревание Оствальда	1646
1976	солнечная бленда	2218
1977	соосаждение	0490
1978	соотношение "включено-выключено" для прерывистых дуг	1618
1979	соотношение давление-напряжение-ток	1844
1980	соотношение массового распределения	1393
1981	сопряженные одновременные техники	0515
1982	сопутствующие вещества (компоненты)	0438
1983	сорбция	2238
1984	составная часть (составляющая)	0448
1985	спад радиоактивности	1952
1986	спаренный промежуток (зазор)	2395
1987	спектр	2298
1988	спектр возбужденной флуоресценции	0841
1989	спектр возбужденной фосфоресценции	0842
1990	спектр флуоресценции с синхронным возбуждением	2389
1991	спектр фосфоресценции с синхронным возбуждением	2390
1992	спектр излучения	1937
1993	спектр радиации	2299
1994	спектр Шпольского	2194
1995	спектральная интенсивность	2271
1996	спектральная светимость	2276
1997	спектральная ширина полосы	2268
1998	спектральная яркость черного тела	2277
1999	спектральные помехи	2272
2000	спектральные приборы	2290
2001	спектральные свойства	2287
2002	спектральные характеристики	2269
2003	спектральный анализ	2282
2004	спектральный буфер	2284
2005	спектральный носитель	2285
2006	спектральный период	2275
2007	спектральный разбавитель	2286
2008	спектральный фон	2267
2009	спектрограмма	2288
2010	спектрограф	2289

2011	спектрография	2291
2012	спектрометр	2292
2013	спектрометр с двойной синхронизацией луча	0700
2014	спектрометр с дифракцией на кристалле	0530
2015	спектрометрия	2294
2016	спектрометрия ионного рассеяния (ИРС)	1219
2017	спектроскоп	2296
2018	спектроскопия	2297
2019	спектроскопия комбинационного рассеяния (рамановская)	1986
2020	спектроскопия обратного рассеяния (СОР)	0199
2021	спектроскопия потерь энергии высокого разрешения (СПЭВР)	1046
2022	спектроскопия потерь энергии электронов	0757
2023	спектроскопия потерь энергии отраженных электронов (СПЭОЭ)	2028
2024	спектроскопия потерь энергии проникающих электронов (СПЭПЭ)	2501
2025	спектроскопия с разрешением во времени	2471
2026	спектрофотометрическое титрование	2295
2027	спектры возбуждения	0839
2028	спонтанное деление	2305
2029	способность к лучеиспусканию	1915
2030	способность к формированию отклика	2072
2031	способность к эмиссионному излучению	1917
2032	спрямленная кривая титрования	1317
2033	специфические помехи	2262
2034	сравнительное значение стандарта pH	2026
2035	среда	1424
2036	среднее значение	0184
2037	среднее время жизни	0187
2038	среднее давление газа	0186
2039	среднее значение	1407
2040	среднее число лигандов	0188
2041	средний коэффициент ионной активности	1410
2042	средний коэффициент активности	1408
2043	средний период жизни	1411
2044	средняя линейная область	1412
2045	средняя мощность	1426
2046	средняя область масс	1413
2047	средняя промежуточная скорость газа-носителя	1409
2048	средняя скорость потока	0185
2049	стабилизированные дуги	2318
2050	стабилизированный источник тока	2319
2051	стабилизирующие кольца	2317
2052	стандартизованный раствор	2324
2053	стандартизованный раствор титранта	2325
2054	стандартная (сравнительная) интенсивность	2022
2055	стандартное вещество сравнения	2327
2056	стандартное отклонение	2322
2057	стандартное отклонение отсчета	2323
2058	стандартные образцы для анализа	0098
2059	стандартные вещества	2330
2060	стандартный наблюдатель	2326
2061	стандартный раствор	2329
2062	стандартный раствор сравнения	2328
2063	стандарты pH	1711
2064	старение	0054
2065	старт	2333
2066	стартовая линия	2335
2067	стартовая точка	2336
2068	статический вес	2343
2069	стационарная фаза	2340
2070	стационарное излучение	2342
2071	стационарное состояние	2344
2072	стеклянный электрод	0992
2073	степень диссоциации	0573
2074	степень извлечения	2007
2075	степень ионизации	0574
2076	степень оттитрованности	0575
2077	стехиометрическая конечная точка	2352
2078	стехиометрия	2353
2079	стоксова люминесценция	2354
2080	столб дуги	0133
2081	столкновения	0400
2082	стриппинг (реэкстракция)	2359
2083	ступенчатое элюирование (вымывание)	2348
2084	стробированный луч	2101
2085	строение дуговой плазмы	0452
2086	ступенчатая переходная флуоресценция	2350
2087	ступенчатый сектор (участок)	2347
2088	ступень	2345

2089	сублимация	2363
2090	субмикроанализ	2364
2091	субследовый анализ	2366
2092	суженный тлеющий разряд	0451
2093	сумма (аккумуляция)	0538
2094	сухой аэрозоль	0709
2095	сфера Ульбрихта	2532
2096	сфокусированное (фокальное) пятно	0927
2097	схема зарядки	0343
2098	схема индикации перегрузки	1662
2099	схема поджига	1074
2100	схема поджигающего устройства	1077
2101	схема распада	0572
2102	схема совпадений	0389
2103	схема с сосредоточенными постоянными	1377
2104	сцинтиллирующий материал	2129
2105	сцинтиллятор	2134
2106	сцинтиллятор детектор	2135
2107	сцинтилляционный детектор	2132
2108	сцинтилляционный спектрометр	2133
2109	сцинтилляционный счетчик	2131
2110	сцинтилляция	2130
2111	счет	0505
2112	счетная трубка (счетчик)	0508
2113	счетчик Гейгера-Мюллера	0980
2114	счетчик импульсов	1989
2115	счетчик радиации	1929
2116	тандемная масс-спектрометрия	2396
2117	таст-интервал	2400
2118	таст-полярография	2401
2119	твердая подложка	2226
2120	твердая проба	2220
2121	твердотельные лазеры	2225
2122	твердотельный детектор	2224
2123	твердый раствор	2221
2124	темновое анодное пространство	0112
2125	темновое катодное пространство	0312
2126	темновое пространство Астона	0143
2127	темновое пространство Крукса	0523
2128	темновое пространство Фарадея	0872
2129	темновое пространство Гитторфа	1048
2130	темновой ток	0559
2131	температура возбуждения	0840
2132	температура газа	0978
2133	температура излучения	1916
2134	температура инжекции	1108
2135	температура ионизации	1211
2136	температура кипения	0250
2137	температура колонки	0411
2138	температура плазмы	1779
2139	температура пламени	0899
2140	температура разделения	2186
2141	температура удерживания	2080
2142	температурные эффекты	2402
2143	тензаметрия	2407
2144	теоретическая конечная точка	2410
2145	теоретическая удельная емкость	2413
2146	теоретический объем удерживания	2412
2147	теоретическое число тарелок	2411
2148	тепловое излучение	2427
2149	тепловое термическое равновесие	2420
2150	тепловой источник	2428
2151	тепловые нейтроны	2425
2152	тепловые процессы	2426
2153	теплопроводность	2417
2154	термическая ионизация	2423
2155	термический анализ	2415
2156	термический анализ частиц	2450
2157	термическое деление	2422
2158	термическое разложение	2419
2159	термическое старение	2414
2160	термическое упаривание	2421
2161	термоаналитические методы	2430
2162	термовесы	2431
2163	термогравиметрическая кривая	2441
2164	термогравиметрический анализ	2440
2165	термогравиметрия	2442
2166	термограмма	2438

2167 термография 2439
2168 термодилатометрия 2433
2169 термодинамическая температура 2436
2170 термодинамические константы 2434
2171 термодинамическое равновесие 2435
2172 термолюминесценция 2443
2173 термомагнетометрия 2444
2174 термометрическая индикация конечной точки 2447
2175 термометрическое титрование 2448
2176 термомеханический анализ 2445
2177 термомеханометрия 2446
2178 термомеханометрия с образцом на (металлической) оплетке 2481
2179 термомикроскопия 2449
2180 термопарометрический анализ 2457
2181 термоптометрия 2453
2182 терморефрактометрия 2454
2183 термосониметрия 2455
2184 термоспектрометрия 2456
2185 термостолбик 2452
2186 термофотометрия 2451
2187 термохимические реакции 2432
2188 термоэлектрометрия 2437
2189 тестирование путем разбавления 0644
2190 тиристорное управление (тиристорная регулировка) 2466
2191 титр 2477
2192 титрант 2473
2193 титрование 2474
2194 титрометрический анализ 2478
2195 титрометрия 2480
2196 тлеющий разряд 0995
2197 тлеющий разряд в магнитном поле 1382
2198 тлеющий разряд, усиленный магнитным полем 0253
2199 ток 0541
2200 ток вершины (пика) 2369
2201 ток двойного слоя 0696
2202 ток заряжения 0344
2203 ток пика 1687
2204 токонесущая плазма дуги 0542
2205 толщина реакционного слоя 2458
2206 тонкая пленка 2459
2207 тонкопленочный электрод 2460
2208 тонкослойная хроматография 2461
2209 топливо (горючее) 0949
2210 тормозное излучение 0266
2211 точка кипения 0249
2212 точка покоя 2073
2213 точка эквивалентности 0818
2214 точность весов 1831
2215 точность взвешивания 1830
2216 точность индикации 1832
2217 точность счета 0512
2218 транспортировка (перемещение) 2507
2219 транспортные помехи 2508
2220 трехщелевая горелка 2464
2221 трехэлектродная конфигурация 2462
2222 трехэлектродная плазма 2463
2223 триплет-триплетное поглощение 2515
2224 тройники (Т-образные трубки) 2393
2225 туман 1468
2226 турбидиметрическая индикация конечной точки 2523
2227 турбидиметрическое титрование 2524
2228 турбидиметрия 2525
2229 турбулентное пламя 2526
2230 турбулентный поток 2527
2231 тушение люминисценции 1368
2232 угловая дисперсия 0101
2233 угловой коэффициент отклика 2070
2234 угол Брэгга 0256
2235 угол высвечивания 0242
2236 угол дивергенции (расхождения) 0100
2237 угол Комптона 0427
2238 угол наклона фотопластинки 2467
2239 уголковый распылитель 0102
2240 удельная активность 2260
2241 удельная ионизация 2263
2242 удельное выгорание 2261
2243 удельный объем удерживания 2264
2244 удерживаемый объем 1050

2245 удерживание 2077
2246 указанный заданный объем 0600
2247 улавливание 2510
2248 улетучивание 2567
2249 ультразвуковый распылитель 2535
2250 ультрамикроанализ 2534
2251 ультраследовый анализ 2536
2252 ультрафиолетовая фотоэлектронная спектроскопия (УФФЭС) 2537
2253 умножающаяся реакция 0077
2254 упругое рассеяние 0732
2255 упругое столкновение 0731
2256 уравнение Брэгга 0257
2257 уравнение Дебая-Хюккеля 0567
2258 уравнение Дернера-Хоскинса 0685
2259 уравнение Ломакина-Шайбе 1354
2260 уравнение Нернста 1548
2261 уравнение Сэнда 2112
2262 уровни возбуждения 0832
2263 уровни усиления 0807
2264 усиленный выход тлеющего разряда 0254
2265 усиленный полый катод 0252
2266 устройство в виде длинной трубки 1355
2267 устройство для градуировки по интенсивности 1132
2268 уширение (линий) 0269
2269 уширение за счет самопоглощения 2160
2270 уширение за счет столкновений 0401
2271 уширение Лоренца 1356
2272 уширение Штарка 2332
2273 фазовая флуориметрия 1714
2274 фазовый анализ 1713
2275 фазовое титрование 1716
2276 фактор, зависящий от линейного преобразователя энергии 1308
2277 фактор извлечения 2008
2278 фактор конверсии (обращения) 0485
2279 фактор обогащения 0809
2280 фактор отражения 2029
2281 фактор поглощения (абсорбции) 0015
2282 фактор разделения 2184
2283 фактор самопоглощения 2162
2284 фактор титрометрического превращения 2479
2285 фактор удерживания (R_f) 2076
2286 фактор Фано 0866
2287 фарадеевский выпрямленный ток 0869
2288 фарадеевский демодуляционный ток 0868
2289 фарадеевский ток 0867
2290 фарадеевское выпрямление высокого уровня 1040
2291 физическая величина 1757
2292 физическая помеха 1756
2293 физическая сорбция 1758
2294 физическое равновесие 1755
2295 физическое старение 1754
2296 фиксированная частица 0893
2297 фиксированные ионы 0894
2298 фильтр 0883
2299 флоккуляция 0904
2300 флотирующая частота 0903
2301 флуоресцентный анализ 0909
2302 флуоресцентный спектрометр 0913
2303 флуоресценция 0908
2304 флуоресценция при прямой линии (прямом переходе) 0653
2305 флуориметрическое детектирование конечной точки 0916
2306 флуориметрическое титрование 0917
2307 флуориметрия с тушением излучения 1912
2308 фокусирующее зеркало 0930
2309 фокусирующие линзы 0929
2310 фокусное расстояние 0926
2311 фон 0193,2546
2312 фон пламени 0896
2313 фон холостого опыта 0236
2314 фоновое излучение 0196
2315 фоновый масс-спектр 0195
2316 фоновый электролит 2377
2317 форма кратера 0521
2318 форма линии 1320
2319 форма контура спектральной линии 2273
2320 формальность 0933
2321 фосфоресцентный анализ 1719
2322 фосфоресцентный спектрометр 1721

2323	фосфоресценция	1718
2324	фосфориметрия	1722
2325	фосфороскопы	1723
2326	фотографическая прозрачность	1738
2327	фотографический параметр	1736
2328	фотодетектор	1725
2329	фотоионизационная спектрометрия	1740
2330	фотоионизация	1739
2331	фотометрическая индикация конечной точки	1741
2332	фотометрическое титрование	1742
2333	фотометрия	1743
2334	фотометрия с фотографической регистрацией	1737
2335	фотон	1745
2336	фотонная активация	1746
2337	фотонная эмиссионная спектроскопия	1748
2338	фотонное излучение	1752
2339	фотонный поток	1750
2340	фотонный эмиссионный спектр	1747
2341	фотопик	1753
2342	фототок	1724
2343	фотоумножитель	1744
2344	фотоэлектрический пик	1728
2345	фотоэлектрический эффект	1727
2346	фотоэлектрон	1729
2347	фотоэлектронная рентгеновская спектроскопия (ФЭРС)	2612
2348	фотоэлектронная спектроскопия (ФЭС)	1731
2349	фотоэлектронный спектр	1730
2350	фотоэмиссионная спектроскопия	1733
2351	фрагментарный ион	0937
2352	фракционная дистилляция	0936
2353	фронт растворителя	2236
2354	фронтальная поверхность	0948
2355	фронтальная хроматография	0946
2356	функция кислотности Гаммета	1017
2357	функция контура спектральной линии	1319
2358	функция распределения	1681
2359	функция Фойта	2565
2360	характеристики колонки	0410
2361	характеристики кристалла	0527
2362	характеристическая концентрация	0334
2363	характеристическая масса	0335
2364	характеристическое рентгеновское излучение	0337
2365	характеристический потенциал	0336
2366	хелатометрическое титрование	0347
2367	хемиионизация	0359
2368	хемилюминесцентный анализ	0361
2369	хемилюминесцентный индикатор	0362
2370	хемилюминесценция	0360
2371	хемосорбция	0363
2372	химическая ионизация	0355
2373	химическая помеха	0354
2374	химические пламена	0353
2375	химические процессы типа "хозяин-гость"	1060
2376	химический актинометр	0348
2377	химический выход	0358
2378	химический дозиметр	0350
2379	химический изотопный обмен	0356
2380	химический сдвиг	0357
2381	химический эквивалент	0352
2382	химическое равновесие	0351
2383	химическое старение	0349
2384	холодные нейтроны	0394
2385	холодный полый катод	0393
2386	холостое распыление	0239
2387	холостое титрование	0241
2388	холостой опыт	0235
2389	холостой опыт по индикатору	1091
2390	холостой опыт по растворителю	2233
2391	хроматограмма	0365
2392	хроматограф	0366
2393	хроматографировать	0367
2394	хроматография	0368
2395	хроматография с програмируемым изменением температуры	2403
2396	хроматография с програмируемым потоком	0906
2397	хроноамперометрия	0369
2398	хроноамперометрия с двойным ступенчатым изменением потенциала	0697
2399	хронокулонометрическая константа	0370
2400	хронокулонометрия	0371

2401	хронокулонометрия с двойным ступенчатым изменением потенциала	0698
2402	хронокулонометрия со ступенчатым изменением потенциала	1808
2403	хронопотенциометрическая константа	0372
2404	хронопотенциометрическое титрование	0374
2405	хронопотенциометрия	0375
2406	хронопотенциометрия с обращением тока	0545
2407	хронопотенциометрия с переменным напряжением	0070
2408	хронопотенциометрия с прерыванием тока	0543
2409	хронопотенциометрия с программируемым током	1863
2410	цементизация	0325
2411	цепочка столкновений	0402
2412	цепь распада	0569
2413	цикл топлива	0950
2414	циклическая вольтамперометрия с треугольной потенциальной разверткой	0552
2415	циклическая полярография с треугольной потенциальной разверткой	0553
2416	циклическая хронопотенциометрия	0550
2417	циклическая хронопотенциометрия со ступенчатым изменением тока	0551
2418	циклотрон	0554
2419	цилиндрическая форма	0557
2420	цилиндрический коллектор	0556
2421	цифро-аналоговый преобразователь (ЦАП)	0641
2422	частичная постоянная распада	1665
2423	частичное выпаривание	1666
2424	частота	0943
2425	частота искры	2247
2426	частота модуляций	0945
2427	частота модуляций	1483
2428	частота умножения	0944
2429	чашечный электрод	0539
2430	чашка Делвиса	0579
2431	чашка для введения пробы	2110
2432	черное тело	0233
2433	четверть-волновые линии	1908
2434	число лигандов	1288
2435	число нейтронов	1562
2436	число нуклонов	1608
2437	число эффективных теоретических тарелок	0722
2438	чистая комната	0378
2439	чистая поверхность	0379
2440	чистить	0380
2441	чистое время удерживания	1551
2442	чистый объем удерживания	1552
2443	чистый станок	0377
2444	чувствительность	2176
2445	чувствительный объем детектора	2175
2446	чувствительность входной системы	2179
2447	чувствительность ионного источника	2180
2448	чувствительность при заданной нагрузке	2177
2449	чувствительность прямого зондирования	2178
2450	ширина	2593
2451	ширина входной щели	0813
2452	ширина линии	1322
2453	ширина пика	1696
2454	ширина пика на половине высоты	1697
2455	ширина полосы	0205
2456	ширина светового пучка	2594
2457	ширина щели	2215
2458	ширина эмиссионной линии на половине высоты	1010
2459	шкала энергий	0801
2460	щелевые горелки	2216
2461	щель	2213
2462	эквивалент	0819
2463	эквивалент дозы	0690
2464	эквивалентная высота теоретической тарелки	1026
2465	эквивалентная высота эффективной теоретической тарелки	1025
2466	экзотерма	0846
2467	экзотермические реакции	0848
2468	экзотермический пик	0847
2469	экзоэнергетические условия	0845
2470	экранированное пламя	2191
2471	экранированный	2137
2472	экранированный индикатор	2136
2473	экранированный электрический разряд	2190
2474	экраны	2192
2475	эксимерная люминесценция	0827
2476	экспозиция	0851
2477	экспозиция излучения	1920
2478	экстрагент	0858

2479 экстрагировать 0856
2480 экстрагируемость 0857
2481 экстракт 0855
2482 экстракционный индикатор 0862
2483 экстракция 0859
2484 экстракция растворителем 2235
2485 электрическая дуга 0733
2486 электрическая проводимость 0734
2487 электрические искры 0739
2488 электрический параметр 0737
2489 электрический разряд 0735
2490 электрический разряд при низком давлении 1363
2491 электрический ток 0740
2492 электроактивное вещество 0741
2493 электровозбуждение 0736
2494 электрогравиметрия 0749
2495 электрогравиметрия при контролируемом потенциале 0475
2496 электрография 0748
2497 электрод-проба 2102
2498 электрод с гетерогенной мембраной 1029
2499 электрод с жесткой матрицей 2085
2500 электрод с подвижным носителем 1475
2501 электрод с ферментативным субстратом 0814
2502 электрод сравнения 2018
2503 электрод типа вакуумной чашки 2548
2504 электродвижущая сила (ЭДС) 0752
2505 электродная резонансная полостная плазма 0745
2506 электролит 0750
2507 электромагнитное излучение 0751
2508 электронная микроскопия 0761
2509 электронная микроскопия для химического анализа (ЭСХА) 0767
2510 электронная температура 0768
2511 электронное давление 0763
2512 электронное размножение 0762
2513 электронно-зондовый микроанализ (ЭЗМА) 0764
2514 электронно-зондовый рентгеновский микроанализ (ЭЗРМА) 0765
2515 электронно-оптические прерыватели (затворы) 0769
2516 электронные спектры 0766
2517 электронный захват 0753
2518 электроразделение 0772
2519 электроразделение при контролируемом потенциале 0476
2520 электроразделение при контролируемом потенциале 0773
2521 электроструя 0774
2522 электрофорез 0770
2523 электрофоретограмма 0771
2524 электроярозия 0747
2525 элемент 0775
2526 элемент изображения 1766
2527 элемент сравнения 2019
2528 элементарная частица 0777
2529 элементный анализ 0776
2530 эллиптическая поляризация 0778
2531 элюат 0779
2532 элюент 0780
2533 элюирующая хроматография 0783
2534 эманационный термический анализ 0787
2535 эмиссионные спектры 0790
2536 эмиссионный монохроматор 0788
2537 эмиссионный профиль 0789
2538 эмиссионный спектр с флуоресцентным возбуждением 1720
2539 эмульсионная градуировочная кривая 0791
2540 эмульсионная градуировочная функция 0792
2541 эндотерма 0793
2542 эндотермический пик 0794
2543 энергетические уровни 0799
2544 энергетический пик горючего компонента 0953
2545 энергетический порог (барьер) 0802
2546 энергия возбуждения 0830
2547 энергия излучения 1918
2548 энергия импульса 1881
2549 энергия ионизации 1206
2550 энергия накачки 1887
2551 энергия оже-пика 0171
2552 энергия поглощения (абсорбции) 0014
2553 энергия радиации 1931
2554 энергия резонанса 2064
2555 энергия связи 0229
2556 энергия эжекции 0730

2557	энергия электрона	0756
2558	энтальпиметрическое титрование	0811
2559	эпикадмиевые нейтроны	0815
2560	эпитермические нейтроны	0816
2561	эффект выгорания	0280
2562	эффект Зеемана	2616
2563	эффект Комптона	0428
2564	эффект Мессбауэра	1517
2565	эффект Сцилларда-Чалмерса	2392
2566	эффект тяжелых атомов	1024
2567	эффект Черенкова	0327
2568	эффективная бесконечная толщина	0721
2569	эффективная запирающая энергия кадмия	0719
2570	эффективная энергия флуоресценции	1817
2571	эффективное термическое поперечное сечение	0723
2572	эффективность	0724
2573	эффективность атомизации	0162
2574	эффективность встроенного детектора	1172
2575	эффективность детектора	0608
2576	эффективность детектирования	0604
2577	эффективность источника	2241
2578	эффективность распыления	1535
2589	эффективность светового воздействия	1373
2580	эффективность совмещения	0516
2581	эффективность счета	0509
2582	эффективность флуоресценции	0910
2583	эффективный эквивалент дозы	0720
2584	эффекты самопоглощения	2161
2585	явление загрязнения при осаждении	0455
2586	ядерная реакция	1599
2587	ядерная система	1601
2588	ядерная химия	1589
2589	ядерная частица	1598
2590	ядерное превращение	1604
2591	ядерное топливо	1593
2592	ядерные изобары	1596
2593	ядерные изомеры	1597
2594	ядерный активационный анализ	1588
2595	ядерный распад	1590
2596	ядерный расход	1605
2597	ядерный реактор	1600
2598	ядерный синтез	1594
2599	ядерный трек	1602
2600	ядрообразование	1606
2601	ярлык (этикетка)	0882
2602	ячейка Керра	1246
2603	ячейка Поккелса	1785

ADDENDA

A01	атомно-абсорбционная спектроскопия	A02
A02	атомно-флуоресцентная спектроскопия	A04
A03	атомно-эмиссионная спектроскопия	A03
A04	внутренний электрод сравнения	A07
A05	молекулярная люминесцентная спектроскопия	A08
A06	оптико-акустические затворы	A01
A07	оптическая эмиссионная спектроскопия	A09
A08	пламенная спектроскопия	A06
A09	рентгеновская флуоресценция спектроскопия	A10
A10	чернение	A05

Chinese

0001	阿斯吞(Aston)暗区	a si tun an qu	0143
0002	爱因斯坦(Einstein)透射概率	ai yin si tan tou she gai lu	0729
0003	爱因斯坦自发射系数	ai yin si tan zi fa she xi shu	0728
0004	氨基多羧酸	an ji duo suo suan	0072
0005	暗电流	an dian liu	0559
0006	螯合滴定	ao he di ding	0347
0007	奥斯脱瓦尔(Ostwald)熟化	ao si tuo wa er shu hua	1646
0008	靶(恩)	ba (en)	0206
0009	靶(子),目标	ba (zi), mu biao	2399
0010	摆[振]动火花	bai [zhen] dong huo hua	1641
0011	(色谱)斑,斑点	(se pu) ban, ban dian	2306
0012	(旋转)板电极	(xuan zhuan) ban dian ji	1783
0013	半波电位	ban bo dian wei	1016
0014	半导体	ban dao ti	2168
0015	半导体表面	ban dao ti biao mian	2167
0016	PIN 半导体检波器	PIN ban dao ti jian bo qi	1762
0017	半导体检测器	ban dao ti jian ce qi	2169
0018	半峰电位	ban feng dian wei	1012
0019	半高峰宽	ban gao feng kuan	1697
0020	半痕量分析	ban hen liang fen xi	1434
0021	半厚度[强度减半厚度]	ban hou du	1013
0022	半积分极谱(法)	ban ji fen ji pu (fa)	2170
0023	半宽度[极大值半处的全宽度](FWHM)	ban kuan du	0955
0024	半 Q 开关	ban Q kai guan	2173
0025	半衰期	ban shuai qi	1011
0026	半透明(反射)镜	ban tou ming (fan she) jing	2174
0027	半微量	ban wei liang	2171
0028	半微量称量	ban wei liang cheng liang	1433
0029	半微量进[试]样	ban wei liang jin [shi] yang	2172
0030	半值层	ban zhi ceng	1014
0031	半值厚	ban zhi hou	1015
0032	(每)半周火花数	(mei) ban zhou huo hua shu	2255
0033	包藏,吸着,吸留	bao cang, xi zhuo, xi liu	1614
0034	包合[含],夹杂物	bao he [han], jia za wu	1087
0035	饱和(度)	bao he (du)	2115
0036	饱和激光变色开关	bao he ji guang bian se kai guan	2113
0037	饱和坪	bao he ping	2117
0038	饱和腔	bao he qiang	2116
0039	饱和溶液	bao he rong ye	2114
0040	保护火焰	bao hu huo yan	2191

0041	保留(值),记忆	bao liu (zhi), ji yi	2077
0042	保留参数	bao liu can shu	2079
0043	保留时间	bao liu shi jian	2081
0044	保留体积	bao liu ti ji	2082
0045	保留温度	bao liu wen du	2080
0046	保留指数	bao liu zhi shu	2078
0047	杯(形)电极	bei xing dian ji	0539
0048	贝茨-古根海姆(Bates-Guggenheim)常规	bei ci-gu gen hai mu chang gui	0213
0049	贝克-萨姆普森-塞德尔(Baker-Sampson-Seidel)变换	bei ke-sa mu pu sen-sai de er bian huan	0202
0050	贝克勒耳(Bq)[放射性强度单位]	bei ke le er	0215
0051	倍频	bei pin	0944
0052	倍增法	bei zeng fa	1522
0053	倍增中子	bei zeng zhong zi	1561
0054	本底[背景]	ben di [bei jing]	0193,2546
0055	本底放射	ben di fang she	0196
0056	本底校正	ben di jiao zheng	0194
0057	本底质谱图	ben di zhi pu tu	0195
0058	本生灯	ben sheng deng	0279
0059	本征光峰效率	ben zheng guang feng xiao lu	1175
0060	本征全能(输出)峰效率	ben zheng quan neng (shu chu) feng xiao lu	1174
0061	本征频率	ben zheng pin lu	0727
0062	泵功率	beng gong lu	1890
0063	泵唧[压载水]系统	beng ji [ya zai shui] xi tong	1885
0064	泵浦[抽运]灯	beng pu [chou yun] deng	1888
0065	泵浦能(量)	beng pu neng (liang)	1887
0066	泵浦[抽运]谐振腔	beng pu [chou yun] xie zhen qiang	1886
0067	泵浦[抽运]循环[终结]	beng pu [chou yun] xun huan [zhong jie]	1889
0068	比保留体积	bi bao liu ti ji	2264
0069	比电离	bi dian li	2263
0070	比活度[性]	bi huo du [xing]	2260
0071	比较溶液	bi jiao rong ye	0416
0072	比例[正比]计数管	bi li [zheng bi] ji shu guan	1868
0073	比燃耗	bi ran hao	2261
0074	比色[较] 器	bi se [jiao] qi	0415
0075	比浊滴定法	bi zhuo di ding fa	2524
0076	比浊法	bi zhuo fa	2525
0077	比浊终点	bi zhuo zhong dian	2523
0078	边壁稳定电弧	bian bi wen ding dian hu	2582
0079	编(制)程序	bian (zhi) cheng xu	1860
0080	扁焰燃烧器	bian yan ran shao qi	2216
0081	变化,方差	bian hua,fang cha	2558
0082	变换,转变	bian huan, zhuan bian	2492
0083	变换常数	bian huan chang shu	2493
0084	变[转]换电子	bian [zhuan] huan dian zi	0484
0085	变宽	bian kuan	0269
0086	变异系数,相对标准偏差	bian yi xi shu, xiang dui biao zhun pian cha	0384
0087	标度零点	biao du ling dian	2618
0088	标记器	biao ji qi	1387
0089	标准参考溶液	biao zhun can kao rong ye	2328

0090	标准参考物	biao zhun can kao wu	2327
0091	标准滴定溶液	biao zhun di ding rong ye	2325
0092	标准观测者	biao zhun guan ce zhe	2326
0093	标准减量法	biao zhun jian liang fa	2331
0094	标准偏差	biao zhun pian cha	2322
0095	标准容量	biao zhun rong liang	0600
0096	(使合)标准[标定]溶液	(shi he) biao zhun [biao ding] rong ye	2324
0097	标准溶液	biao zhun rong ye	2329
0098	标准温度	biao zhun wen du	1587
0099	标准[基准]物	biao zhun [ji zhun] wu	2330
0100	标准物添加法	biao zhun wu tian jia fa	2321
0101	表观能量	biao guan neng liang	0127
0102	表观浓度	biao guan nong du	0125
0103	表观 pH 值	biao guan pH zhi	0126
0104	表观温度	biao guan wen du	0129
0105	表面	biao mian	2381
0106	表面反应	biao mian fan ying	2387
0107	表面分析	biao mian fen xi	2382
0108	表面复盖度	biao mian fu gai du	2386
0109	表面富集	biao mian fu ji	2384
0110	表面火花[电介质击穿]	biao mian huo hua [dian jie zhi ji chuan]	2388
0111	表面势[位]垒半导体检测计	biao mian shi [wei] lei ban dao ti jian ce ji	2383
0112	表面污染	biao mian wu ran	2385
0113	并联[平行]点火	bing lian [ping xing] dian huo	1660
0114	并行-并行镜面对称构象	bing xing-bing xing jing mian dui cheng gou xiang	1661
0115	波埃尔斯马(Boersma)型装置	bo ai er si ma xing zhuang zhi	0248
0116	波长分[色]散	bo chang fen [se] san	2585
0117	波长调制	bo chang tiao zhi	2586
0118	波长位形[对称构象]	bo chang wei xing [dui cheng gou xiang]	2584
0119	波道效应	bo dao xiao ying	0333
0120	波高	bo gao	2583
0121	波数	bo shu	2587
0122	玻璃电极	bo li dian ji	0992
0123	玻璃电极的零点	bo li dian ji de ling dian	2617
0124	玻璃电极内参比电极	bo li dian ji nei can bi dian ji	1161
0125	玻璃电极内充液	bo li dian ji nei chong ye	1159
0126	玻璃电极误差	bo li dian ji wu cha	0993
0127	薄层色谱(法)	bo ceng se pu (fa)	2461
0128	薄膜	bo mo	2459
0129	薄膜电极	bo mo dian ji	2460
0130	补充电极	bu chong dian ji	2375
0131	捕集	bu ji	2510
0132	捕集剂	bu ji ji	0396
0133	部分衰变常数	bu fen shuai bian chang shu	1665
0134	部分蒸发(作用)	bu fen zheng fa (zuo yong)	1666
0135	不导电粉末	bu dao dian fen mo	1573
0136	不对称电势	bu dui cheng dian shi	0144
0137	不规则照射脉冲	bu gui ze zhao she mai chong	1225
0138	不均匀材料	bu jun yun cai liao	1580
0139	不均匀磁场	bu jun yun ci chang	1579

0140	不能分辨的带状光谱	bu neng fen bian de dai zhuang guang pu	2544
0141	不清晰的带状光谱	bu qing xi de dai zhuang guang pu	2545
0142	布喇格(Bragg)方程	bu la ge fang cheng	0257
0143	布喇格角	bu la ge jiao	0256
0144	步,阶	bu, jie	2345
0145	参比材料	can bi cai liao	2024
0146	参比电极	can bi dian ji	2018
0147	参比溶液	can bi rong ye	2025
0148	参比稳定器	can bi wen ding qi	2021
0149	参比[起读]线	can bi [qi du] xian	2023
0150	参比元素	can bi yuan su	2019
0151	参比元素技术	can bi yuan su ji shu	2020
0152	α参数	α can shu	0062
0153	残余电流	can yu dian liu	2051
0154	残余气体	can yu qi ti	2052
0155	残余液体界面	can yu ye ti jie mian	2054
0156	残余液体界面电位	can yu ye ti jie mian dian wei	2055
0157	(展开)槽饱和	(zhan kai) cao bao he	0331
0158	操作 pH 池	cao zuo pH chi	1623
0159	操作 pH 标尺[度]	cao zuo pH biao chi [du]	1624
0160	操作 pH 标准	cao zuo pH biao zhun	1625
0161	测定发光量子产率[额]	ce ding fa guang liang zi chan lu [e]	1417
0162	测定磷光激发	ce ding lin guang ji fa	1416
0163	(被)测定强度	(bei)ce ding qiang du	1415
0164	测定(极)限	ce ding (ji) xian	1298
0165	测定荧光激发	ce ding ying guang ji fa	1414
0166	测量[定],计量	ce liang [ding], ji liang	1418
0167	测微光度计	ce wei guang du ji	1452
0168	测温滴定法	ce wen di ding fa	2448
0169	测温终点	ce wen zhong dian	2447
0170	层[片]流	ceng [pian] liu	1253
0171	层流火焰	ceng liu huo yan	1252
0172	(惰性气体)层流屏蔽[护套]	(duo xing qi ti) ceng liu ping bi [hu tao]	1254
0173	层平衡	ceng ping heng	1283
0174	差动泵送	cha dong beng song	0627
0175	差热分析	cha re fen xi	0629
0176	差热分析结合热重分析法	cha re fen xi jie he re zhong fen xi fa	0598,0599
0177	差(示)热分析图	cha (shi) re fen xi tu	2438
0178	差热重量分析	cha re zhong liang fen xi	0632
0179	差示电流分析法	cha shi dian liu fen xi fa	0618
0180	差示电位滴定	cha shi dian wei di ding	0624
0181	差示电位分析法	cha shi dian wei fen xi fa	0625
0182	差示[微分]伏安法	cha shi [wei fen] fu an fa	0634
0183	差示极谱法	cha shi ji pu fa	0623
0184	差示检测器	cha shi jian ce qi	0621
0185	差示脉冲极谱	cha shi mai chong ji pu	0626
0186	差示热膨胀测量法	cha shi re peng zhang ce liang fa	0622
0187	差示扫描量热法	cha shi sao miao liang re fa	0628
0188	差示色谱法	cha shi se pu fa	0620
0189	差示色谱图	cha shi se pu tu	0619

0190	差示温谱图	cha shi wen pu tu	0631
0191	长幅余摆线质谱仪	chang fu yu bai xian zhi pu yi	1865
0192	长管装置	chang guan zhuang zhi	1355
0193	常数，恒量	chang shu, heng liang	0444
0194	常压微波感应等离子体	chang ya wei bo gan ying deng li zi ti	0145
0195	场致电离	chang zhi dian li	0878
0196	超镉中子	chao ge zhong zi	0815
0197	超痕量分析	chao hen liang fen xi	2536
0198	超临界阻尼放电	chao lin jie zu ni fang dian	1651
0199	超热发光	chao re fa guang	2380
0200	超热中子	chao re zhong zi	0816
0201	超声雾化器	chao sheng wu hua qi	2535
0202	超微量分析	chao wei liang fen xi	2534
0203	沉淀	chen dian	1824
0204	沉淀滴定	chen dian di ding	1827
0205	沉淀物	chen dian wu	1823
0206	沉淀污染现象	chen dian wu ran xian xiang	0455
0207	沉淀杂质	chen dian za zhi	0454
0208	沉淀指示剂	chen dian zhi shi ji	1826
0209	成分，元[组]件	cheng fen, yuan [zu] jian	0425
0210	成[构，组]分	cheng [gou, zu] fen	0448
0211	成核作用，晶核形成	cheng he zuo yong, jing he xing cheng	1606
0212	成熟	cheng shu	2086
0213	称量[重]法	cheng liang [zhong] fa	2589
0214	称重精度	cheng zhong jing du	1831
0215	程序	cheng xu	1859
0216	程序变电流计时电位分析(法)	cheng xu bian dian lui ji shi dian wei fen xi (fa)	1863
0217	程序变流色谱(法)	cheng xu bian liu se pu (fa)	0906
0218	程序[工作，次序，时间，进度]表	cheng xu [gong zuo, ci xu, shi jian, jin du] biao	1861
0219	程序升温色谱(法)	cheng xu sheng wen se pu (fa)	2403
0220	炽热辐射	chi re fu she	1084
0221	炽热体	chi re ti	1083
0222	充电电流	chong dian dian liu	0344
0223	充电电路	chong dian dian lu	0343
0224	充电电阻(器)	chong dian dian zu (qi)	0345
0225	充电时间常数	chong dian shi jian chang shu	0346
0226	冲洗	chong xi	0918
0227	重排分子离子	chong pai fen zi li zi	2001
0228	重排离子	chong pai li zi	2002
0229	出发点	chu fa dian	2336
0230	出峰电势，起始电位	chu feng dian shi, qi shi dian wei	0128
0231	初级离子选择性电极	chu ji li zi xuan ze xing dian ji	1848
0232	初级燃烧	chu ji ran shao	1845
0233	储能电路	chu neng dian lu	2397
0234	触发的稳定间隔[隙]	chu fa de wen ding jian ge [xi]	2514
0235	出口压力隙间速度	chu kou ya li xi jian su du	1167
0236	传质量控制电解质速度常数	chuan zhi liang kong zhi dian jie zhi su du chang shu	1398
0237	γ 串级[级联]	γ chuan ji [ji lian]	0957
0238	串联点火	chuan lian dian huo	2189
0239	串联质谱计	chuan lian zhi pu ji	2396

0240	串列[连]式间隙	chuan lie [lian] shi jian xi	2395
0241	遄流	chuan liu	2527
0242	遄流焰	chuan liu yan	2526
0243	磁场	ci chang	1381
0244	磁场辉光放电	ci chang hui guang fang dian	1382
0245	磁偏转	ci pian zhuan	1380
0246	磁稳定电弧	ci wen ding dian hu	1379
0247	磁性彭宁(Penning)效应	ci xing peng ning xiao ying	1383
0248	磁子	ci zi	1384
0249	次级放电	ci ji fang dian	2142
0250	次级辐射	ci ji fu she	2149
0251	次级燃烧	ci ji ran shao	2141
0252	催化电流	cui hua dian liu	0310
0253	带电粒子	dai dian li zi	0339
0254	带电粒子活化	dai dian li zi huo hua	0338
0255	带(状)光谱	dai (zhuang) guang pu	0204
0256	带间区[范围]	dian jian qu [fan wei]	1170
0257	单(通)道脉冲高[幅]度分析器[仪]	dan (tong) dao mai chong gao [fu] du fen xi qi [yi]	2204
0258	单电离元素	dan dian li yuan su	2211
0259	单分子层	dan fen zi ceng	1513
0260	单分子层载量	dan fen zi ceng zai liang	1514
0261	单线态-单线态吸收	dan xian tai-dan xian tai xi shou	2210
0262	单官能离子交换剂	dan guan neng li zi jiao huan ji	1512
0263	单光束分光计	dan guang shu fen guang ji	2203
0264	单核二元配合物	dan he er yuan pei he wu	1515
0265	单核三元配合物	dan he san yuan pei he wu	1516
0266	单(电)极微波等离子体	dan (dian) ji wei bo deng li zi ti	1616
0267	单交替	dan jiao ti	1509
0268	单聚焦质谱仪	dan ju jiao zhi pu yi	2207
0269	单脉冲[冲量]	dan mai chong [chong liang]	2208
0270	单扫描示波极谱法	dan sao miao shi bo ji pu fa	2209
0271	单色辐射	dan se fu she	1510
0272	单色器[仪]	dan se qi [yi]	1511
0273	单色器[镜](波)通量	dan se qi [jing] (bo) tong liang	0925
0274	单色指示剂	dan se zhi shi ji	1615
0275	单逃逸峰	dan tao yi feng	2206
0276	单向放电	dan xiang fang dian	2541
0277	单向火花	dan xiang huo hua	2542
0278	单元素灯	dan yuan su deng	2205
0279	当量	dang liang	0819
0280	当量点	dang liang dian	0818
0281	导电率	dao dian lu	0440
0282	导[函]数	dao [han] shu	0587
0283	导数电位[势]滴定	dao shu dian wei [shi] di ding	0592
0284	导数分光光度法	dao shu fen guang guang du fa	0594
0285	导数伏安法	dao shu fu an fa	0597
0286	导数极谱法	dao shu ji pu fa	0591
0287	导数计时电位分析法	dao shu ji shi dian wei fen xi fa	0588
0288	导数脉冲极谱法	dao shu mai chong ji pu fa	0593
0289	导数膨胀计测定法	dao shu peng zhang ji ce ding fa	0589

0290	导数热分析	dao shu re fen xi	0595
0291	导数热重量分析法	dao shu re zhong liang fen xi fa	0596
0292	导数终点	dao shu zhong dian	0590
0293	德拜-休克耳(Debye-Hückel)方程式	de bai-xiu ke er fang cheng shi	0567
0294	德尔韦斯(Delves)杯	de er wei si bei	0579
0295	灯[灯泡]	deng [deng pao]	1255
0296	等板高度	deng ban gao du	1026
0297	等待时间曲线	deng dai shi jian qu xian	2470
0298	等电位点	deng dian wei dian	1230
0299	等离子体	deng li zi ti	1773
0300	等离子体[区]	deng li zi ti [qu]	1772
0301	等离子体参量	deng li zi ti can liang	1778
0302	等离子体吹管	deng li zi ti chui guan	1780
0303	等离子体负载	deng li zi ti fu zai	1777
0304	等离子体观测带	deng li zi ti guan ce dai	1613
0305	等离子体喷注	deng li zi ti pen zhu	1776
0306	等离子体起始	deng li zi ti qi shi	1775
0307	等离子体[区]温度	deng li zi ti [qu] wen du	1779
0308	等离子体阻抗	deng li zi ti zu kang	1774
0309	等温环境差热分析	deng wen huan jing cha re fen xi	0630
0310	等温线	deng wen xian	1231
0311	等温重量变化测定	deng wen zhong liang bian hua ce ding	1232
0312	低电流电弧	di dian liu dian hu	1357
0313	低电压火花	di dian ya huo hua	1365
0314	低能电子衍射(LEED)	di neng dian zi yan she	1359
0315	低能碰撞	di neng peng zhuang	1358
0316	低损耗	di sun hao	1360
0317	低压放电	di ya fang dian	1363
0318	低压放电灯	di ya fang dian deng	1362
0319	低压弧光灯	di ya hu guang deng	1361
0320	低压微波感应等离子体	di ya wei bo gan ying deng li zi ti	1364
0321	滴定(法)	di ding (fa)	2474
0322	滴定碘法	di ding dian fa	1181
0323	滴定度	di ding du	0575
0324	滴定法	di ding fa	2480
0325	滴定分析	di ding fen xi	2478
0326	滴定分析换算因子	di ding fen xi huan suan yin zi	2479
0327	滴定剂	di ding ji	2473
0328	滴定率	di ding lu	2477
0329	滴定曲线	di ding qu xian	2469
0330	滴定误差	di ding wu cha	2476
0331	滴定浓度[含量]	di ding nong du [han liang]	1285
0332	滴发生器	di fa sheng qi	0704
0333	滴汞电极	di gong dian ji	0707
0334	滴液电极库仑分析法	di ye dian ji ku lun fen xi fa	0706
0335	碘滴定法	dian di ding fa	1178
0336	碘量滴定法	dian liang di ding fa	1180
0337	碘量(滴定)法,碘滴定法[碘为滴定剂]	dian liang (di ding) fa, dian di ding fa	1179
0338	点滴雾化器	dian di wu hua qi	0705
0339	点对点对称构象	dian dui dian dui cheng gou xiang	1787

0340	点对平面对称构象	dian dui ping mian dui cheng gou xiang	1786
0341	点火	dian huo	1073
0342	点火火花	dian huo huo hua	1076
0343	点火器线路	dian huo qi xian lu	1077
0344	点火交流电弧	dian huo jiao liu dian hu	1072
0345	点火脉冲	dian huo mai chong	1075
0346	点火线路	dian huo xian lu	1074
0347	电参数	dian can shu	0737
0348	电磁辐射	dian ci fu she	0751
0349	电导滴定法	dian dao di ding fa	0441
0350	电导分析法	dian dao fen xi fa	0442
0351	电导率[性]	dian dao lu [xing]	0734
0352	电动喷雾	dian dong pen wu	0774
0353	电动势	dian dong shi	0752
0354	电放电	dian fang dian	0735
0355	电分离	dian fen li	0772
0356	电-光快门	dian-guang kuai men	0769
0357	电荷传递	dian he chuan di	0342
0358	电荷交换	dian he jiao huan	0340
0359	电荷跃极谱法	dian he yue ji pu fa	0341
0360	电弧	dian hu	0131,0733
0361	电弧的电压-时间关系	dian hu de dian ya-shi jian guan xi	2572
0362	电弧等离子体压缩	dian hu deng li zi ti ya suo	0452
0363	电弧爬升	dian hu pa sheng	0381
0364	电弧(炉内)气氛	dian hu (lu nei) qi fen	0132
0365	电弧柱	dian hu zhu	0133
0366	电火花	dian huo hua	0739
0367	电火隙	dian huo xi	2248
0368	电活性物质	dian huo xing wu zhi	0741
0369	电激发[兴奋]	dian ji fa [xing fen]	0736
0370	电极-溶液界面	dian ji-rong ye jie mian	0746
0371	电极谐振腔等离子体	dian ji xie zhen qiang deng li zi ti	0745
0372	电解质[液]	dian jie zhi [ye]	0750
0373	电浸蚀	dian jin shi	0747
0374	电离,离子化(作用)	dian li, li zi hua (zuo yong)	1200
0375	电离度	dian li du	0574
0376	电离[离子]辐射	dian li [li zi] fu she	1213
0377	电离干扰	dian li gan rao	1207
0378	电离	dian li gai lu	1210
0379	电离截面	dian li jie mian	0526
0380	电离能	dian li neng	1206
0381	电离室	dian li shi	1203
0382	电离势	dian li shi	1209
0383	电离温度	dian li wen du	1211
0384	电离系数	dian li xi shu	1204
0385	电离效率曲线	dian li xiao lu qu xian	1205
0386	电离[指,磁]针	dian li [zhi, ci] zhen	1208
0387	电流	dian liu	0740
0388	电流滴定法	dian liu di ding fa	0074
0389	电流分析法	dian liu fen xi fa	0075

0390	电流扫描极谱法	dian liu sao miao ji pu fa	0546
0391	电流时间曲线	dian liu shi jian qu xian	0547
0392	电流终点检测	dian liu zhong dian jian ce	0073
0393	电谱法	dian pu fa	0748
0394	电桥溶液	dian qiao rong ye	0267
0395	电容耦合等离子体	dian rong ou he deng li zi ti	0290
0396	电容器电压	dian rong qi dian ya	0292
0397	电容微波等离子体	dian rong wei bo deng li zi ti	0291
0398	电容电流	dian rong dian liu	0294
0399	电势[位]	dian shi [wei]	1807
0400	电势测定终点	dian shi ce ding zhong dian	1809
0401	电势[位]滴定法	dian shi [wei] di ding fa	1811
0402	电势[位]分析法	dian shi [wei] fen xi fa	1813
0403	电势[位]控制电分离	dian shi [wei] kong zhi dian fen li	0773
0404	电势[位]选择性系数	dian shi [wei] xuan ze xing xi shu	1810
0405	电位跃计时库仑法	dian wei yue ji shi ku lun fa	1808
0406	电位重量滴定法	dian wei zhong liang di ding fa	1812
0407	电压,伏特数	dian ya, fu te shu	2570
0408	电压-电流[伏安]特性	dian ya-dian liu [fu an] te xing	2571
0409	电泳	dian yong	0770
0410	电泳图	dian yong tu	0771
0411	电源参数	dian yuan can shu	0738
0412	电源电压	dian yuan dian ya	1385
0413	电晕放电	dian yun fang dian	0270
0414	电重量分析法	dian zhong liang fen xi fa	0749
0415	电子倍增作用	dian zi bei zeng zuo yong	0762
0416	电子动能	dian zi dong neng	0760
0417	电子对产生	dian zi dui chan sheng	1658
0418	电子对衰减系数	dian zi dui shuai jian xi shu	1657
0419	电子俘获	dian zi fu huo	0753
0420	电子俘获检测器	dian zi fu huo jian ce qi	0754
0421	电子交换聚合物	dian zi jiao huan ju he wu	2014
0422	电子能量	dian zi neng liang	0756
0423	电子能量传递过程	dian zi neng liang chuan di guo cheng	0758
0424	电子能量损耗能谱(法)	dian zi neng liang sun hao neng pu (fa)	0757
0425	电子能谱	dian zi neng pu	0766
0426	电子碰撞离子化	dian zi peng zhuang li zi hua	0759
0427	电子探针 X 射线微量分析(EXPMA)	dian zi tan zhen X she xian wei liang fen xi	0765
0428	电子探针微量分析(EPMA)	dian zi tan zhen wei liang fen xi	0764
0429	电子温度	dian zi wen du	0768
0430	电子显微镜	dian zi xian wei jing	0761
0431	电子压	dian zi ya	0763
0432	电子衍射	dian zi yan she	0755
0433	电子直线加速器	dian zi zhi xian jia su qi	1305
0434	电阻器	dian zu qi	2057
0435	叠加	die jia	2371
0436	顶替色谱法	ding ti se pu fa	0670
0437	定标[校正]电路,(脉冲)计数电路	ding biao [jiao zheng] dian lu, (mai chong) ji shu dian lu	2120
0438	定标器	ding biao qi	2119
0439	定量差热分析	ding liang cha re fen xi	1899

0440	定量分析	ding liang fen xi	1898
0441	定量(极)限	ding liang (ji) xian	1299
0442	定率计	ding lu ji	1989
0443	定性分析	ding xing fen xi	1896
0444	动力补偿[驱动]差示扫描量热法	dong li bu chang [qu dong] cha shi sao miao liang re fa	1816
0445	动力电流	dong li dian liu	1248
0446	动力(学)体系	dong li (xue) ti xi	0715
0447	动力学分析(法)	dong li xue fen xi (fa)	1247
0448	动量传递	dong liang chuan di	1508
0449	动态差示量热法	dong tai cha shi liang re fa	0713
0450	动态场致质谱计	dong tai chang zhi zhi pu ji	0714
0451	动态热解重量分析	dong tai re jie zhong liang fen xi	0716
0452	动态热力学分析法	dong tai re li xue fen xi fa	0717
0453	杜阿尼-亨特(Duane-Hunt)短波长极限	du a ni-heng te duan bo chang ji xian	0710
0454	读出率,清晰度	du chu lu, qing xi du	1997
0455	读数	du shu	1998
0456	短路放电	duan lu fang dian	2193
0457	断续器,切光器	duan xu qi,qie guang qi	0364
0458	堆积效应	dui ji xiao ying	1760
0459	对电极	dui dian ji	0506
0460	对流	dui liu	0481
0461	对流计时安培[电流]分析法	dui liu ji shi an pei [dian liu] fen xi fa	0482
0462	对流计时库仑分析法	dui liu ji shi ku lun fen xi fa	0483
0463	对数滴定曲线	dui shu di ding qu xian	1353
0464	对数分配系数	dui shu fen pei xi shu	1352
0465	对数衰减[减量]	dui shu shuai jian [jian liang]	1351
0466	对应于基峰强度	dui ying yu ji feng qiang du	1137
0467	对照滴定	dui zhao di ding	0480
0468	对照谱带	dui zhao pu dai	0463
0469	对照[参比]强度	dui zhao [can bi] qiang du	2022
0470	多槽燃烧器	duo cao ran shao qi	1524
0471	多(分子)层	duo (fen zi) ceng	1520
0472	多道分析器	duo dao fen xi qi	1518
0473	多道脉冲高度分析器	duo dao mai chong gao du fen xi qi	1519
0474	多功能离子交换剂	duo gong neng li zi jiao huan ji	1796
0475	多核[环]配合物	duo he [huan] pei he wu	1797
0476	多孔电极	duo kong dian ji	1799
0477	多路[程,通]方法	duo lu [cheng, tong] fang fa	1521
0478	多内尔-奥斯金斯(Doerner-Hoskins 方程	duo nei er-ao si jin si fang cheng	0685
0479	多普勒(Doppler)(谱线)变宽	duo pu le (pu xian) bian kuan	0686
0480	多普勒谱线半宽度	duo pu le pu xian ban kuan du	0687
0481	多普勒位移	duo pu le wei yi	0688
0482	多扫描极谱法	duo sao miao ji pu fa	1525
0483	多色光度术[学]	duo se guang du shu [xue]	1028
0484	多色仪	duo se yi	1795
0485	多[通]用[性能]火花源	duo [tong] yong [xing neng] huo hua yuan	1523
0486	多原子的	duo yuan zi de	1794
0487	惰性电解质	duo xing dian jie zhi	1094
0488	惰性放电气氛	duo xing fang dian qi fen	1103
0489	俄歇(Auger)电子	e xie dian zi	0168

0490	俄歇电子产额	e xie dian zi chan e	0170,0173
0491	俄歇电子能谱学	e xie dian zi neng pu xue	0169
0492	俄歇峰(值)能(量)	e xie feng (zhi) neng (liang)	0171
0493	俄歇能谱	e xie neng pu	0172
0494	俄歇效应	e xie xiao ying	0167
0495	二次电光效应	er ci dian guang xiao ying	1894
0496	二次电子	er ci dian zi	2144
0497	二次电子倍增器	er ci dian zi bei zeng qi	2143
0498	二次电子产额	er ci dian zi chan e	2148
0499	二次离子	er ci li zi	2146
0500	二次离子质谱仪(SIMS)	er ci li zi zhi pu yi	2147
0501	二次荧光	er ci ying guang	2145
0502	二极管-阵列检测器	er ji guan-zhen lie jian ce qi	0645
0503	二阶导数电位[势]滴定	er jie dao shu dian wei [shi] di ding	2152
0504	二色指示剂	er se zhi shi ji	2529
0505	二元碰撞	er yuan peng zhuang	0227
0506	发光	fa guang	1366
0507	发光参数	fa guang can shu	1367
0508	发光层	fa guang ceng	1375
0509	发光猝灭	fa guang cui mie	1368
0510	发光分光计	fa guang fen guang ji	1369
0511	发光光度法	fa guang guang du fa	1370
0512	发光光谱学	fa guang guang pu xue	1371
0513	发光量子产额	fa guang liang zi chan e	1906
0514	发光量	fa guang liang	1376
0515	发光量子效率	fa guang liang zi xiao lu	1903
0516	发光能量[输出当量]产额	fa guang neng liang [shu chu dang liang] chan e	0805
0517	发光线偏振	fa guang xian pian zhen	1311
0518	发光效力	fa guang xiao li	1372
0519	发光效率	fa guang xiao lu	1373
0520	(接种)发[点]火弧	(jie zhong) fa [dian] huo hu	2154
0521	发散角	fa san jiao	0100
0522	发色团间无辐射跃迁	fa se tuan jian wu fu she yue qian	1140
0523	发色团内无辐射传递	fa se tuan nei wu fu she chuan di	1171
0524	发射光单色器	fa she guang dan se qi	0788
0525	发射光谱	fa she guang pu	0790
0526	发射光谱分析法	fa she guang pu fen xi fa	1628
0527	发射轮廓	fa she lun kuo	0789
0528	发射线谱的半强度宽度	fa she xian pu de ban qiang du kuan du	1010
0529	法拉弟(Faraday)暗区	fa la di an qu	0872
0530	法拉弟常数	fa la di chang shu	0870
0531	法拉弟电流	fa la di dian liu	0867
0532	法拉弟集电极	fa la di ji dian ji	0871
0533	法拉弟调解电流	fa la di tiao jie dian liu	0868
0534	法拉弟整流电流	fa la di zheng liu dian liu	0869
0535	法诺(Fano)因子	fa nuo yin zi	0866
0536	反常辉光放电	fan chang hui guang fang dian	0001
0537	反冲	fan chong	2005
0538	反萃取,反提取	fan cui qu, fan ti qu	0190,0191
0539	反萃[提]取(法)	fan cui [ti] qu (fa)	0192

0540	反导数电位滴定(法)	fan dao shu dian wei di ding (fa)	1177
0541	反滴定	fan di ding	0201
0542	反滴定法	fan di ding fa	0200
0543	反康普顿(Compton)γ射线	fan kang pu dun γ she xian	0119
0544	反康普顿(散射)γ射线谱仪	fan kang pu dun (san she) γ she xian pu yi	0120
0545	反馈系统	fan kui xi tong	0875
0546	反粒子	fan li zi	0121
0547	反(向)散射	fan (xiang) san she	0197
0548	反散射能谱法(BSS)	fan san she neng pu fa	0199
0549	反闪	fan shan	0900
0550	反射比[本领]	fan she bi [ben ling]	2027
0551	反射电子能量损耗能谱法(REELS)	fan she dian zi neng liang sun hao neng pu fa	2028
0552	反射高能电子衍射(RHEED)	fan she gao neng dian zi yan she	2030
0553	反射系数	fan she xi shu	2029
0554	反斯托克斯(Stokes)荧光	fan si tuo ke si ying guang	0122
0555	反向散射电子	fan xiang san she dian zi	0198
0556	反相色谱(法)	fan xiang se pu (fa)	2084
0557	反应层	fan ying ceng	1996
0558	反应层厚度	fan ying ceng hou du	2458
0559	反应间距	fan ying jian ju	1995
0560	反载体	fan zai ti	1049
0561	范围	fan wei	1988
0562	方波电流	fang bo dian lui	2313
0563	方波极谱法	fang bo ji pu fa	2314
0564	放,排,放电	fang, pai, fang dian	0655
0565	放大反应	fang da fan ying	0077
0566	放电	fang dian	0735
0567	放电常压	fang dian chang ya	0656
0568	放电电流	fang dian dian liu	0658
0569	放电电路	fang dian dian lu	0657
0570	放电极谱法	fang dian ji pu fa	0660
0571	放电气体	fang dian qi ti	0659
0572	放电真空	fang dian zhen kong	0661
0573	放(射)化(学)纯度	fang (she) hua (xue) chun du	1968
0574	放(射)化(学)分离	fang (she) hua (xue) fen li	1969
0575	放能条件	fang neng tiao jian	0845
0576	放热	fang re	0846
0577	放热反应	fang re fan ying	0848
0578	放热峰	fang re feng	0847
0579	放射半衰期	fang she ban shuai qi	1953
0580	放射测量终点	fang she ce liang zhong dian	1978
0581	放射滴定	fang she di ding	1979
0582	放射分析化学	fang she fen xi hua xue	1966
0583	放射分析提纯[纯化]	fang she fen xi ti chun [chun hua]	1967
0584	放射化学	fang she hua xu	1971
0585	放射化学产额	fang she hua xue chan e	1970
0586	放射平衡	fang she ping heng	1951
0587	放射色谱仪	fang she se pu yi	1972
0588	放射(性)衰变	fang she (xing) shuai bian	1949
0589	放射(性)系[族]	fang she (xing) xi [zu]	1961

0590	放射性[学]	fang she xing [xue]	1965
0591	放射性标记	fang she xing biao ji	1956
0592	放射性材料[物质]	fang she xing cai liao [wu zhi]	1957
0593	放射性测定年代	fang she xing ce ding nian dai	1948
0594	放射性落尘	fang she xing luo chen	1952
0595	放射性的	fang she xing de	1944
0596	放射性废物	fang she xing fei wu	1964
0597	放射性核	fang she xing he	1959
0598	放射性核素	fang she xing he su	1960,1980
0599	放射性核素纯度	fang she xing he su chun du	1981
0600	放射性胶体	fang she xing jiao ti	1973
0601	放射性年令(测定)	fang she xing nian ling (ce ding)	1945
0602	放射性排出物[废水]	fang she xing pai chu wu [fei shui]	1950
0603	放射性平均寿命	fang she xing ping jun shou ming	1958
0604	放射性试剂	fang she xing shi ji	1982
0605	放射性示踪物	fang she xing shi zong wu	1963
0606	放射性衰变链	fang she xing shuai bian lian	1946
0607	放射性同位素	fang she xing tong wei su	1955, 1976
0608	放射性物质	fang she xing wu zhi	1962
0609	放射指示剂	fang she zhi shi ji	1954
0610	放射自显影	fang she zi xian ying	0179
0611	方式	fang shi	1479
0612	非迭加电流	fei die jia dian liu	1569
0613	非法拉弟导纳	fei fa la di dao na	1578
0614	非光谱干扰	fei guang pu gan rao	1582
0615	非晶性电极	fei jing xing dian ji	1574
0616	非均质性,各向异性	fei jun zhi xing, ge xiang yi xing	0107
0617	非离散连续辐射	fei li san lian xu fu she	1577
0618	非水(溶液)滴定	fei shui (rong ye) di ding	1570
0619	非弹性碰撞	fei tan xing peng zhuang	1101
0620	非弹性散射	fei tan xing san she	1102
0621	非特异[种类]干扰	fei te yi [zhong lei] gan rao	1581
0622	非相干电磁辐射	fei xiang gan dian ci fu she	1571
0623	非相干辐射脉冲	fei xiang gan fu she mai chong	1088
0624	非相干光源	fei xiang gan guang yuan	1572
0625	沸点	fei dian	0249
0626	沸腾温度	fei teng wen du	0250
0627	废气分析	fei qi fen xi	0725
0628	废气监测	fei qi jian ce	0726
0629	分辨[解],离析,溶解	fen bian [jie], li xi, rong jie	2058
0630	分辨 10% 谷值	fen bian 10% gu zhi	2059
0631	分辨率[能力]	fen bian lu [neng li]	2060
0632	分辨时间	fen bian shi jian	2061
0633	分辨时间校正	fen bian shi jian jiao zheng	2062
0634	分别溶液法	fen bie rong ye fa	2183
0635	分段线荧光	fen duan xian ying guang	2350
0636	分光化学缓冲剂	fen guang hua xue huan chong ji	2284
0637	分光光度滴定	fen guang guang du di ding	2295
0638	分光计,光谱仪	fen guang ji, guang pu yi	2292
0639	分光计分辨率	fen guang ji fen bian lu	2293

0640	分光镜[仪,器]	fen guang jing [yi, qi]	2296
0641	分解,溶解	fen jie, rong jie	1620
0642	分阶洗脱	fen jie xi tuo	2348
0643	分开的火焰	fei kai de huo yan	2182
0644	分离过程	fen li guo cheng	2185
0645	分离温度	fen li wen du	2186
0646	分离系数	fen li xi shu	2184
0647	α—分裂	α—fen lie	0059
0648	分馏	fen liu	0936
0649	分配[布]	fen pei [bu]	0678
0650	(萃取)分配比	(cui qu) fen pei bi	0684
0651	分配常数	fen pei chang shu	0680,1680
0652	分配函数	fen pei han shu	1681
0653	分配[布](定)律	fen pei [bu] (ding) lu	0683
0654	分配平衡	fen pei ping heng	0681
0655	分配平衡常数	fen pei ping heng chang shu	0682
0656	分配色谱法	fen pei se pu fa	1678
0657	(不溶混液间)分配物	(bu rong hun ye jian) fen pei wu	0677
0658	分配系数	fen pei xi shu	0679,0860,1679
0659	分散(体),色散,离差	fen san (ti), se san, li cha	0667
0660	(流动注射分析)(FIA)分散度	(lui dong zhu she fen xi) fen san du	0668
0661	分析标准	fen xi biao zhun	0098
0662	分析电极	fen xi dian ji	0090
0663	分析电子显微镜法	fen xi dian zi xian wei jing fa	0091
0664	分析放射化学	fen xi fang she hua xue	0095
0665	分析功能	fen xi gong neng	0093
0666	分析加入法	fen xi jia ru fa	0084
0667	分析间隙	fen xi jian xi	0094
0668	分析校正	fen xi jiao zheng	0086
0669	分析校正函数	fen xi jiao zheng han shu	0087
0670	分析结果	fen xi jie guo	0096
0671	分析评定曲线	fen xi ping ding qu xian	0092
0672	分析曲线	fen xi qu xian	0088
0673	分析曲线法	fen xi qu xian fa	0089
0674	分析天平	fen xi tian ping	0085
0675	(被)分析物	(bei) fen xi wu	0080
0676	分析物加入法	fen xi wu jia ru fa	0081
0677	(被)分析物离子	(bei) fen xi wu li zi	0083
0678	(被)分析物原子	(bei) fen xi wu yuan zi	0082
0679	分析系统	fen xi xi tong	0099
0680	分析样品	fen xi yang pin	0097
0681	分析元素	fen xi yuan su	0079
0682	分支比	fen zhi bi	0260
0683	分支(衰变)分额	fen zhi (shuai bian) fen e	0259
0684	分支衰变	fen zhi shuai bian	0258
0685	分子	fen zi	1507
0686	分子法	fen zi fa	1503
0687	分子发光光谱法	fen zi fa guang guang pu fa	1502
0688	分子发射光谱法	fen zi fa she guang pu fa	1499
0689	分子发射光谱学	fen zi fa she guang pu xue	1500

0690	分子辐照[射]物	fen zi fu zhao [she] wu	1504
0691	分子光谱法	fen zi guang pu fa	1505
0692	分子光谱谱带	fen zi guang pu pu dai	1497
0693	分子光谱学	fen zi guang pu xue	1506
0694	分子间无辐射跃迁	fen zi jian wu fu she yue qian	1152
0695	分子离子	fen zi li zi	1501
0696	分子阳离子	fen zi yang li zi	1498
0697	分子阴离子	fen zi yin li zi	1496
0698	分子吸收光谱测定(法)	fen zi xi shou guang pu ce ding (fa)	1494
0699	分子吸收光谱法[学]	fen zi xi shou guang pu fa [xue]	1495
0700	粉末技术	fen mo ji shu	1815
0701	粉末样品	fen mo yang pin	1814
0702	封闭空心阴极灯	feng bi kong xin yin ji deng	2140
0703	峰(值)	feng (zhi)	1682
0704	峰[谷](值),顶点,尖	feng [gu] (zhi), ding dian, jian	0124
0705	峰[顶点]电流	feng [ding dian] dian liu	2369
0706	峰电位	feng dian wei	2370
0707	峰顶点,峰最大值	feng ding dian, feng zui da zhi	1693
0708	峰分辨度	feng fen bian du	1694
0709	峰高	feng gao	1691
0710	峰基线	feng ji xian	1685
0711	峰宽	feng kuan	1696
0712	峰面积	feng mian ji	1684
0713	峰强度	feng qiang du	1692
0714	峰值电流	feng zhi dian liu	1687
0715	峰值[最大]电容电压	feng zhi [zui da] dian rong dian ya	1686
0716	峰值分析	feng zhi fen xi	1683
0717	峰值[最大]火花电流	feng zhi [zui da] huo hua dian liu	1695
0718	峰值拟合法	feng zhi ni he fa	1690
0719	峰值[最大]洗脱[流出]容积	feng zhi [zui da] xi tuo [liu chu] rong ji	1689
0720	峰值[最大]衍射系数	feng zhi [zui da] yan she xi shu	1688
0721	伏安测量法	fu an ce liang fa	2575
0722	伏安常数	fu an chang shu	2574
0723	伏安(法)分析	fu an (fa) fen xi	2573
0724	伏安谱(图)	fu an pu (tu)	2576
0725	伏特安倍谱(图)	fu te an bei pu (tu)	2577
0726	俘获	fu huo	0299
0727	俘获γ辐射	fu huo γ fu she	0301
0728	俘获截面	fu huo jie mian	0300
0729	符号规定[常规]	fu hao gui ding [chang gui]	2197
0730	符合电路	fu he dian lu	0389
0731	符合分解(分辨)时间	fu he fen jie (fen bian) shi jian	0391
0732	符合计数	fu he ji shu	0390
0733	辐解作用	fu jie zuo yong	1977
0734	辐[放]射(线),放射物,照射(作用)	fu [fang] she (xian), fang she wu, zhao she (zuo yong)	1927
0735	辐[照]射,辐射(率,密度),光亮度	fu [zhao] she, fu she (lu, mi du), guang liang du	1915
0736	γ辐射[射线]	γ fu she [she xian]	0959
0737	X—辐射[射线]	X—fu she [she xian]	2598
0738	(光)辐射发射率[系数]	(guang) fu she fa she lu [xi shu]	1917
0739	辐射分光计	fu she fen guang ji	1936

0740	辐射复合	fu she fu he	1942
0741	辐射俘获	fu she fu huo	1938
0742	γ 辐射[线]俘获	γ fu she [xian] fu huo	0960
0743	(光)辐射功率[流量]/立体角	(guang) fu she gong lu [liu liang]/li ti jiao	1925
0744	辐射过程	fu she guo cheng	1941
0745	辐射[放射]化学	fu she [fang she] hua xue	1928
0746	辐射激发	fu she ji fa	1940
0747	辐射计数器[管]	fu she ji shu qi [guan]	1929
0748	辐射冷却	fu she leng que	1947
0749	(光)辐射粒子	(guang) fu she li zi	1923
0750	(光)辐射量	(guang) fu she liang	1926
0751	辐射量	fu she liang	1934
0752	辐射脉冲	fu she mai chong	1933
0753	(光)辐射能	(guang) fu she neng	1918
0754	辐射能	fu she neng	1931
0755	(光) 辐射能曝露(量)	(guang) fu she neng bao lu (liang)	1920
0756	(光)辐射能(量)密度	(guang) fu she neng (liang) mi du	1919
0757	(光)辐射谱	(guang) fu she pu	1937
0758	辐射谱	fu she pu	2299
0759	(光)辐射强度	(guang) fu she qiang du	1135
0760	辐射强度	fu she qiang du	1922
0761	辐射去激发[活]	fu she qu ji fa [huo]	1939
0762	辐射热测量器	fu she re ce liang qi	0251
0763	辐射散射	fu she san she	2126
0764	辐射探[检]测器	fu she tan [jian] ce qi	1930
0765	(光)辐射通量	(guang) fu she tong liang	1921
0766	(光)辐射(表面)突起[羽状]处	(guang) fu she (biao mian) tu qi [yu zhuang] chu	1924
0767	辐[照]射温度	fu [zhao] she wen du	1916
0768	γ 辐射吸收	γ fu she xi shou	0016
0769	辐射源	fu she yuan	1935
0770	辐射跃迁	fu she yue qian	1943
0771	辐照,照射	fu zhao, zhao she	1224
0772	辐照度[率],辐照通量密度	fu zhao du [lu], fu zhao tong liang mi du	1223
0773	辅助电花隙	fu zhu dian hua xi	0183
0774	辅助电极	fu zhu dian ji	0182
0775	辅助放电	fu zhu fang dian	0181
0776	辅助放电	fu zhu fang dian	2374
0777	附聚(作用)	fu ju (zuo yong)	0051
0778	附随物质	fu sui wu zhi	0438
0779	负电带,阴极区	fu dian dai, yin ji qu	1542
0780	负电子	fu dian zi	1543
0781	负离子质谱	fu li zi zhi pu	1541
0782	负载循环	fu zai xun huan	0711
0783	复弧	fu hu	0691
0784	副标准	fu biao zhun	2150
0785	副标准溶液	fu biao zhun rong ye	2151
0786	富集,浓缩	fu ji, nong suo	0808
0787	富集因子	fu ji yin zi	0809
0788	富油的火焰	fu you de huo yan	0952
0789	傅里叶(Fourier)变换质谱仪	fu li ye bian huan zhi pu yi	0935

0790	盖革—谬勒(Geiger—Müller)计数管	gai ge—miu le ji shu guan	0980
0791	盖革—谬勒区域	gai ge—miu le qu yu	0981
0792	盖革—谬勒阈[极限](值)	gai ge—miu le yu [ji xian] (zhi)	0982
0793	盖斯勒(Geissler)灯	gai si le deng	0983
0794	改进活性固体	gai jin huo xing gu ti	1480
0795	改进能斯脱(Nernst)方程	gai jin neng si tuo fang cheng	1481
0796	改性[进,良]剂,改性物,调节物	gai xing [jin, liang] ji, gai xing wu, tiao ji wu	1482
0797	感光板斜置	gan guang ban xie zhi	2467
0798	感光强度测定	gan guang qiang du ce ding	1735
0799	感光[照相]乳剂校正[定标]	gan guang [zhao xiang] ru ji xiao zheng [ding biao]	1734
0800	感光透射比	gan guang tou she bi	1738
0801	感生放射性	gan sheng fang she xing	1096
0802	感应耦合等离子体	gan ying ou he deng li zi ti	1099
0803	感应圈	gan ying quan	1098
0804	干气溶胶	gan qi rong jiao	0709
0805	干扰,干涉,噪声	gan rao, gan she, zao sheng	1144
0806	干扰谱线	gan rao pu xian	1147
0807	干扰曲线	gan rao qu xian	1145
0808	干扰物	gan rao wu	1146
0809	干扰物质	gan rao wu zhi	1148
0810	高重复火花	gao chong fu huo hua	1045
0811	高重复率预发火花	gao chong fu lu yu fa huo hua	1044
0812	高低射标法	gao di she biao fa	0255
0813	高电平法拉弟整流	gao dian ping fa la di zheng liu	1040
0814	高(饱和)反射率	gao (bao he) fan she lu	1043
0815	高分辨率能量损耗能谱法(HRELS)	gao fen bian lu neng liang sun hao neng pu fa	1046
0816	高频滴定法	gao pin di ding fa	1038
0817	高频电导滴定法	gao pin dian dao di ding fa	1034
0818	高频电导分析法	gao pin dian dao fen xi fa	1035
0819	高频电导分析终点	gao pin dian dao fen xi zhong dian	1033
0820	高频短路电容器	gao pin duan lu dian rong qi	1036
0821	高频升压变压器	gao pin sheng ya bian ya qi	1037
0822	高强度空心阴极	gao qiang du kong xin yin ji	1039
0823	高压电弧	gao ya dian hu	1041
0824	高压电火花	gao ya dian huo hua	1047
0825	高压氙灯	gao ya xian deng	1042
0826	高(次)谐波交流极谱法	gao (ci) xie bo jiao liu ji pu fa	1031
0827	格拉赫(Gerlach) 同调[匀称]线	ge la he tong tiao [yun cheng] xian	0989
0828	镉切断有效能(量)	gng qie duan you xiao neng (liang)	0719
0829	戈瑞(gray)[吸收剂量单位]	ge rui	1005
0830	工作电极	gong zuo dian ji	2595
0831	工作气	gong zuo qi	2596
0832	公[标]称线性流量	gong [biao] cheng xian xing liu liang	1568
0833	汞池电极	gong chi dian ji	1431
0834	汞滴[滴下]时间	gong di [di xia] shi jian	0708
0835	共沉淀(作用)	gong chen dian (zuo yong)	0490
0836	共存元素效应	gong cun yuan su xiao ying	1142
0837	共离子	gong li zi	0392
0838	共振分光仪	gong zhen fen guang yi	2067
0839	共振加宽	gong zhen jia kuan	2063

0840	共振能	gong zhen neng	2064
0841	共[谐]振腔	gong [xie] zhen qiang	2068
0842	共振线	gong zhen xian	2065
0843	共振中子	gong zhen zhong zi	2066
0844	固定辐射	gu ding fu she	2342
0845	固定基体[矩阵]电极	gu ding ji ti [ju zhen] dian ji	2085
0846	固定离子	gu ding li zi	0894
0847	固定频率	gu ding pin lu	0893
0848	固定相	gu ding xiang	2339
0849	固定相部分	gu ding xiang bu fen	2340
0850	固定相体积	gu ding xiang ti ji	2341
0851	固(态)溶液[体]	gu (tai) rong ye [ti]	2221
0852	固态溶液样品	gu tai rong ye yang pin	2222
0853	固态激光器	gu tai ji guang qi	2225
0854	固态检测器	gu tai jian ce qi	2224
0855	固态晶格	gu tai jing ge	2223
0856	固体试样	gu ti shi yang	2220
0857	鼓风[吹入]方法	gu feng [cui ru] fang fa	0246
0858	T(形)管	T (xing) guan	2393
0859	观测[值]	guan ce [zhi]	1611
0860	观察高度	guan ce gao du	1027,1612
0861	管装置	guan zhuang zhi	2517
0862	光泵	guang beng	1632
0863	光传导	guang chuan dao	1626
0864	光电倍增管	guang dian bei zeng guan	1744
0865	光电发射能谱法	guang dian fa she neng pu fa	1733
0866	光电检测[波]器	guang dian jian ce [bo] qi	1725
0867	光电流	guang dian liu	1724
0868	光电衰减系数	guang dian shuai jian xi shu	1726
0869	光电效应	guang dian xiao ying	1727
0870	光电子	guang dian zi	1729
0871	光电子产额[率]	guang dian zi chan e [lu]	1732
0872	光电子峰	guang dian zi feng	1753
0873	光电子能谱	guang dian zi neng pu	1730
0874	光电子能谱法(PES)	guang dian zi neng pu fa	1731
0875	光度滴定	guang du di ding	1742
0876	光度法终点	guang du fa zhong dian	1741
0877	光度学[法]	guang du xue [fa]	1743
0878	光量	guang liang	1633
0879	光密度	guang mi du	1627
0880	光密度计	guang mi du ji	0583
0881	光[波]谱,谱	guang [bo] pu, pu	2298
0882	光谱背景[本底]	guang pu bei jing [ben di]	2267
0883	光谱测定(法)	guang pu ce ding (fa)	2294
0884	光谱带宽	guang pu dai kuan	2268
0885	光谱法[摄谱方法]	guang pu fa	2291
0886	光谱辐射	guang pu fu she	2276
0887	光谱辐射量	guang pu fu she liang	2279
0888	光谱辐射能(量)密度	guang pu fu she neng (liang) mi du	2278
0889	光谱干扰	guang pu gan rao	2272

0890	光谱化学分析	guang pu hua xue fen xi	2282
0891	光谱化学特性	guang pu hua xue te xing	2287
0892	光谱化学稀释剂	guang pu hua xue xi shi ji	2286
0893	光谱化学应用	guang pu hua xue ying yong	2283
0894	光谱化学载体	guang pu hua xue zai ti	2285
0895	光谱化学准确度	guang pu hua xue zhun que du	2281
0896	光谱级	guang pu ji	1637
0897	光谱连续	guang pu lian xu	2270,0890
0898	光谱强度	guang pu qiang du	2271
0899	光谱特征	guang pu te zheng	2269
0900	光谱图	guang pu tu	2288
0901	光谱响应曲线	guang pu xiang ying qu xian	2280
0902	光谱学[测量]	guang pu xue [ce liang]	2297
0903	光谱仪,摄谱仪	guang pu yi, she pu yi	2289
0904	光谱仪器	guang pu yi qi	2290
0905	光谱周期	guang pu zhou qi	2275
0906	光弱等离子体	guang ruo deng li zi ti	1631
0907	光束宽度	guang shu kuan du	2594
0908	光调制	guang tiao zhi	1289
0909	光通量	guang tong liang	1374
0910	光信号调制	guang xin hao tiao zhi	1634
0911	光学系统,光具组	guang xue xi tong, guang ju zu	1635
0912	光学纤维	guang xue xian wei	1629
0913	光致电离	guang zhi dian li	1739
0914	光致电离能谱法	guang zhi dian li neng pu fa	1740
0915	光子	guang zi	1745
0916	光子发射产额	guang zi fa she chan e	1749
0917	光子发射能谱法	guang zi fa she neng pu fa	1748
0918	光子发射谱	guang zi fa she pu	1747
0919	光子辐射	guang zi fu she	1752
0920	光子活化	guang zi huo hua	1746
0921	光子通量	guang zi tong liang	1750
0922	光子通量密度	guang zi tong liang mi du	1751
0923	规度[当量浓度]	gui du [dang liang nong du]	1585
0924	规度[当量]溶液	gui du [dang liang] rong ye	1586
0925	硅光电二极管	gui guang dian er ji guan	2198
0926	过饱和	guo bao he	2373
0927	过饱和溶液	guo bao he rong ye	2372
0928	过渡时间	guo du shi jian	2500
0929	哈梅特(Hammett)酸标指示剂	ha mei te suan biao zhi shi ji	1018
0930	哈梅特酸度函数	ha mei te suan du han shu	1017
0931	焓测量终点	han ce liang zhong dian	0810
0932	毫克当量清晰度	hao ke dang liang qing xi du	1464
0933	毫(克量)库仑分析法	hao (ke liang) ku lun fen xi fa	1463
0934	耗液速度	hao ye su du	1991
0935	(原子)核反应	(yuan zi) he fan ying	1599
0936	核(子)反应堆	he (zi) fan ying dui	1600
0937	(原子)核化学	(yuan zi) he hua xue	1589
0938	核活化分析	he huo hua fen xi	1588
0939	核径迹	he jing ji	1602

0940	核径迹检测计	he jing ji jian ce ji	1603
0941	核晶形成速度	he jing xing cheng su du	1992
0942	核聚变	he ju bian	1594
0943	核聚变反应	he ju bian fan ying	1595
0944	核粒，核粒子	he li, he li zi	1598
0945	核裂变	he lie bian	1591
0946	核裂变反应	he lie bian fan ying	1592
0947	核燃料	he ran liao	1593
0948	核衰变	he shuai bian	1590
0949	核素	he su	1609
0950	核素质量	he su zhi liang	1610
0951	核同量异位素	he tong liang yi wei su	1596
0952	核系统	he xi tong	1601
0953	核跃迁	he yue qian	1605
0954	(原子)核转变	(yuan zi) he zhuan bian	1604
0955	核(单)子	he (dan) zi	1607
0956	核子数	he zi shu	1608
0957	黑体	hei ti	0233
0958	黑体的光谱辐射	hei ti de guang pu fu she	2277
0959	黑体辐射	hei ti fu she	0234
0960	痕量，示踪	hen liang, shi zong	2487
0961	痕量分析	hen liang fen xi	2488
0962	痕量构分	hen liang gou fen	2489
0963	恒电流	heng dian liu	0445
0964	恒电压	heng dian ya	0447
0965	恒温电弧	heng wen dian hu	0446
0966	横向电磁方式	heng xiang dian ci fang shi	2509
0967	横向静电场分析器	heng xiang jing dian chang fen xi qi	1914
0968	横向扩散干扰	heng xiang kuo san gan rao	1281
0969	宏观[粗量]截面	hong guan [cu liang] jie mian	1378
0970	红移	hong yi	2016
0971	后沉淀	hou chen dian	1806
0972	π一弧度磁场检测器	π—hu du ci chang jian ce qi	1763
0973	π/2一弧度磁场检测器	π/2—hu du ci chang jian ce qi	1764
0974	π/3一弧度磁场检测器	π/3—hu du ci chang jian ce qi	1765
0975	弧光灯	hu guang deng	0135
0976	弧坑深度	hu keng shen du	0519
0977	弧坑形状	hu keng xing zhuang	0521
0978	弧坑直径	hu keng zhi jing	0520
0979	弧隙	hu xi	0134
0980	(电)弧线	(dian) hu xian	0137
0981	化学产率[额]	hua xue chan lu [e]	0358
0982	化学当量	hua xue dang liang	0352
0983	化学电离	hua xue dian li	0355,0359
0984	化学发光	hua xue fa guang	0360
0985	化学发光分析	hua xue fa guang fen xi	0361
0986	化学发光指示剂	hua xue fa guang zhi shi ji	0362
0987	化学分析电子能谱法(ESCA)	hua xue fen xi dian zi neng pu fa	0767
0988	化学干扰	hua xue gan rao	0354
0989	化学光量计	hua xue guang liang ji	0348

0990	化学火焰	hua xue huo yan	0353
0991	化学计量(法,学)	hua xue ji liang (fa, xue)	2353
0992	化学剂量计	hua xue ji liang ji	0350
0993	化学计量终点	hua xue ji liang zhong dian	2352
0994	化学老化	hua xue lao hua	0349
0995	化学平衡	hua xue ping heng	0351
0996	化学位移	hua xue wei yi	0357
0997	化学吸附	hua xue xi fu	0363
0998	环阳[正]极	huan yang [zheng] ji	0111
0999	缓冲	huan chong	0272
1000	缓冲电弧	huan chong dian hu	0275
1001	缓冲剂[垫]	huan chong ji [dian]	0271
1002	缓冲剂加入法	huan chong ji jia ru fa	0273
1003	缓冲容量[能力]	huan chong rong liang [neng li]	0274
1004	缓冲指数	huan chong zhi shu	0276
1005	缓发荧光	huan fa ying guang	0577
1006	缓发中子	huan fa zhong zi	0578
1007	缓和[减速]剂	huan he [jian su] ji	1478
1008	挥发(作用)	hui fa (zuo yong)	2567
1009	挥发速率	hui fa su lu	2568
1010	挥发性改进剂	hui fa xing gai jin ji	2569
1011	辉光放电	hui guang fang dian	0995
1012	辉纹	hui wen	2358
1013	回流雾化器	hui liu wu hua qi	2031
1014	回收百分率	hui shou bai fen lu	1700
1015	回收率	hui shou lu	2007
1016	(已知分析物量)回收试验	(yi zhi fen xi wu liang) hui shou shi yan	2009
1017	回收系数	hui shou xi shu	2008
1018	回旋共振质谱计	hui xuan gong zhen zhi pu ji	0555
1019	回旋积分线性扫描伏安法	hui xuan ji fen xian xing sao miao fu an fa	0486
1020	回旋加速器	hui xuan jia su qi	0554
1021	混合金属配合物	hun he jin shu pei he wu	1473
1022	混合配(体络)合物,杂配物	hun he pei (ti lou) he wu, za pei wu	1469,1472
1023	混合溶液法	hun he rong ye fa	1474
1024	混合指示剂	hun he zhi shi ji	1471
1025	混晶	hun jing	1470
1026	活度	huo du	0037
1027	活度浓度	huo du nong du	0039
1028	活度系数	huo du xi shu	0038
1029	活化(作用)	huo hua (zuo yong)	0032
1030	活化分析	huo hua fen xi	0033
1031	活化截面	huo hua jie mian	0034
1032	活性介质	huo xing jie zhi	0035
1033	(放射性核素)活性增长曲线	(fang she xing he su) huo xing zheng chang qu xian	1007
1034	火花重复率	huo hua chong fu lu	2253
1035	火花点火	huo hua dian huo	2250
1036	火花发生器	huo hua fa sheng qi	2249
1037	火花放电	huo hua fang dian	2246
1038	火花交叉激发	huo hua jiao cha ji fa	2245
1039	火花频率	huo hua pin lu	2247

1040	火花平衡	huo hua ping heng	2251
1041	火花室[腔]	huo hua shi [qiang]	2244
1042	火花源电离	huo hua yuan dian li	2254
1043	(发)火[电]花载体[架,台]	(fa) huo [dian] hua zai ti [jia,tai]	2256
1044	火焰	huo yan	0895
1045	火焰本底	huo yan ben di	0896
1046	火焰光谱法	huo yan guang pu fa	0898
1047	火焰几何形干扰	huo yan ji he xing gan rao	0897
1048	火焰温度	huo yan wen du	0899
1049	击穿电压	ji chuan dian ya	0263
1050	击穿时间	ji chuan shi jian	0262
1051	击穿跳动	ji chuan tiao dong	0261
1052	机械化	ji xie hua	1422,1423
1053	机械夹带	ji xie jia dai	1419
1054	机械闸,开闭器	ji xie zha, kai bi qi	1420
1055	机制[构,理]	ji zhi [gou, li]	1421
1056	积分标准	ji fen biao zhun	1127
1057	积分反射系数	ji fen fan she xi shu	1126
1058	积分球	ji fen qiu	1128
1059	积分色谱图	ji fen se pu tu	1124
1060	积分吸收	ji fen xi shou	1123
1061	积分型检测器	ji fen xing jian ce qi	1125
1062	基本参比标准	ji ben can bi biao zhun	1851
1063	基本粒子	ji ben li zi	0777
1064	基本 pH 标准	ji ben pH biao zhun	1849
1065	基本气体	ji ben qi ti	1856
1066	基底电解质	ji di dian jie zhi	0208
1067	基峰	ji feng	0211
1068	基态	ji tai	1006
1069	基体[质,块]	ji ti [zhi, kuai]	1400
1070	基体改性物	ji ti gai xing wu	1403
1071	基体效应	ji ti xiao ying	1401
1072	基线	ji xian	0209
1073	基质[主]材料	ji zhi [zhu] cai liao	1061
1074	基(本标)准	ji (ben biao) zhun	1853
1075	基准[参比]光束	ji zhun [can bi] guang shu	2017
1076	基(本标)准溶液	ji (ben biao) zhun rong ye	1854
1077	基准物(质)	ji zhun wu (zhi)	1855
1078	激发	ji fa	0828
1079	激发单色器	ji fa dan se qi	0835
1080	激发电位[势]	ji fa dian wei [shi]	0836
1081	激发一发射光谱	ji fa—fa she guang pu	0829
1082	激发干扰	ji fa gan rao	0831
1083	激发光谱	ji fa guang pu	0839
1084	激发能级	ji fa neng ji	0832
1085	激发[厉]机制[构]	ji fa [li] ji zhi [gou]	0833
1086	激发磷光光谱	ji fa lin guang guang pu	0842
1087	激发模式	ji fa mo shi	0834
1088	激发能	ji fa neng	0830
1089	激发能量	ji fa neng liang	0844

1090	激发区域	ji fa qu yu	0837
1091	激发[受激]态	ji fa [shou ji] tai	0843
1092	激发温度	ji fa wen du	0840
1093	激发荧光光谱	ji fa ying guang guang pu	0841
1094	激发源	ji fa yuan	0838
1095	激光(器)	ji guang (qi)	1256
1096	激光分析	ji guang fen xi	1259
1097	激光腐蚀	ji guang fu shi	1264
1098	激光活性物质	ji guang huo xing wu zhi	1258
1099	激光尖峰	ji guang jian feng	1279
1100	激光镜	ji guang jing	1269
1101	激光卷流	ji guang juan liu	1272
1102	激光喇曼(Raman)微量分析(LRMA)	ji guang la man wei liang fen xi	1275
1103	激光脉冲	ji guang mai chong	1274
1104	激光区域分析	ji guang qu yu fen xi	1265
1105	激光射程	ji guang she cheng	1277
1106	激光生成汽雾	ji guang sheng cheng qi wu	1273
1107	激光输出	ji guang shu chu	1270
1108	激光输出能量	ji guang shu chu neng liang	1271
1109	激光束	ji guang shu	1262
1110	激光束电离	ji guang shu dian li	1263
1111	激光微区发射光谱法(LAMES)	ji guang wei qu fa she guang pu fa	1266
1112	激光微区质量分析法(LAMMS)	ji guang wei qu zhi liang fen xi fa	1267
1113	激光微探针分析	ji guang wei tan zhen fen xi	1268
1114	激光原子化器	ji guang yuan zi hua qi	1261
1115	激光谐振器	ji guang xie zhen qi	1276
1116	激光阈值	ji guang yu zhi	1280
1117	激光源	ji guang yuan	1278
1118	激光原子化与激发	ji guang yuan zi hua yu ji fa	1260
1119	激光作用	ji guang zuo yong	1257
1120	激态原子[分子]发光	ji tai yuan zi [fen zi] fa guang	0827
1121	极化作用	ji hua zuo yong	1789
1122	极谱滴定(法)	ji pu di ding (fa)	1793
1123	极谱法	ji pu fa	1792
1124	极限催化电流	ji xian cui hua dian liu	1292
1125	极限电流	ji xian dian liu	1293
1126	极限动力电流	ji xian dong li dian liu	1295
1127	极限焦点直径	ji xian jiao dian zhi jing	2533
1128	极限扩散电流	ji xian kuo san dian liu	1294
1129	极限迁移电流	ji xian qian yi dian liu	1296
1130	极限吸附电流	ji xian xi fu dian liu	1291
1131	极性	ji xing	1788
1132	急[骤]冷气	ji [zhou] leng qi	1910
1133	几何量	ji he liang	0986
1134	几何衰减	ji he shuai jian	0987
1135	几何因子[数]	ji he yin zi [shu]	0988
1136	计时安培分析法,计时电流法	ji shi an pei fen xi fa, ji shi dian liu fa	0369
1137	计时电位常数	ji shi dian wei chang shu	0372
1138	计时电位滴定法	ji shi dian wei di ding fa	0374
1139	计时电位分析法	ji shi dian wei fen xi fa	0375

1140	计时电位(分析)终点	ji shi dian wei (fen xi) zhong dian	0373
1141	计时库仑常数	ji shi ku lun chang shu	0370
1142	计时库仑分析法	ji shi ku lun fen xi fa	0371
1143	计数[算]标准偏差	ji shu [suan] biao zhun pian cha	2323
1144	计数管	ji shu guan	0508
1145	计数几何条件	ji shu ji he tiao jian	0510
1146	计数精度	ji shu jing du	0512
1147	计数率	ji shu lu	0513,0514
1148	计数损失	ji shu sun shi	0511
1149	计数效率	ji shu xiao lu	0509
1150	计算[数]	ji suan [shu]	0505
1151	记录,资料	ji lu, zi liao	2006
1152	剂[药]量	ji [yao] liang	0689
1153	剂量当量	ji liang dang liang	0690
1154	加和性	jia he xing	0041
1155	加热速率	jia re su lu	1023
1156	(化合,原子)价电子	(hua he, yuan zi) jia dian zi	2551
1157	鉴别[频,相]器	jian bie [pin, xiang] qi	0664
1158	检测查定	jian ce cha ding	0603
1159	检测功率[能力]	jian ce gong lu [neng li]	1818
1160	检测计的灵敏区	jian ce ji de ling min qu	2175
1161	检测[波]器	jian ce [bo] qi	0606
1162	检测器分辨率	jian ce qi fen bian lu	0609
1163	检测器空载时间	jian ce qi kong zai shi jian	0607
1164	检测器内禀效率	jian ce qi nei bing xiao lu	1172
1165	检测器效率	jian ce qi xiao lu	0608
1166	检测(极)限	jian ce (ji) xian	0605,1297
1167	检测效率	jian ce xiao lu	0604
1168	检定参考物质	jian ding can kao wu zhi	0329
1169	碱	jian	0207
1170	碱度[性]	jian du [xing]	0212
1171	碱化	jian hua	0058
1172	碱量滴定(法)	jian liang di ding (fa)	0056,0057
1173	(激光)尖峰功率	(ji guang) jian feng gong lu	2300
1174	减活化(能力)方法	jian huo hua (neng li) fang fa	0562
1175	减色移(动)	jian se yi (dong)	1069
1176	间接滴定	jian jie di ding	1095
1177	间隙	jian xi	0964
1178	间隙电阻器	jian xi dian zu qi	0965
1179	间歇电弧	jian xie dian hu	1164
1180	间歇方法[程序]	jian xie fang fa [cheng xu]	0662
1181	间歇同时测定技术	jian xie tong shi ce ding ji shu	0663
1182	溅射	jian she	2310
1183	溅射产额	jian she chan e	2312
1184	溅射电离	jian she dian li	1202
1185	溅射(速)率	jian she (su) lu	2311
1186	交变电压	jiao bian dian ya	0068
1187	交变电压极谱法	jiao bian dian ya ji pu fa	0071
1188	交变电压计时电位分析法	jiao bian dian ya ji shi dian wei fen xi fa	0070
1189	交变电压振幅	jiao bian dian ya zhen fu	0069

1190	交流电弧	jiao liu dian hu	0065
1191	交流极谱法	jiao liu ji pu fa	0067
1192	交流计时电位分析法	jiao liu ji shi dian wei fen xi fa	0066
1193	交流振幅	jiao liu zhen fu	0064
1194	胶片剂量计	jiao pian ji liang ji	0882
1195	焦班[点]	jiao ban [dian]	0927
1196	焦班直径	jiao ban zhi jing	0928
1197	焦点长度,焦距	jiao dian chang du, jiao ju	0926
1198	角喷雾器	jiao pen wu qi	0102
1199	角色散	jiao se san	0101
1200	校正保留时间	jiao zheng bao liu shi jian	0499
1201	校正保留体积	jiao zheng bao liu ti ji	0500
1202	校正发光量子产额	jiao zheng fa guang liang zi chan e	0498
1203	校正发射光谱	jiao zheng fa she guang pu	0493
1204	校正激发光谱	jiao zheng ji fa guang pu	0494
1205	校正冷光发射偏振光谱	jiao zheng leng guang fa she pian zhen guang pu	0496
1206	校正冷光激发偏振光谱	jiao zheng leng guang ji fa pian zhen guang pu	0497
1207	校正磷光衰变时间	jiao zheng lin guang shuai bian shi jian	0492
1208	校正强度	jiao zheng qiang du	0495
1209	校正[准]曲线	jiao zheng [zhun] qu xian	0287
1210	校正荧光衰变时间	jiao zheng ying guang shuai bian shi jian	0491
1211	解(掩)蔽	jie (yan) bi	0580
1212	解调极谱法	jie tiao ji pu fa	0582
1213	介电测试法	jie dian ce shi fa	0616
1214	介电常数	jie dian chang shu	0614
1215	介电法滴定	jie dian fa di ding	0615
1216	介质[体],媒质	jie zhi [ti], mei zhi	1424
1217	结果	jie guo	2075
1218	结合能	jie he neng	0229
1219	截留杂质	jie liu za zhi	0453
1220	截面	jie mian	0525
1221	界面	jie mian	1143
1222	阶梯极谱法	jie ti ji pu fa	2320
1223	阶梯扇	jie ti shan	2347
1224	金箔探测器	jin bo tan ce qi	0932
1225	金属一离子配合物的稳定常数	jin shu—li zi pei he wu de wen ding chang shu	2316
1226	金属(灯)丝原子化器	jin shu (deng) si yuan zi hua qi	1435
1227	金属荧光指示剂	jin shu ying guang zhi shi ji	1437
1228	金属蒸气电弧	jin shu zheng qi dian hu	1438
1229	金属蒸气弧光灯	jin shu zheng qi hu guang deng	1439
1230	金属置换沉淀	jin shu zhi huan chen dian	0325
1231	金属(显色)指示剂	jin shu (xian se) zhi shi ji	1436
1232	近似单色光度法	jin si dan se guang du fa	1909
1233	进样间隔[循环]	jin yang jian ge [xun huan]	2105
1234	进样器	jin yang qi	2104
1235	精度	jing du	1828
1236	精密零点电位法	jing mi ling dian dian wei fa	1829
1237	(结)晶格(子)	(jie) jing ge (zi)	0531
1238	晶态离子选择电极	jing tai li zi xuan ze dian ji	0532
1239	晶体控制振荡器	jing ti kong zhi zhen dang qi	0528

1240	晶体特性	jing ti te xing	0527
1241	晶体衍射	jing ti yan she	0529
1242	晶体衍射分光计	jing ti yan she fen guang ji	0530
1243	净保留时间	jing bao liu shi jian	1551
1244	净保留体积	jing bao liu ti ji	1552
1245	净法拉弟电流	jing fa la di dian liu	1550
1246	净化，提纯	jing hua, ti chun	0380
1247	镜象组态	jing xiang zu tai	1467
1248	静电场质谱计	jing dian chang zhi pu ji	2337
1249	静态电极伏安法	jing tai dian ji fu an fa	2338
1250	久期平衡	jiu qi ping heng	2153
1251	居里[放射性强度单位＝3.7x10^{10}次衰变/秒]	ju li	0540
1252	局部挥发百分率	ju bu hui fa bai fen lu	1348
1253	局部气体温度	ju bu qi ti wen du	1349
1254	局部去溶剂化百分率	ju bu qu rong ji hua bai fen lu	1347
1255	局部热平衡	ju bu re ping heng	1350
1256	局部雾化百分率	ju bu wu hua bai fen lu	1346
1257	局部雾[原子]化效率	ju bu wu [yuan zi] hua xiao lu	1345
1258	矩阵匹配	ju zhen pi pei	1402
1259	(核)聚变	ju bian	0956
1260	聚集(作用)	ju ji (zuo yong)	0053
1261	聚集体	ju ji ti	0052
1262	聚焦反射镜	ju jiao fan she jing	0930
1263	聚焦(透)镜	ju jiao (tou) jing	0929
1264	巨脉冲	ju mai chong	0991
1265	绝对光峰效率	jue dui guang feng xiao lu	0004
1266	绝对计数	jue dui ji shu	0002
1267	绝对清洁[纯化]工作台	jue dui qing jie [chun hua] gong zuo tai	0377
1268	绝对清洁面	jue dui qing jie mian	0379
1269	绝对清洁室	jue dui qing jie shi	0378
1270	绝对全能量峰效率	jue dui quan neng liang feng xiao lu	0003
1271	绝对温度	jue dui wen du	0005
1272	均相膜电极	jun xiang mo dian ji	1057
1273	均相溶液中的沉淀作用	jun xiang rong ye zhong de chen dian zuo yong	1825
1274	均匀磁场	jun yun ci chang	1056
1275	均匀分配系数	jun yun fen pei xi shu	1055
1276	卡尔韦特(Calvet)型装置	ka er wei te xing zhuang zhi	0289
1277	卡罗塞克(Kalousek)极谱法	ka luo sai ke ji pu fa	1245
1278	开端空心阴极	kai duan kong xin yin ji	1619
1279	Q(光亮)开关	Q (guang liang) kai guan	1892
1280	Q 开关操作	Q kai guan cao zuo	1893
1281	开口管色谱法	kai kou guan se pu fa	1621
1282	康普顿(Compton)电子	kang pu dun dian zi	0429
1283	康普顿角	kang pu dun jiao	0427
1284	康普顿散射	kang pu dun san she	0430
1285	康普顿效应	kang pu dun xiao ying	0428
1286	抗衡[相反]离子	kang heng [xiang fan] li zi	0507
1287	科泽尔(Kossel)法	ke ze er fa	1249
1288	可靠的支撑体	ke kao de zhi cheng ti	2226

1289	可活化材料	ke huo hua cai liao	0031
1290	可控硅控制	ke kong gui kong zhi	2466
1291	可裂变的	ke lie bian de	0888
1292	可逆线性分散	ke ni xian xing fen san	2004
1293	可漂白(变色)物质	ke piao bai (bian se) wu zhi	0244
1294	可调整火花	ke tiao zheng huo hua	2212
1295	可转换材料	ke zhuan huan cai liao	0876
1296	可转换核素	ke zhuan huan he su	0877
1297	刻度扩展	ke du kuo zhan	2118
1298	刻度[分离,分割]值	ke du [fen li, fen ge] zhi	2552
1299	克尔(Kerr)盒	ke er he	1246
1300	克鲁克斯(Crookes)暗区	ke lu ke si an qu	0523
1301	克式浓度	ke shi nong du	0933
1302	(激光放射)坑,放电痕,喷口	(ji guang fang she) keng, fang dian hen, pen kou	0518
1303	空白	kong bai	0235
1304	空白本底	kong bai ben di	0236
1305	空白测量	kong bai ce liang	0238
1306	空白滴定	kong bai di ding	0241
1307	空白校正	kong bai jiao zheng	0237
1308	空白溶液	kong bai rong ye	0240
1309	空白散射[扩散]	kong bai san she [kuo san]	0239
1310	空间电荷	kong jian dian he	2242
1311	空间分辨[判定]技术	kong jian fen bian [pan ding] ji	2243
1312	空间分辨率	kong jian fen bian lu	2258
1313	空间分布干扰	kong jian fen bu gan rao	2257
1314	空间稳定	kong jian wen ding	2259
1315	空气峰	kong qi feng	0055
1316	空心电极	kong xin dian ji	1054
1317	空心阴极灯	kong xin yin ji deng	1052
1318	空心阴极放电	kong xin yin ji fang dian	1051
1319	空心阴极源	kong xin yin ji yuan	1053
1320	空载显示	kong zai xian shi	1567
1321	孔径光阑直径	kong jing guang lan zhi jing	0123
1322	控制波形电花	kong zhi bo xing dian hua	0478
1323	控制电流电位[势]滴定法	kong zhi dian liu dian wei [shi] di ding fa	0468
1324	控制电流电位[势]分析法	kong zhi dian liu dian wei [shi] fen xi fa	0469
1325	控制电流库仑法	kong zhi dian liu ku lun fa	0467
1326	控制电势[位]电分离	kong zhi dian shi [wei] dian fen li	0476, 0773
1327	控制电势电重量分析法	kong zhi dian shi dian zhong liang fen xi fa	0475
1328	控制电势极谱法	kong zhi dian shi ji pu fa	0477
1329	控制电势库仑滴定法	kong zhi dian shi ku lun di ding fa	0473
1330	控制电势库仑法	kong zhi dian shi ku lun fa	0474
1331	控制高压火花发生器	kong zhi gao ya huo hua fa sheng qi	0472
1332	控制火花隙	kong zhi huo hua xi	0479
1333	控制火花时间	kong zhi huo hua shi jian	0470
1334	控制间隙	kong zhi jian xi	0464
1335	控制交流电弧	kong zhi jiao liu dian hu	0465
1336	控制流速雾化器	kong zhi liu su wu hua qi	0471
1337	控制气氛	kong zhi qi fen	0466
1338	库仑滴定	ku lun di ding	0503

1339	库仑分析法	ku lun fen xi fa	0504
1340	块电路	kuai dian lu	1377
1341	快速扫描[飞行时间]质谱仪	kuai su sao miao [fei xing shi jian] zhi pu yi	2469
1342	快原子轰击(FAB)	kuai yuan zi hong ji	0873
1343	快中子	kuai zhong zi	0874
1344	宽度	kuan du	2593
1345	扩散,漫射	kuo san, man she	0637
1346	扩散电流	kuo san dian liu	0639
1347	扩散电流常数	kuo san dian liu chang shu	0640
1348	扩散接界半导体检测器	kuo san jie jie ban dao ti jian ce qi	0636
1349	扩散系数	kuo san xi shu	0638
1350	拉德(rad)[=100 尔格/克]	la de	1913
1351	喇曼(Raman)辐射	la man fu she	1984
1352	喇曼光谱学	la man guang pu xue	1986
1353	喇曼散射	la man san she	1985
1354	蓝移	lan yi	0247
1355	老化,陈化	lao hua, chen hua	0054
1356	雷利(Rayleigh) 散射	lei li san she	1993
1357	雷姆	lei mu	2047
1358	累积分布	lei ji fen bu	0533
1359	累积裂变[裂解]产额	lei ji lie bian [lie jie] chan e	0534
1360	累积生成常数	lei ji sheng cheng chang shu	0535
1361	累积稳定常数	lei ji wen ding chang shu	0537
1362	累积质子化常数	lei ji zhi zi hua chang shu	0536
1363	累积总数	lei ji zong shu	0538
1364	类弧光[电弧]放电	lei hu guang [dian hu] fang dian	0136
1365	类火花(样)放电	lei huo hua (yang) fang dian	2252
1366	冷空心阴极	leng kong xin yin ji	0393
1367	冷空心阴极灯	leng kong xin yin ji deng	0487
1368	冷却曲线	leng que qu xian	0488
1369	冷却速率曲线	leng que su lu qu xian	0489
1370	冷中子	leng zhong zi	0394
1371	离合[开闭]比	li he [kai bi] bi	1618
1372	离解常数	li jie chang shu	0672
1373	离解电位	li jie dian wei	0674
1374	离解度	li jie du	0573
1375	离解干扰	li jie gan rao	0673
1376	离解作用	li jie zuo yong	0671
1377	离析气体分析法(EGA)	li xi qi ti fen xi fa	0825
1378	离析气体检测法(EGD)	li xi qi ti jian ce fa	0826
1379	离子	li zi	1182
1380	离子一分子反应	li zi—fen zi fan ying	1216
1381	离子光谱	li zi guang pu	1196
1382	离子光[波]谱线	li zi guang [bo] pu xian	1197
1383	离子轨道长度	li zi gui dao chang du	1195
1384	离子轰击	li zi hong ji	1183
1385	离子缓冲	li zi huan chong	1201
1386	离子回旋共振质谱计	li zi hui xuan gong zhen zhi pu ji	1186
1387	离子活度系数	li zi huo du xi shu	1192
1388	离子交换(作用)	li zi jiao huan (zuo yong)	1187

1389	离子交换等温线	li zi jiao huan deng wen xian	1189
1390	离子交换剂	li zi jiao huan ji	1191
1391	离子交换剂的体积溶胀比率	li zi jiao huan ji de ti ji rong zhang bi lu	2580
1392	离子交换膜	li zi jiao huan mo	1190
1393	离子交换色谱法	li zi jiao huan se pu fa	1188
1394	离子介质	li zi jie zhi	1194
1395	离子(电)流	li zi (dian) liu	1185
1396	离子强度	li zi qiang du	1198
1397	离子强度调节缓冲剂	li zi qiang du tiao jie huan chong ji	1199
1398	离子散射能谱分析法(ISS)	li zi san she neng pu fen xi fa	1219
1399	离子色谱(法)	li zi se pu (fa)	1184
1400	离子探针微区分析(IPMA)	li zi tan zhen wei qu fen xi	1218
1401	离子特效性[选择性]电极	li zi te xiao xing [xuan ze xing] dian ji	1222
1402	离子线	li zi xian	1193
1403	离子显微镜检查法,离子显微术	li zi xian wei jing jian cha fa, li zi xian wei shu	1215
1404	离子选择(性)电极	li zi xuan ze (xing) dian ji	1220
1405	离子源	li zi yuan	1221
1406	离子源的灵敏度	li zi yuan de ling min du	2180
1407	理论(塔)板数目	li lun (ta) ban shu mu	2411
1408	理论保留体积	li lun bao liu ti ji	2412
1409	理论交换容量[能力]	li lun ji huan rong liang [neng li]	2413
1410	理论终点	li lun zhong dia	2410
1411	α粒(子)	α li (zi)	0063
1412	粒度分析	li du fen xi	1676
1413	β粒子	β li zi	0220
1414	粒子大小[细度]	li zi da xiao [xi du]	1675
1415	粒子大小分布	li zi da xiao fen bu	1677
1416	粒子辐射	li zi fu she	1674
1417	粒子轰击	li zi hong ji	1668
1418	粒子激发[诱导]X射线发射光谱学(PIXES)	li zi ji fa [you dao] X she xian fa she guang pu xue	1673,1767
1419	粒子检测器	li zi jian ce qi	1669
1420	粒子数反转,布居反转	li zi shu fan zhuan, bu ju fan zhuan	1798
1421	粒子通量	li zi tong liang	1670
1422	粒子通量密度	li zi tong liang mi du	1671
1423	粒子吸收	li zi xi shou	0017
1424	粒子湮没	li zi yan mo	1667
1425	粒子诱导[激发]X射线发射光谱分析法	li zi you dao [ji fa] X she xian fa she guang pu fen xi fa	1672
1426	立体角	li ti jiao	2219
1427	连式裂变产额	lian shi lie bian chan e	0330
1428	连续波(CW)	lian xu bo	0460
1429	连续波(CW)操作	lian xu bo cao zuo	0461
1430	连续法	lian xu fa	0457
1431	连续辐射	lian xu fu she	0458
1432	连续光谱	lian xu guang pu	0462
1433	连续进料[供给]	lian xu jin liao [gong gei]	0459
1434	连续流动分析	lian xu liu dong fen xi	0456
1435	量	liang	1900
1436	量焓滴定	liang han di ding	0811
1437	量热法	liang re fa	0288

1438	两性溶剂	liang xing rong ji	0076
1439	γ量子	γ liang zi	0958
1440	量子产额[率]	liang zi chan e [lu]	1904
1441	量子计数器	liang zi ji shu qi	1901
1442	裂变,裂解[质谱]	lie bian, lie jie [zhi pu]	0887
1443	裂变产额	lie bian chan e	0892
1444	裂变产物	lie bian chan wu	0891
1445	(易)裂变的(物质)	(yi) lie bian de (wu zhi)	0885
1446	裂变核素	lie bian he su	0886
1447	裂变碎片	lie bian sui pian	0889
1448	裂变中子	lie bian zhong zi	0890
1449	β一裂解	β—lie jie	0218
1450	临界阻尼放电	lin jie zu ni fang dian	0522
1451	磷光(现象)	lin guang (xian xiang)	1718
1452	磷光分光计	lin guang fen guang ji	1721
1453	磷光分析	lin guang fen xi	1719
1454	磷光光度法	lin guang guang du fa	1722
1455	磷光[镜]计,磷光测定器	lin guang [jing] ji, lin guang ce ding qi	1723
1456	磷光激发发射光谱	lin guang ji fa fa she guang pu	1720
1457	磷光增强分析	lin guang zeng qiang fen xi	0806
1458	灵敏度[性]	ling min du [xing]	2176
1459	菱镜基体	ling jing ji ti	1857
1460	流,电流	liu, dian liu	0541
1461	流出[排流]速度	liu chu [pai liu] su du	1648
1462	流动相	liu dong xiang	1476
1463	流动载体电极	liu dong zai ti dian ji	1475
1464	流动注射分析(FIA)	liu dong zhu she fen xi	0905
1465	流率[量,速],比移值	liu lu [liang, su], bi yi zhi	0907
1466	流体动力(学)伏安法	liu ti dong li (xue) fu an fa	1066
1467	漏过容量[能力]	lou guo rong liang [neng li]	0264
1468	漏过体积	lou guo ti ji	0265
1469	卢瑟福(Rutherford)反散射(RBS)	lu se fu fan san she	2095
1470	滤出	lu chu	0884
1471	滤光片[器]	lu guang pian [qi]	1630
1472	(辐射)滤器	(fu she) lu qi	0883
1473	伦琴(Roentgen)(单位)	lun qin (dan wei)	2089
1474	洛伦兹(Lorentz)变宽	luo lun zi bian kuan	1356
1475	洛马金一谢贝(Lomakin—Scheibe)方程	luo ma jin—xie bei fang cheng	1354
1476	麻痹[间歇]电路	ma bi [jian xie] dian lu	1662
1477	马陶克一赫洛格(Mattauch—Herzog)几何	ma tao ke—he luo zi ge ji he	1404
1478	麦克斯韦一波兹曼(Maxwell—Boltzmann)定律	mai ke si wei—bo zi man ding lu	1405
1479	麦克拉佛特(McLafferty)重排	mai ke la fo te zhong pai	1406
1480	脉冲放大器	mai chong fang da qi	1874
1481	脉冲放电灯[管]	mai chong fang dian deng [guan]	1878
1482	脉冲幅度分布	mai chong fu du fen bu	1876
1483	脉冲幅度分析器	mai chong fu du fen xi qi	1875
1484	脉冲辐度选择器	mai chong fu du xuan ze qi	1877
1485	脉冲高度分析器	mai chong gao du fen xi qi	1882
1486	脉冲功率	mai chong gong lu	1884
1487	脉冲激光(器电)源	mai chong ji guang (qi dian) yuan	1879

1488	脉冲极谱(法)	mai chong ji pu (fa)	1883
1489	脉冲宽度,脉冲持续时间	mai chong kuan du, mai chong chi xu shi jian	1880
1490	脉冲能量	mai chong neng liang	1881
1491	漫射[杂散]光	man she [za san] guang	2356
1492	慢[低速]中子	man [di su] zhong zi	2217
1493	毛细管	mao xi guan	0298
1494	毛细管电弧	mao xi guan dian hu	0296
1495	毛细管电极	mao xi guan dian ji	0297
1496	(空心)毛细管柱	(kong xin) mao xi guan zhu	1622
1497	酶(底物)电极	mei (di wu) dian ji	0814
1498	梅(Mie)散射[分散]	mei san she [fen san]	1460
1499	免疫测定法	mian yi ce ding fa	1080
1500	敏化发光	min hua fa guang	2181
1501	摩尔(克分子)电离能	mo er (ke fen zi) dian li neng	1492
1502	(重量)摩尔(浓度)活度	(zhong liang) mo er (nong du) huo du	1486
1503	(体积)摩尔(浓度)活度	(ti ji) mo er (nong du) huo du	1490
1504	(体积)摩尔活度系数	(ti ji) mo er huo du xi shu	1491
1505	摩尔活度系数	mo er huo du xi shu	1487
1506	(体积)摩尔浓度	(ti ji) mo er nong du	1493
1507	摩尔吸光系数	mo er xi guang xi shu	1489
1508	(隔)膜	(ge) mo	1430
1509	模拟[仿真]技术	mo ni [fang zhen] ji shu	2199
1510	模拟数字变换器(ADC)	mo ni shu zi bian huan qi	0078
1511	模式	mo shi	1479
1512	母离子	mu li zi	1664
1513	母体离子	mu ti li zi	1835
1514	目视指示剂	mu shi zhi shi ji	2564
1515	目视终点	mu shi zhong dian	2563
1516	穆斯鲍尔(Mössbauer)效应	mu si bao er xiao ying	1517
1517	纳[纤,毫微]痕量	na [xian,hao wei] hen liang	1527
1518	内标,内标物[准]	nei biao, nei biao wu [zhun]	1163
1519	内禀效率	nei bing xiao lu	1173
1520	内参比线	nei can bi xian	1162
1521	内充溶液	nei chong rong ye	1158
1522	内[初级燃烧]带[区]	nei [chu ji ran shao] dai [qu]	1113
1523	内电解(法)	nei dian jie (fa)	1157
1524	内电重量分析法	nei dian zhong liang fen xi fa	1156
1525	内轫致辐射	nei ren zhi fu she	1112
1526	内吸收[光]度	nei xi shou [guang] du	1153
1527	内消旋	nei xiao xuan	1432
1528	内振荡振幅	nei zhen dang zhen fu	1160
1529	内转换[化]	nei zhuan huan [hua]	1154
1530	内转换系数	nei zhuan huan xi shu	1155
1531	内锥焰区	nei zhui yan qu	1141
1532	能级	neng ji	0799
1533	能量标度	neng liang biao du	0801
1534	能量分辨(率)	neng liang fen bian (lu)	0800
1535	能量分散	neng liang fen san	0796
1536	能量分散 X—射线荧光分析	neng liang fen san X—she xian ying guang fen xi	0797
1537	能量阈值	neng liang yu zhi	0802

1538	能量跃迁[转变]	neng liang yue qian [zhuan bian]	0804
1539	能量转移	neng liang zhuan yi	0803
1540	能斯脱(Nernst)方程	neng si tuo fang cheng	1548
1541	能斯脱响应	neng si tuo xiang ying	1549
1542	能通量密度	neng tong liang mi du	0798
1543	尼尔—约翰逊(Nier—Johnson)几何构造	ni er—yue han xun ji he gou zao	1565
1544	拟出程序并按程序工作	ni chu cheng xu bing an cheng xu gong zuo	1862
1545	逆冷却曲线(速)率	ni leng que qu xian (su) lu	1176
1546	逆向电流计时电位分析法	ni xiang dian liu ji shi dian wei fen xi fa	0545
1547	逆向注射燃烧器	ni xiang zhu she ran shao qi	2083
1548	凝胶过滤	ning jiao guo lu	0984
1549	凝胶渗透色谱(法)	ning jiao shen tou se pu (fa)	0985
1550	凝结[聚],絮[胶]凝	ning jie [ju], xu [jiao] ning	0382
1551	扭辫分析	niu bian fen xi	2481
1552	浓度	nong du	0431
1553	浓度比	nong du bi	0436
1554	浓度常数	nong du chang shu	0432
1555	浓度[萃取]分配比	nong du [cui qu] fen pei bi	0433
1556	浓度平衡常数	nong du ping heng chang shu	0434
1557	浓度指数	nong du zhi shu	0435
1558	钕玻璃	nu bo li	1544
1559	耦合器	ou he qi	0517
1560	耦合效率	ou he xiao lu	0516
1561	偶联同时测定技术	ou lian tong shi ce ding ji shu	0515
1562	旁通注射[进样]器	pang tong zhu she [jin yang] qi	0286
1563	配[络]合(作用)	pei [luo] he (zuo yong)	0420
1564	配合[络合]滴定	pei he [luo he] di ding	0422
1565	配合[络合]滴定法	pei he [luo he] di ding fa	0423
1566	配合生成平衡	pei he sheng cheng ping heng	0421
1567	配[络]阴离子	pei [luo] yin li zi	0419
1568	配位(体)数	pei wei (ti) shu	1288
1569	配位体	pei wei ti	1286
1570	配位酮[氨羧配位[络合]剂]	pei wei tong [an suo pei wei [luo he] ji]	0418,0424
1571	喷[排,挤]出能(量)	pen [pai, ji] chu neng (liang)	0730
1572	喷气限制孔	pen qi xian zhi kong	0975
1573	喷雾放电	pen wu fang dian	2308
1574	喷雾[洒]器,喷头[嘴]	pen wu [sa] qi, pen tou [zui]	2309
1575	喷雾室	pen wu shi	2307
1576	彭宁(Penning)气体混合物	peng ning qi ti hun he wu	1698
1577	膨胀计测定法	peng zhang ji ce ding fa	0642
1578	碰撞	peng zhuang	0400
1579	碰撞半宽度	peng zhuang ban kuan du	0405
1580	碰撞过程	peng zhuang guo cheng	0406
1581	碰撞截面	peng zhuang jie mian	0403
1582	碰撞链	peng zhuang lian	0402
1583	碰撞位移	peng zhuang wei yi	0407
1584	碰撞消激发	peng zhuang xiao ji fa	0404
1585	碰撞展宽	peng zhuang zhan kuan	0401
1586	pH 标准	pH biao zhun	1711
1587	pH 标准参比值	pH biao zhun can bi zhi	2026

1588	pH 操作标度	pH cao zuo biao du	1710
1589	pH 测量	pH ce liang	1709
1590	pH 滴定	pH di ding	1712
1591	pH 电极	pH dian ji	1706
1592	pH 判读	pH pan du	1708
1593	pH 实际测量	pH shi ji ce liang	1819
1594	pH 指示剂	pH zhi shi ji	1707
1595	皮克痕量	pi ke hen liang	1759
1596	匹配电容器	pi pei dian rong qi	1399
1597	偏[误]差	pian [wu] cha	0611
1598	偏压[流]	pian ya [liu]	0224
1599	偏压线性脉冲放大器	pian ya xian xing mai chong fang da qi	0225
1600	偏振化[极化]的	pian zhen hua [ji hua] de	1790
1601	漂移	piao yi	0702
1602	漂移频率	piao yi ping lu	0903
1603	品质因子	pin zhi yin zi	1897
1604	频率	pin lu	0943
1605	频率调制	pin lu tiao zhi	0945
1606	平板阴极辉光放电源	ping ban yin ji hui guang fang dian yuan	1769
1607	平衡常数	ping heng chang shu	0817
1608	平衡[休止]点	ping heng [xiu zhi] dian	2073
1609	(宜多元素)平衡[折衷]条件	(yi duo yuan su) ping heng[zhe zhong] tiao jian	0426
1610	平均	ping jun	0184
1611	平均(值),中数,中项	ping jun (zhi), zhong shu, zhong xiang	1407
1612	平均活度系数	ping jun huo du xi shu	1408
1613	平均离子活度系数	ping jun li zi huo du xi shu	1410
1614	平均流量[流速率]	ping jun liu liang [liu su lu]	0185
1615	平均配位体数	ping jun pei wei ti shu	0188
1616	平均气压	ping jun qi ya	0186
1617	平均寿命	ping jun shou ming	0187,1411
1618	平均线性范围[直线射程]	ping jun xian xing fan wei [zhi xian she cheng]	1412
1619	平均载气隙间速度	ping jun zai qi xi jian su du	1409
1620	平均质量线性范围	ping jun zhi liang xian xing fan wei	1413
1621	平面偏振化(作用)[极化]	ping mian pian zhen hua (zuo yong) [ji hua]	1770
1622	平面偏振化的	ping mian pian zhen hua de	1771
1623	平行校正	ping xing jiao zheng	0397
1624	坪,平顶[稳,坦][常指曲线]	ping, ping ding [wen, tan]	1782
1625	屏蔽	ping bi	2137
1626	屏蔽放电	ping bi fang dian	2190
1627	普克尔斯 (Pockels) 盒	pu ke er si he	1785
1628	普朗克(Planck)定律	pu lang ke ding lu	1768
1629	谱带宽度	pu dai kuan du	0205
1630	谱尾	pu wei	2394
1631	谱线轮廓函数	pu xian lun kuo han shu	1319
1632	谱线强度	pu xian qiang du	1136
1633	(光)谱线位移	(guang) pu xian wei yi	1321,2274
1634	谱线形状	pu xian xing zhuang	1320,2273
1635	谱线展宽参数	pu xian zhan kuan can shu	1318
1636	γ 谱仪	γ pu yi	0962
1637	赫尔特—德里菲尔德 (Hurter—Driffield) 曲线	he er te—de li fei er de qu xian	1065

1638	齐拉—查曼斯(Szilard—Chalmers)效应	qi la—cha man si xiao ying	2392
1639	(色谱)起点,起动点	(se pu) qi dian, qi dong dian	2333
1640	起动,出发线	qi dong, chu fa xian	2334
1641	起偏振器	qi pian zhen qi	1791
1642	起始[初]电流	qi shi [chu] dian liu	1104
1643	起始[初]温度	qi shi [chu] wen du	1105
1644	起(动)线	qi (dong) xian	2335
1645	契伦科夫(Cerenkov)辐射	qi lun ke fu fu she	0328
1646	契伦科夫检波器	qi lun ke fu jian bo qi	0326
1647	契伦科夫效应	qi lun ke fu xiao ying	0327
1648	气动雾化器	qi dong wu hua qi	1784
1649	气固色谱法	qi gu se pu fa	0976
1650	气溶胶	qi rong jiao	0049
1651	气溶胶载气	qi rong jiao zai qi	0050
1652	气体(电离)放大	qi ti (dian li) fang da	0966
1653	气体放电	qi ti fang dian	0968
1654	气体激光,气体光激射器	qi ti ji guang, qi ti guang ji she qi	0971
1655	(燃)气体温度	(ran) qi ti wen du	0978
1656	气体形态[形成]	qi ti xing tai [xing cheng]	0969
1657	气体滞留容积	qi ti zhi liu rong ji	0970
1658	气体正比检测器	qi ti zheng bi jian ce qi	0974
1659	气—液色谱法	qi—ye se pu fa	0972
1660	气相色谱法	qi xiang se pu fa	0967
1661	(蒸)汽化(作用)	(zheng) qi hua (zuo yong)	2553
1662	汽相干扰	qi xiang gan rao	2556
1663	(蒸)气云	(zheng) qi yun	2554
1664	迁移,移动	qian yi, yi dong	1461
1665	迁移电流	qian yi dian liu	1462
1666	迁移干扰	qian yi gan rao	2508
1667	前表面,前置面	qian biao mian, qian zhi mian	0948
1668	前体,产物母体	qian ti, chan wu mu ti	1834
1669	前体同位素	qian ti tong wei su	1836
1670	前置滤光器效应	qian zhi lu guang qi xiao ying	1837
1671	腔式喷雾器	qiang shi pen wu qi	0332
1672	强电流电弧	qiang dian liu dian hu	1030
1673	强度	qiang du	1129
1674	强度分路	qiang du fen lu	1130
1675	强度分路比率	qiang du fen lu bi lu	1131
1676	强度校准器	qiang du jiao zhun qi	1132
1677	强度调制光束	qiang du tiao zhi guang shu	1133
1678	强度调制空心阴极灯	qiang du tiao zhi kong xin yin ji deng	1134
1679	桥式配体配[络]合物	qiao shi pei ti pei [luo] he wu	1287
1680	切光器	qie guang qi	0364
1681	清除(的)	qing chu (de)	2128
1682	清除剂	qing chu ji	2127
1683	氢离子浓度	qing li zi nong du	1705
1684	氢气电极	qing qi dian ji	1067
1685	氰分子带	qing fen zi dai	0549
1686	球形电弧	qiu xing dian hu	0994
1687	曲线校正器[板]	qu xian jiao zheng qi [ban]	0548

1688	取样杯	qu yang bei	2110
1689	取样环	qu yang huan	2111
1690	取样舟	qu yang zhou	2109
1691	去极化剂,起偏振镜	qu ji hua ji, qi pian zhen jing	0584
1692	去溶剂化(作用)	qu rong ji hua (zuo yong)	0602
1693	去溶剂化部分	qu rong ji hua bu fen	0601
1694	全能峰	quan neng feng	0953
1695	燃耗	ran hao	0284
1696	燃耗率	ran hao lu	0285
1697	燃料	ran liao	0949
1698	燃料循环	ran liao xun huan	0950
1699	燃烧	ran shao	0414
1700	燃烧点	ran shao dian	0281
1701	燃烧电压	ran shao dian ya	0283
1702	燃烧速度	ran shao su du	0282
1703	染料激光器	ran liao ji guang qi	0712
1704	热磁测定(法)	re ci ce ding (fa)	2444
1705	热导率,导热(性)系数	re dao lu, dao re (xing) xi shu	2417
1706	热导式检测器	re dao shi jian ce qi	2418
1707	热点火电弧	re dian huo dian hu	2424
1708	热电测量(法)	re dian ce liang (fa)	2437
1709	热[温差]电堆	re [wen cha] dian dui	2452
1710	热电检测计	re dian jian ce ji	1891
1711	热电离	re dian li	2423
1712	热动[力(学)]平衡	re dong [li (xue)] ping heng	2435
1713	热分析	re fen xi	2415
1714	热分析方法	re fen xi fang fa	2430
1715	热光度法	re guang du fa	2451
1716	热光(特性)分析法	re guang (te xing) fen xi fa	2453
1717	热光谱测定	re guang pu ce ding	2456
1718	热过程	re guo cheng	2426
1719	热化学反应	re hua xue fan ying	2432
1720	热分解(作用)	re fen jie (zuo yong)	2419
1721	热解重量分析	re jie zhong liang fen xi	2440
1722	热解重量分析曲线	re jie zhong liang fen xi qu xian	2441
1723	热空心阴极	re kong xin yin ji	1064
1724	热老化,加热老陈	re lao hua, jia re lao chen	2414
1725	热力学测量(法)	re li xue ce liang (fa)	2446
1726	热力学常数	re li xue chang shu	2434
1727	热力学分析	re li xue fen xi	2445
1728	热力学温度	re li xue wen du	2436
1729	热流[通量]差分扫描量热法	re liu [tong liang] cha fen sao miao liang re fa	1020
1730	热能辐射	re neng fu she	2427
1731	热膨胀分析法	re peng zhang fen xi fa	2433
1732	热平衡	re ping heng	2420
1733	热曲线	re qu xian	1022
1734	热曲线测定	re qu xian ce ding	1021
1735	热声测量(法)	re sheng ce liang (fa)	2455
1736	热声透射测定法	re sheng tou she ce ding fa	2429
1737	热室[放射性小室]	re shi[fang she xing xiao shi]	1063

1738	热丝色谱法	re si se pu fa	0879
1739	热天平	re tian ping	2431
1740	热微粒分析	re wei li fen xi	2450
1741	热显微术	re xian wei shu	2449
1742	热(中子)源	re (zhong zi) yuan	2428
1743	热原子	re yuan zi	1062
1744	热折射法	re zhe she fa	2454
1745	热蒸发	re zheng fa	2421
1746	热蒸汽测量分析	re zheng qi ce liang fen xi	2457
1747	热致发光	re zhi fa guang	2443
1748	热重量分析法	re zhong liang fen xi fa	2442
1749	热[慢]中子	re [man] zhong zi	2425
1750	热中子裂变	re zhong zi lie bian	2422
1751	热柱	re zhu	2416
1752	人工放射(性)	ren gong fang she (xing)	0138
1753	韧致辐射	ren zhi fu she	0266
1754	日光盲式光电倍增器	ri guang mang shi guang dian bei zeng qi	2218
1755	(体积)容量	(ti ji) rong liang	2578
1756	容量因子	rong liang yin zi	0295
1757	溶出,反萃取,解吸	rong chu, fan cui qu, jie xi	2359
1758	溶出(极谱)分析法	rong chu (ji pu) fen xi fa	2360
1759	溶出溶液,反提取剂	rong chu rong ye, fan ti qu ji	2361
1760	溶度积	rong du ji	2227
1761	溶剂	rong ji	2232
1762	溶剂萃取[抽提]	rong ji cui qu [chou ti]	2235
1763	溶剂空白	rong ji kong bai	2233
1764	(展开)溶剂前沿	(zhan kai) rong ji qian yan	2236
1765	溶剂效应	rong ji xiao ying	2234
1766	溶剂移动距离	rong ji yi dong ju li	2237
1767	溶解	rong jie	0675
1768	溶液,溶解(作用)	rong ye, rong jie (zuo yong)	2230
1769	溶液平衡	rong ye ping heng	2231
1770	溶质气化干扰	rong zhi qi hua gan rao	2229
1771	溶质一溶剂相互作用	rong zhi—rong ji xiang hu zuo yong	2228
1772	熔化	rong hua	0956
1773	乳化校准函数	ru hua jiao zhun han shu	0792
1774	乳化校准曲线	ru hua jiao zhun qu xian	0791
1775	入口(管)系统的灵敏度	ru kou (guan) xi tong de ling min du	2179
1776	入口狭缝高度	ru kou xia feng gao du	0812
1777	入口狭缝宽度	ru kou xia feng kuan du	0813
1778	入射幅射	ru she fu she	1086
1779	入射功率	ru she gong lu	1085
1780	软化水[订正,原为去离子水]	ruan hua shui [ding zheng, yuan wei qu li zi shui]	0576,0581
1781	软片模糊,灰雾	ruan pian mo hu, hui wu	0931
1782	弱电位梯度	ruo dian wei ti du	2588
1783	撒哈一俄戈特(Saha—Eggert)定律	sa ha—e ge te ding lu	2096
1784	塞曼(Zeeman)背景校正	sai man bei jing xiao zheng	2615
1785	塞曼效应	sai man xiao ying	2616
1786	三槽燃烧器	san cao ran shao qi	2464

1787	三电极等离子体	san dian ji deng li zi ti	2463
1788	三电极构型	san dian ji gou xing	2462
1789	三角波伏安法	san jiao bo fu an fa	2513
1790	三角波极谱(法)	san jiao bo ji pu (fa)	2512
1791	三棱镜的有效底长	san leng jing de you xiao di chang	1284
1792	三线态—三线态吸收	san xian tai—san xian tai xi shou	2515
1793	散射,分散	san she, fen san	2124
1794	散射物,漫[分]散	san she wu, man [fen] san	2125
1795	桑德(Sand)方程	sang de fang cheng	2112
1796	扫描电子显微镜(SEM)	sao miao dian zi xian wei jing	2121
1797	扫描激光分析	sao miao ji guang fen xi	2122
1798	扫描透射式电子显微镜	sao miao tou she shi dian zi xian wei jing	2123
1799	色[层析]谱	se [ceng xi] pu	0365
1800	色谱法	se pu fa	0368
1801	色谱仪	se pu yi	0366
1802	色(谱)区[带,层]	se (pu) qu [dia, ceng]	2619
1803	筛子电极	shai zi dian ji	2196
1804	闪光灯	shan guang deng	0902
1805	闪光荧光分析(法)	shan guang ying guang fen xi (fa)	0901
1806	闪烁(现象)	shan shuo (xian xiang)	2130
1807	闪烁材料	shan shuo cai liao	2129
1808	闪烁光谱测定法	shan shuo guang pu ce ding fa	2133
1809	闪烁剂[器,体]	shan shuo ji [qi,	2134
1810	闪烁计数器	shan shuo ji shu qi	2131
1811	闪烁检测器	shan shuo jian ce qi	2132
1812	闪烁体检测计	shan shuo ti jian ce ji	2135
1813	闪耀波长	shan yao bo chang	0243
1814	闪耀角	shan yao jiao	0242
1815	上升时间	shang sheng shi jian	2087
1816	上升速度	shang sheng shu du	2088
1817	烧除效应	shao chu xiao ying	0280
1818	(较)少[小]量组[成,构]分	(jiao) shao [xiao] liang zu [cheng, gou] fen	1466
1819	射频等离子体	she pin deng li zi ti	1974
1820	射频极谱法	she pin ji pu fa	1975
1821	射气热分析	she qi re fen xi	0787
1822	γ 射线	γ she xian	0961
1823	X— 射线测量	X—she xian ce liang	2611
1824	X— 射线电子发生	X—she xian dian zi fa sheng	2607
1825	X— 射线发射光谱测定法	X—she xian fa she guang pu ce ding fa	2602
1826	X— 射线发射光谱学	X—she xian fa she guang xue	2603
1827	X— 射线发生	X—she xian fa sheng	2606
1828	X— 射线分析	X—she xian fen xi	2599
1829	X— 射线光电子能谱法 (XPS)	X—she xian guang dian zi neng pu fa	2612
1830	γ 射线光谱	γ she xian guang pu	0963
1831	X— 射线光子	X—she xian guang zi	2613
1832	X— 射线光子发生	X—she xian guang zi fa sheng	2608
1833	X— 射线谱	X—she xian pu	2614
1834	X— 射线强度	X—she xian qiang du	2610
1835	X— 射线数据判读	X—she xian shu ju pan du	2600
1836	X— 射线逃逸峰	X—she xian tao yi feng	2604

1837	X— 射线衍射	X—she xian yan she	2601
1838	X— 射线荧光	X—she xian ying guang	2605
1839	X— 射线正离子发生	X—she xian zheng li zi fa sheng	2609
1840	摄入率	she ru lu	2547
1841	摄影参数	she ying can shu	1736
1842	深度分辨	shen du fen bian	0586
1843	深度轮廓[剖面]	shen du lun kuo [pou mian]	0585
1844	渗透色谱法	shen tou se pu fa	1703
1845	升华(作用)	sheng hua (zuo yong)	2363
1846	声(频)冲击波筛子系统	sheng (pin) chong ji bo shai zi xi tong	0030
1847	剩余(宇宙线)电离	sheng yu (yu zhou xian) dian li	2053
1848	实时	shi shi	2000
1849	实用比能力	shi yong bi neng li	1821
1850	实用分辨	shi yong fen bian	1820
1851	石墨棒炉	shi mo bang lu	1000
1852	石墨杯原子化器	shi mo bei yuan zi hua qi	0999
1853	石墨管炉	shi mo guan lu	1002
1854	石墨火花	shi mo huo hua	1001
1855	时间常数	shi jian chang shu	2468
1856	时间分辨	shi jian fen bian	2404
1857	时间分辨光谱(法)	shi jian fen bian guang pu (fa)	2471
1858	时间分辨技术	shi jian fen bian ji shu	2472
1859	时间跨距[空号]	shi jian kua ju [kong hao]	2405
1860	时间稳定	shi jian wen ding	2406
1861	史泊尔斯基(Shpol′ skii)(光)波谱	shi bo er si ji (guang) bo pu	2194
1862	示踪	shi zhong	1251
1863	示波法	shi bo fa	1644
1864	示波极谱法	shi bo ji pu fa	1645
1865	示踪物	shi zong wu	2490
1866	示踪原子,标记	shi zong yuan zi, biao ji	1250
1867	试验表面	shi yan biao mian	0849
1868	试验电极	shi yan dian ji	2409
1869	试[抽,送]样	shi [chou, song] yang	2100
1870	试样电极	shi yang dian ji	2102
1871	试样光束	shi yang guang shu	2101
1872	试样重量分极	shi yang zhong liang fen ji	2108
1873	(较大)试样(夹持)装置[台,座]	(jiao da) shi yang (jia chi) zhuang zhi [tai, zuo]	0245
1874	(差热分析)试[样]品座组合	(cha re fen xi) shi [yang] pin zuo zu he	2266
1875	释放[防粘]剂	shi fang [fang nian] ji	2045
1876	释放机制	shi fang ji zhi	2046
1877	释放装置[剂,器]	shi fang zhuang zhi [ji, qi]	2044
1878	释热[燃料]元件	shi re [ran liao] yuan jian	0951
1879	收集(品)	shou ji (pin)	0395
1880	收集剂[器]	shou ji ji [qi]	0396
1881	收缩效应	shou suo xiao ying	1761
1882	受激发射跃迁概率	shou ji fa she yue qian gai lu	2498
1883	(样品)输入[入口]系统	(yang pin) shu ru [ru kou] xi tong	1111
1884	数据,基线	shu ju, ji xian	0560
1885	数字模拟转换器(DAC)	shu zi mo ni zhuan huan qi	0641
1886	束缚	shu fu	0228

1887	树脂基体	shu zhi ji ti	2056
1888	刷形放电	shua xing fang dian	0270
1889	衰变[化]	shuai bian [hua]	0568
1890	α 衰变	α shuai bian	0061
1891	β 衰变	β shuai bian	0219
1892	衰变常数	shuai bian chang shu	0570
1893	衰变链	shuai bian lian	0569
1894	衰变曲线	shuai bian qu xian	0571
1895	衰变图式	shuai bian tu shi	0572
1896	衰减,稀释	shuai jian, xi shi	0165
1897	衰减系数	shuai jian xi shu	0166
1898	双安培滴定	shuang an pei di ding	0222
1899	双安培分析法	shuang an pei fen xi fa	0223
1900	双安培终点检测	shuang an pei zhong dian jian ce	0221
1901	双层电流	shuang ceng dian liu	0696
1902	双电势[位]分析法	shuang dian shi [wei] fen xi fa	0232
1903	双电位[势]滴定	shuang dian wei [shi] di ding	0231
1904	双电位跃计时安培[电流]分析法	shuang dian wei yue ji shi an pei [dian liu] fen xi fa	0697
1905	双电位跃计时库仑滴定法	shuang dian wei yue ji shi ku lun di ding fa	0698
1906	双调极谱法	shuang tiao ji pu fa	0701
1907	双官能离子交换剂	shuang guan neng li zi jiao huan ji	0226
1908	双光谱光束分光计	shuang guang pu guang shu fen guang ji	0699
1909	双光束分光计	shuang guang shu fen guang ji	0692
1910	双光束系统	shuang guang shu xi tong	0693
1911	双核配合物	shuang he pei he wu	0230
1912	双接界参比电极电桥溶液	shuang jie jie can bi dian ji dian qiao rong ye	0268
1913	双聚焦质谱计	shuang ju jiao zhi pu ji	0695
1914	双逃脱峰	shuang tao tuo feng	0694
1915	双同步光束分光光度计	shuang tong bu guang shu fen guang guang du ji	0700
1916	双雾化器	shuang wu hua qi	2528
1917	双向色谱法	shuang xiang se pu fa	2530
1918	双原子的	shuang yuan zi de	0613
1919	水平[卧式]电极	shui ping [wo shi] dian ji	1059
1920	瞬发符合	shun fa fu he	1866
1921	瞬发中子	shun fa zhong zi	1867
1922	瞬时测量[定]	shun shi ce liang [ding]	1118
1923	瞬时电流	shun shi dian liu	1116
1924	瞬时流动速率	shun shi liu dong su lu	1117
1925	顺磁化合物	shun shi hua he wu	1663
1926	斯塔克(Stark)变宽	si ta ke bian kuan	2332
1927	斯特恩—福尔默(Stern—Volmer)定律	si te en—fu er mo ding lu	2351
1928	斯托克斯(Stokes)荧光	si tuo ke si ying guang	2354
1929	死[空耗]时间	si [kong hao] shi jian	0564
1930	死时间校正	si shi jian jiao zheng	0565
1931	死体积	si ti ji	0566
1932	死停终点	si ting zhong dian	0563
1933	四分之一波长谱	si fen zhi yi bo chang xian	1908
1934	四分之一跃迁时间电位	si fen zhi yi yue qian shi jian dian wei	1907
1935	四极质量分析仪	si ji zhi liang fen xi yi	1895
1936	素烧杯电极	su shao bei dian ji	1800

1937	酸度常数	suan du chang shu	0028
1938	酸度函数	suan du han shu	0029
1939	酸碱滴定	suan jian di ding	0025
1940	酸碱相互作用	suan jian xiang hu zuo yong	0024
1941	酸碱指示剂	suan jian zhi shi ji	0023
1942	酸量滴定(法)	suan liang di ding (fa)	0026,0027
1943	随机重合	sui ji chong he	1987
1944	碎片离子	sui pian li zi	0937
1945	塔斯脱(Tast)极谱(法)	ta si te ji pu (fa)	2401
1946	塔斯特间隔[距,歇][周期]	ta si te jian ge [ju, xie] [zhou qi]	2400
1947	弹性碰撞	tan xing peng zhuang	0731
1948	弹性散射	tan xing san she	0732
1949	碳棒炉[加热器]	tan bang lu [jia re qi]	0304
1950	碳杯原子化器	tan bei yuan zi hua qi	0302
1951	碳灯丝原子化器	tan deng si yuan zi hua qi	0303
1952	唐森德(Townsend)第一系数	tang sen de di yi xi shu	2486
1953	忒斯拉(Tesla)线圈[磁通量单位]	te si la xian quan	2408
1954	特性干扰	te xing gan rao	2262
1955	特性[固有,波]阻抗	te xing [gu you, bo] zu kang	1530
1956	特征电位	te zheng dian wei	0336
1957	特征 X—辐射	te zheng X—fu she	0337
1958	特征浓度	te zheng nong du	0334
1959	特征质量	te zheng zhi liang	0335
1960	梯度层	ti du ceng	0997
1961	梯度填料	ti du tian liao	0998
1962	梯度洗脱[提]	ti du xi tuo [ti]	0996
1963	梯段[级]高度	ti duan [ji] gao du	2346
1964	提[萃]取	ti [cui] qu	0856
1965	提[萃]取(法)	ti [cui] qu fa	0859
1966	提[萃]取液	ti [cui] qu yei	0855
1967	提取常数	ti qu chang shu	0861
1968	提[萃]取剂	ti [cui] qu ji	0858
1969	提[萃]取系数	ti [cui] qu xi shu	0860
1970	提取性	ti qu xing	0857
1971	提[萃]取指示剂	ti [cui] qu zhi shi ji	0862
1972	提余液	ti yu ye	1983
1973	体积分布系数	ti ji fen bu xi shu	2579
1974	体积流速[流量]	ti ji liu su [liu liang]	2581
1975	天平,平衡	tian ping, ping heng	0203
1976	天平精度	tian ping jing du	1830
1977	天然放射性	tian ran fang she xing	1533
1978	天[自]然丰度	tian [zi] ran feng du	1528
1979	天然同位素丰度	tian ran tong wei su feng du	1531
1980	添加剂	tian jia ji	0040
1981	填充气体	tian chong qi ti	0880
1982	填充柱	tian chong zhu	1655
1983	调(制)频(率)	tiao (zhi) pin (lu)	1483
1984	调谐短截线	tiao xie duan jie xian	2522
1985	调谐放大器测量系统	tiao xie fang da qi ce liang xi tong	2518
1986	调整保留体积	tiao zheng bao liu ti ji	0042

1987	调制[幅]极谱法	tiao zhi [fu] ji pu fa	1485
1988	停电流计时电位法	ting dian liu ji shi dian wei fa	0543
1989	停留时间	ting liu shi jian	2050
1990	停止本领	ting zhi ben ling	2355
1991	通量	tong liang	0919
1992	通量积分	tong liang ji fen	0922
1993	通量监测器[体]	tong liang jian ce qi [ti]	0923
1994	通量降低	tong liang jiang di	0921
1995	通量密度	tong liang mi du	0920
1996	2200m/s 通量密度	2200m/s tong liang mi du	2531
1997	通量微扰[扰动]	tong liang wei rao [rao dong]	0924
1998	同步激发磷光光谱	tong bu ji fa lin guang guang pu	2390
1999	同步激发荧光光谱	tong bu ji fa ying guang guang pu	2389
2000	同步技术	tong bu ji shu	2202
2001	同步转动间隙	tong bu zhuan dong jian xi	2391
2002	同分异构态	tong fen yi gou tai	1228
2003	同核异构体[物]	tong he yi gou ti [wu]	1597
2004	同量异位质量变异测定	tong liang yi wei zhi liang bian yi ce ding	1226
2005	同时差热分析	tong shi cha re fen xi	2201
2006	同时分析体系	tong shi fen xi ti xi	2200
2007	同位素[原子序数相同]	tong wei su	1234
2008	同位素分离	tong wei su fen li	1243
2009	同位素分析	tong wei su fen xi	1240
2010	同位素丰度	tong wei su feng du	1239
2011	同位素交换(法)	tong wei su jiao huan (fa)	1237
2012	同位素交换化学法	tong wei su jiao huan hua xue fa	0356
2013	同位素离子	tong wei su li zi	1242
2014	同位素稀释(法)	tong wei su xi shi (fa)	1235
2015	同位素稀释分析法	tong wei su xi shi fen xi fa	1236
2016	同位素载体	tong wei su zai ti	1241
2017	同位素指示剂[示踪]	tong wei su zhi shi ji [shi zong]	1244
2018	同心雾化器	tong xin wu hua qi	0437
2019	同质异能核	tong zhi yi neng he	1597
2020	同质异能跃迁	tong zhi yi neng yue qian	1229
2021	同中子异核[荷]素	tong zhong zi yi he [he] su	1233
2022	同轴导波器	tong zhou dao bo qi	0383
2023	统计权重	tong ji quan zhong	2343
2024	统计数值	tong ji shu zhi	2559
2025	透明度	tou ming du	2506
2026	透气膜	tou qi mo	0973
2027	透射比	tou she bi	2505
2028	透射式电子能量损耗(能)谱术[法](TEELS)	tou she shi dian zi neng liang sun hao (neng) pu shu [fa]	2501
2029	透射式电子显微术(TEM)	tou she shi dian zi xian wei shu	2502
2030	透射式高能电子衍射(法)(THEED)	tou she shi gao neng dian zi yan she (fa)	2504
2031	透射[传递]因子	tou she [chuan di] yin zi	2503
2032	投影屏	tou ying ping	1864
2033	图象设备	tu xiang she bei	1079
2034	蜕变常数	tui bian chang shu	0665
2035	蜕变率	tui bian lu	0666
2036	脱挥发器[剂]	tuo hui fa qi [ji]	0612

2037	椭圆偏振	tuo yuan pian zhen	0778
2038	外标准(物)	wai biao zhun (wu)	0854
2039	外带[区]	wai dai [qu]	1647
2040	外加电位	wai jia dian wei	0130
2041	外推开始	wai tui kai shi	0863
2042	外推量程[范围]	wai tui liang cheng [fan wei]	0864
2043	外引燃火花放电	wai yin ran huo hua fang dian	0853
2044	外重原子效应	wai zhong yuan zi xiao ying	0852
2045	完全蒸发[汽化]	wan quan zheng fa [qi hua]	0417
2046	网口灯	wang kou deng	1429
2047	威斯特科特(Westcott)截面	wei si te ke te jie mian	2592
2048	微波	wei bo	1457
2049	微波等离子体	wei bo deng li zi ti	1459
2050	微波感应等离子体	wei bo gan ying deng li zi ti	1458
2051	微处理机,微型信息处理机	wei chu li ji, wei xing xin xi chu li ji	1453
2052	微分[差示](的)	wei fen [cha shi] (de)	0617
2053	微分热重量分析法	wei fen re zhong liang fen xi fa	0633
2054	微观截面	wei guan jie mian	1455
2055	微痕量	wei hen liang	1456
2056	微库仑分析法	wei ku lun fen xi fa	1451
2057	微量称样	wei liang cheng yang	1454
2058	微量分析	wei liang fen xi	1446
2059	微量化学天平	wei liang hua xue tian ping	1448
2060	微量天平	wei liang tian ping	1447
2061	微量组分	wei liang zu fen	1449
2062	微型计算机	wei xing ji suan ji	1450
2063	未极化的	wei ji hua de	2543
2064	温度记录(法)	wen du ji lu (fa)	2439
2065	温度效应	wen du xiao ying	2402
2066	稳定常数	wen ding chang shu	2315
2067	稳定电弧	wen ding dian hu	2318
2068	稳定环	wen ding huan	2317
2069	稳流电源	wen liu dian yuan	2319
2070	稳(定)气(体)电弧	wen (ding) qi (ti) dian hu	0977
2071	稳态	wen tai	2344
2072	沃伊特(Voigt)函数	wo yi te han shu	2565
2073	沃伊特轮廓	wo yi te lun kuo	2566
2074	乌布利希(Ulbricht)球,积算球[积算式光度计的]	wu bu li xi qiu, ji suan qiu	2532
2075	钨一卤(素)灯	wu—lu (su) deng	2521
2076	钨丝灯	wu si deng	2520
2077	无电极放电灯	wu dian ji fang dian deng	0742
2078	无电极激发	wu dian ji ji fa	0743
2079	无电极谐振腔等离子体	wu dian ji xie zhen qiang deng li zi ti	0744
2080	无电流等离子体	wu dian liu deng li zi ti	0544
2081	无辐射跃迁	wu fu she yue qian	1932
2082	无控制高压火花发生器	wu kong zhi gao ya huo hua fa sheng qi	2539
2083	无控制交流电弧	wu kong zhi jiao liu dian hu	2538
2084	无损活化分析	wu sun huo hua fen xi	1576
2085	无载(电)流等离子体	wu zai (dian) liu deng li zi ti	1575

2086	无载体(的)	wu zai ti (de)	0308
2087	物理干扰	wu li gan rao	1756
2088	物理老化	wu li lao hua	1754
2089	物理平衡	wu li ping heng	1755
2090	物理数值[量]	wu li shu zhi [liang]	1757
2091	物理吸附	wu li xi fu	1758
2092	物质分散	wu zhi fen san	0669
2093	(烟)雾	(yan) wu	1468
2094	雾化	wu hua	1534
2095	雾化[喷雾]器	wu hua [pen wu] qi	1536
2096	雾化器一火焰系统	wu hua qi—huo yan xi tong	1537
2097	雾化[喷雾]效率	wu hua [pen wu] xiao lu	1535
2098	误差	wu cha	0820
2099	误差百分数,百分误差	wu cha bai fen shu, bai fen wu cha	1699
2100	西韦特(Sievert)[单位]	xi wei te	2195
2101	析出气体检测[EGD]	xi chu qi ti jian ce	0826
2102	熄[猝]灭	xi [cu] mie	1911
2103	熄[猝]灭荧光分析法	xi [cu] mie ying guang fen xi fa	1912
2104	吸附(作用)	xi fu (zuo yong)	0045
2105	吸附电流	xi fu dian liu	0047
2106	吸附剂	xi fu ji	0036,0044
2107	吸附色谱(法)	xi fu se pu (fa)	0046
2108	(被)吸附物	(bei) xi fu wu	0043
2109	吸附指示剂	xi fu zhi shi ji	0048
2110	吸光系数	xi guang xi shu	0021
2111	(真空)吸气器剂	(zhen kong) xi qi qi [ji]	0990
2112	吸气[入],抽出	xi qi [ru], chou chu	0139
2113	吸气[入,出],抽出	xi qi [ru,chu], chou chu	0140
2114	吸气率	xi qi lu	0141
2115	吸气器	xi qi qi	0142
2116	吸热	xi re	0793
2117	吸热峰	xi re feng	0794
2118	吸入雾化[喷雾]器	xi ru wu hua [pen wu] qi	2367
2119	吸收(作用)	xi shou (zuo yong)	0010
2120	吸收电子	xi shou dian zi	0008
2121	吸收度	xi shou du	0006
2122	吸收光程长度	xi shou guang cheng chang du	0018
2123	吸收剂[器,体]	xi shou ji [qi, ti]	0009
2124	吸收剂量	xi shou ji liang	0007
2125	吸收轮廓	xi shou lun kuo	0019
2126	吸收能量	xi shou neng liang	0014
2127	吸收系数	xi shou xi shu	0011,0021
2128	吸收线半强度宽度	xi shou xian ban qiang du kuan du	1009
2129	吸收性	xi shou xing	0021
2130	吸收(边)沿波长	xi shou (bian) yan bo chang	0013
2131	吸收(边)沿跃比(率)	xi shou (bian) yan yue bi (lu)	0012
2132	吸收因数(率)	xi shou yin shu (lu)	0015
2133	吸收跃迁概率	xi shou yue qian gai lu	0020,2496
2134	吸液速度	xi ye su du	1990
2135	吸着[附](作用)	xi zhuo [fu] (zuo yong)	2238

2136	吸着[附]等温线	xi zhuo [fu] deng wen xian	2239
2137	希托夫(Hittorf) 暗区	xi tuo fu an qu	1048
2138	稀释剂	xi shi ji	0643
2139	稀释试验	xi shi shi yan	0064
2140	稀有气体	xi you qi ti	1566
2141	洗出液[物]	xi chu ye [wu]	0779
2142	洗涤	xi di	2138
2143	洗涤液	xi di ye	2139
2144	洗脱[出,提]	xi tuo [chu, ti]	0781
2145	洗脱带	xi tuo dai	0782
2146	洗脱峰	xi tuo feng	0785
2147	洗脱曲线	xi tuo qu xian	0784
2148	洗脱容[体]积	xi tuo rong [ti] ji	0786
2149	洗脱色谱(法)	xi tuo se pu (fa)	0783
2150	洗脱液剂	xi tuo ye ji	0780
2151	系(列),串(联),级数	xi (lie), chuan (lian), ji shu	2188
2152	系间[线路]交叉	xi jian [xian lu] jiao cha	1169
2153	α[副反应]系数	α [fu fan ying] xi shu	0060
2154	隙间部分	xi jian bu fen	1165
2155	隙间速度	xi jian su du	1166
2156	隙间体积	xi jian ti ji	1168
2157	狭缝	xia feng	2213
2158	狭缝高度	xia feng gao du	2214
2159	狭缝宽度	xia feng kuan du	2215
2160	氙灯	xian deng	2597
2161	显[指]示精度	xian [zhi] shi jing du	1832
2162	显[指]示器	xian [zhi] shi qi	1090
2163	现场[原地]微量X射线衍射	xian cheng [yuan di] wei liang X she xian yan she	1115
2164	限流板	xian liu ban	0443
2165	限制辉光放电	xian zhi hui guang fang dian	2074
2166	线,谱线,长度单位	xian, pu xian, chang du dan wei	1300
2167	线宽度	xian kuan du	1322
2168	线能量转移	xian neng liang zhuan yi	1307
2169	线能量转移相关系数[因子]	xian neng liang zhuan yi xiang guan xi shu [yin zi]	1308
2170	线偏振镜	xian pian zhen jing	1312
2171	线(路)调谐振荡器	xian (lu) tiao xie zhen dang qi	2519
2172	线型	xian xing	1320
2173	线性滴定曲线	xian xing di ding qu xian	1317
2174	线性递减[减量]	xian xing di jian [jian liang]	1303
2175	线性电势[位]扫描	xian xing dian shi [wei] sao miao	1313
2176	线性电势[位]扫描一阳极溶出计时电流[安培]法	xian xing dian shi (wei) sao miao—rong chu ji shi dian liu [an pei] fa	0115
2177	线性分散[色散,离差]	xian xing fen san [se san, li cha]	1304
2178	线性光电吸收系数	xian xing guang dian xi shou xi shu	1310
2179	线性光电效应	xian xing guang dian xiao ying	1306
2180	线性加速质谱仪	xian xing jia su zhi pu yi	1301
2181	线性[直线]流动速率	xian xing [zhi xian] liu dong su lu	1309
2182	线性脉冲放大器	xian xing mai chong fang da qi	1314
2183	线性偏振辐射调制	xian xing pian zhen fu she tiao zhi	1484
2184	线性散射系数	xian xing san she xi shu	1315

2185	线性扫描伏安法	xian xing sao miao fu an fa	1316
2186	线性衰减系数	xian xing shuai jian xi shu	1302
2187	相比率	xiang bi lu	1715
2188	(分)相滴定	(fen) xiang di ding	1716
2189	相对保留(值)	xiang dui bao lui (zhi)	2042
2190	相对标准偏差	xiang dui biao zhun pian cha	2043
2191	相对活度	xiang dui huo du	2036
2192	相对计数	xiang dui ji shu	2039
2193	相对密度	xiang dui mi du	2040
2194	相对生物效用	xiang dui sheng wu xiao yong	2038
2195	相对原子质量	xiang dui yuan zi zhi liang	2037
2196	相对(比较)误差	xiang dui (bi jiao) wu cha	2041
2197	相分析	xiang fen xi	1713
2198	相干单位系统	xiang gan dan wei xi tong	0388
2199	相干电磁辐射	xiang gan dian ci fu she	0385
2200	相干辐射	xiang gan fu she	0386
2201	相干源	xiang gan yuan	0387
2202	相互干扰	xiang hu gan rao	1526
2203	相互作用时间	xiang hu zuo yong shi jian	1138
2204	相敏整流[检波]高谐波交流极谱法	xiang min zheng liu [jian bo] gao xie bo jiao liu ji pu fa	1032
2205	相体积比	xiang ti ji bi	1717
2206	相荧光测定法	xiang ying guang ce ding fa	1714
2207	响应常数	xiang ying chang shu	2069
2208	响应时间	xiang ying shi jian	2071
2209	响应斜率	xiang ying xie lu	2070
2210	响应性,敏感度	xiang ying xing, min gan du	2072
2211	向红移	xiang hong yi	0214
2212	向蓝移,向短波长移	xiang lan yi, xiang duan bo chang yi	1070
2213	象素	xiang su	1766
2214	消[湮]灭	xiao [yan] mie	0108
2215	消[湮]灭(作用),湮没[熄火](现象)	xiao [yan] mie (zuo yong), yan mo [xi huo] (xian xiang)	0109
2216	小(口径)孔,孔板	xiao (kou jing) kong, kong ban	1639
2217	效率	xiao lu	0724
2218	效应[果],结果,影响	xiao ying [guo],jie guo, ying xiang	0718
2219	谐波重叠	xie bo chong die	1019
2220	E—型缓发[迟滞]荧光	E—xing huan fa [chi zhi] ying guang	0822
2221	P—型缓发[迟滞]荧光	P—xing huan fa [chi zhi] ying guang	1873
2222	形成常数	xing cheng chang shu	0934
2223	絮凝(作用)	xu ning (zuo yong)	0904
2224	选定区电子衍射	xuan ding qu dian zi yan she	2155
2225	选通光检测器	xuan tong guang jian ce qi	0979
2226	选通间隔[周期]	xuan tong jian ge [zhou qi]	2362
2227	选择挥发	xuan ze hui fa	2157
2228	选择渗透性	xuan ze shen tou xing	1704
2229	选择(性)系数	xuan ze (xing) xi shu	2158
2230	选择性洗脱	xuan ze xing xi tuo	2156
2231	旋转镜快[闸]门	xuan zhuan jing kuai [zha] men	2091
2232	旋转圆盘快[闸]门	xuan zhuan yuan pan kuai [zha] men	2090
2233	旋转温度	xuan zhuan wen du	2094

2234	循环计时电位分析法	xun huan ji shi dian wei fen xi fa	0550
2235	循环三角波伏安法	xun huan san jiao bo fu an fa	0553
2236	循环三角波极谱法	xun huan san jiao bo ji pu fa	0552
2237	循环电流阶跃计时电位分析法	xun huan dian liu jie yue ji shi dian wei fen xi fa	0551
2238	压力[强]	ya li [qiang]	1841
2239	压力—电压—电流关系	ya li—dian ya—dian liu guan xi	1844
2240	压力梯度校正因子	ya li ti du jiao zheng yin zi	1843
2241	压力效应	ya li xiao ying	1842
2242	压缩弧	ya suo hu	0439
2243	压缩(的)辉光放电	ya suo (de) hui guang fang dian	0451
2244	亚痕量分析	ya hen liang fen xi	2366
2245	亚化学计量同位素稀释分析	ya hua xue ji liang tong wei su xi shi fen xi	2365
2246	亚临界阻尼振荡[动]放电	ya lin jie zu ni zhen dang [dong] fang dian	2540
2247	亚微量分析	ya wei liang fen xi	2364
2248	亚稳	ya wen	1441
2249	亚稳(态)的	ya wen (tai) de	1440
2250	亚稳(态)分解	ya wen (tai) fen jie	1442
2251	亚稳离子峰	ya wen li zi feng	1443
2252	亚稳(定)态	ya wen (ding) tai	1445
2253	亚稳态峰	ya wen tai feng	1444
2254	湮没辐射	yan mo fu she	0110
2255	盐(助)溶	yan (zhu) rong	2097
2256	(加)盐(分)析	(jia) yan (fen) xi	2098
2257	盐析色谱(法)	yan xi se pu (fa)	2099
2258	掩蔽	yan bi	1388,2187
2259	掩蔽剂	yan bi ji	1389
2260	衍射电子	yan she dian zi	0635
2261	阳极暗区	yang ji an qu	0112
2262	阳极辉光	yang ji hui guang	0114
2263	阳极间隙	yang ji jian xi	0113
2264	阳极汽化	yang ji qi hua	0118
2265	阳极溶出伏安法	yang ji rong chu fu an fa	0117
2266	阳极溶出控制电势[位]库仑法	yang ji rong chu kong zhi dian shi [wei] ku lun fa	0116
2267	阳[正]离子	yang [zheng] li zi	1803
2268	阳离子电泳	yang li zi dian yong	0311
2269	阳离子交换	yang li zi jiao huan	0323
2270	阳离子交换剂[器]	yang li zi jiao huan ji [qi]	0324
2271	阳离子效应	yang li zi xiao ying	0322
2272	氧化还原滴定(法)	yang hua huan yuan di ding (fa)	1654,2015
2273	氧化还原离子交换树脂	yang hua huan yuan li zi jiao huan shu zhi	2013
2274	氧化还原指示剂	yang hua huan yuan zhi shi ji	1653,2012
2275	氧化剂	yang hua ji	1652
2276	样品,标本	yang pin, biao ben	2265
2277	样品架,试样座	yang pin jia, shi yang zuo	2103
2278	样品溶液	yang pin rong ye	2107
2279	液固色谱法	ye gu se pu fa	1339
2280	液体固定相	ye ti gu ding xiang	1340
2281	液体激光器	ye ti ji guang qi	1328
2282	液体接界	ye ti jie jie	1326
2283	液(体)接界电势	ye (ti) jie jie dian shi	1327

2284	液体一凝胶色谱(法)	ye ti—ning jiao se pu (fa)	1325
2285	液体闪烁体计数器	ye ti shan shuo ti ji shu qi	1337
2286	液体闪烁体检测计	ye ti shan shuo ti jian ce ji	1338
2287	液体提[萃]取(法)	ye ti ti [cui] qu (fa)	1324
2288	液体体积	ye ti ti ji	1341
2289	液体样品	ye ti yang pin	1335
2290	液体样品注射	ye ti yang pin zhu she	1336
2291	液相	ye xiang	1334
2292	液相色谱(法)	ye xiang se pu (fa)	1323
2293	液一液萃取	ye—ye cui qu	1332
2294	液一液分配	ye—ye fen pei	1330
2295	液一液接界	ye—ye jie jie	1333
2296	液一液平衡	ye—ye ping heng	1331
2297	液一液色谱法	ye—ye se pu fa	1329
2298	一步激发源	yi bu ji fa yuan	1617
2299	一次放电	yi ci fang dian	1846
2300	依尔科维奇(Ilkovič)方程符号常规[规定]	yi er ke wei qi fang cheng fu hao chang gui [gui ding]	1078
2301	仪器活化分析	yi qi huo hua fen xi	1119
2302	仪器显示	yi qi xian shi	1120
2303	移动时间	yi dong shi jian	2511
2304	逸出深度	yi chu shen du	0821
2305	异构重整[形成]	yi gou chong zheng [xing cheng]	1227
2306	异相膜电极	yi xiang mo dian ji	1029
2307	抑制剂	yi zhi ji	2379
2308	因数重量	yin shu zhong liang	0865
2309	阴极暗区	yin ji an qu	0312
2310	阴极层电弧	yin ji ceng dian hu	0316
2311	阴极过程符号常规	yin ji guo cheng fu hao chang gui	0317
2312	阴极辉光	yin ji hui guang	0315,1539
2313	阴极溅射	yin ji jian she	0319
2314	阴极汽化	yin ji qi hua	0321
2315	阴极溶出	yin ji rong chu	0320
2316	阴极射线极谱法	yin ji she xian ji pu fa	0318
2317	阴极势降电压	yin ji shi jiang dian ya	0314
2318	阴极势降区	yin ji shi jiang qu	0313
2319	阴[负]离子	yin [fu] li zi	1540
2320	阴离子交换	yin li zi jiao huan	0104
2321	阴离子交换剂	yin li zi jiao huan ji	0105
2322	阴离子配[络]合物	yin li zi pei [luo] he wu	0106
2323	阴离子效应	yin li zi xiao ying	0103
2324	引发,起始	yin fa, qi shi	1106
2325	隐蔽(作用)	yin bi (zuo yong)	2187
2326	迎头色谱(法)	ying tou se pu (fa)	0946
2327	荧光	ying guang	0908
2328	荧光测定终点	ying guang ce ding zhong dian	0916
2329	荧光产额	ying guang chan e	0915
2330	荧光滴定	ying guang di ding	0917
2331	荧光分光计	ying guang fen guang ji	0913
2332	荧光分析	ying guang fen xi	0909
2333	荧光功率系数	ying guang gong lu xi shu	0911,1817

2334	荧光量子产额	ying guang liang zi chan e	1905
2335	荧光(的)量子效率	ying guang (de) liang zi xiao lu	0912,1902
2336	荧光效率	ying guang xiao lu	0910
2337	荧光(的)总量子效率	ying guang (de) zong liang zi xiao lu	0914,2484
2338	用色谱(法)分析	yong se pu (fa) fen xi	0367
2339	有机(体,物)效应[作用]	you ji (ti, wu) xiao ying [zuo yong]	1638
2340	有限体积	you xian ti ji	1290
2341	有限样品粒度	you xian yang pin li du	2106
2342	有效剂量当量	you xiao ji liang dang liang	0720
2343	有效理论塔板数	you xiao li lun ta ban shu	0722
2344	有效理论塔板等板高度	you xiao li lun ta ban deng ban gao du	1025
2345	有效热截面	you xiao re jie mian	0723
2346	有效无限厚度	you xiao wu xian hou du	0721
2347	诱导反应	you dao fan ying	1097
2348	诱导物,感应器[物,体]	you dao wu, gan ying qi [wu, ti]	1100
2349	宇宙辐射	yu zhou fu she	0502
2350	预电弧时间	yu dian hu shi jian	1822
2351	预电离	yu dian li	1838
2352	预混燃烧器	yu hun ran shao qi	1839
2353	预火花时期	yu huo hua shi qi	1840
2354	阈能	yu neng	2465
2355	元素	yuan su	0775
2356	元素分析	yuan su fen xi	0776
2357	原地[现场]微量分析	yuan di [xian chang] wei liang fen xi	1114
2358	原离子	yuan li zi	1847
2359	原子的同位素	yuan zi de tong wei su	1238
2360	原子发射	yuan zi fa she	0149
2361	原子发射光谱法	yuan zi fa she guang pu fa	0150
2362	原子光谱	yuan zi guang pu	0157
2363	原子光谱基线法	yuan zi guang pu ji xian fa	0210
2364	原子光谱线	yuan zi guang pu xian	0158
2365	原子光谱学	yuan zi guang pu xue	0159
2366	原子轰[撞,打]击	yuan zi hong [zhuang, da] ji	0148
2367	原子[雾]化	yuan zi [wu] hua	0161
2368	原子[雾]化部分	yuan zi [wu] hua bu fen	0163
2369	原子化器	yuan zi hua qi	0164
2370	原子[雾]化效率	yuan zi [wu] hua xiao lu	0162
2371	原子间的碰撞	yuan zi jian de peng zhuang	1139
2372	原子量	yuan zi liang	0160
2373	原子吸收	yuan zi xi shou	0146
2374	原子吸收光谱法	yuan zi xi shou guang pu fa	0147
2375	原子(谱)线	yuan zi (pu) xian	0153
2376	原子序(数)	yuan zi xu (shu)	0156
2377	原子荧光	yuan zi ying guang	0151
2378	原子荧光光谱法	yuan zi ying guang guang pu fa	0152
2379	原子质量	yuan zi zhi liang	0154
2380	原子质量单位	yuan zi zhi liang dan wei	0155
2381	圆偏振(作用)	yuan pian zhen (zuo yong)	0376
2382	圆柱形	yuan zhu xing	0557
2383	圆柱[筒]形收集器	yuan zhu [tong] xing shou ji qi	0556

2384	(离子)源,放射源	(li zi) yuan, fang she yuan	2240
2385	源效率	yuan xiao lu	2241
2386	远程耦合	yuan cheng ou he	2048
2387	跃迁概率	yue qian gai lu	2495
2388	跃迁时间	yue qian shi jian	2499
2389	匀称线	yun cheng xian	1058
2390	运输,移置,运输工具	yun shu, yi zhi, yun shu gong ju	2507
2391	杂散辐射	za san fu she	2357
2392	载荷电阻器	zai he dian zu qi	1343
2393	载流电弧等离子体	zai liu dian hu deng li zi ti	0542
2394	载气	zai qi	0309
2395	载体[气,液]	zai ti [qi, ye]	0305
2396	载体,支座	zai ti, zhi zuo	2376
2397	载体电极	zai ti dian ji	0307
2398	载体蒸馏电弧	zai ti zheng liu dian hu	0306
2399	载重量	zai zhong liang	0293
2400	再(度)沉淀	zai (du) chen dian	2049
2401	再生电流	zai sheng dian liu	2034
2402	再引弧电压	zai yin hu dian ya	2035
2403	增电荷极谱法	zeng dian he ji pu fa	1089
2404	增量,尖峰	zeng liang, jian feng	2301
2405	增强	zeng qiang	0807
2406	增强磁场辉光放电	zeng qiang ci chang hui guang fang din	0253
2407	增强空心阴极	zeng qiang kong xin yin ji	0252
2408	增强输出辉光放电	zeng qiang shu chu hui guang fang dian	0254
2409	增色移(动)	zeng se yi (dong)	1068
2410	展开	zhan kai	0610
2411	张力法	zhang li fa	2407
2412	罩,屏,保护	zhao, ping, bao hu	2192
2413	照[曝]射量,暴光	zhao [pu] she liang, bao guang	0851
2414	照相光度术	zhao xiang guang du shu	1737
2415	遮蔽指示剂	zhe bi zhi shi ji	2136
2416	折光指数,折射率	zhe guang zhi shu, zhe she lu	2033
2417	折射效应	zhe she xiao ying	2032
2418	折衷条件	zhe zhong tiao jian	0426
2419	针阀	zhen fa	1538
2420	真重合[一致]	zhen chong he [yi zhi]	2516
2421	真空杯电极	zhen kong bei dian ji	2548
2422	真空光电管	zhen kong guang dian guan	2549
2423	真空火花	zhen kong huo hua	2550
2424	真实吸收	zhen shi xi shou	1999
2425	振荡[周期性]电流	zhen dang [zhou qi xing] dian liu	1701
2426	振荡反应	zhen dang fan ying	1640
2427	振荡能量跃迁	zhen dang neng liang yue qian	2561
2428	振荡器	zhen dang qi	1642
2429	振动温度	zhen dong wen du	2562
2430	振簧静电计	zhen huang jing dian ji	2560
2431	振子强度	zhen zi qiang du	1643
2432	蒸发(作用)	zheng fa (zuo yong)	0823
2433	蒸发平衡	zheng fa ping heng	0824

2434	蒸馏(作用)	zheng liu (zuo yong)	0676
2435	蒸汽喷射	zheng qi pen she	2555
2436	蒸汽羽烟	zheng qi yu yan	2557
2437	整流交流电弧	zheng liu jiao liu dian hu	2010
2438	整流(电)源	zheng liu (dian) yuan	2011
2439	整体浓度	zheng ti nong du	0277
2440	整体样品	zheng ti yang pin	0278
2441	正常辉光放电	zheng chang hui guang fang dian	1584
2442	正常眼睛	zheng chang yan jing	1583
2443	正电柱,阳极区[塔]	zheng dian zhu, yang ji qu [ta]	1801
2444	正电柱电流密度	zheng dian zhu dian liu mi du	1802
2445	正交电磁场	zheng jiao dian ci chang	0524
2446	正子,阳电子	zheng zi, yang dian zi	1804
2447	支承[垫]板	zhi cheng [dian] ban	2378
2448	支持电解质	zhi chi dian jie zhi	1094,2377
2449	直接裂变产率[额]	zhi jie lie bian chan lu [e]	0651
2450	直接权重校正	zhi jie quan zhong jiao zheng	0501
2451	直接探测[针]	zhi jie tan ce [zhen]	0654
2452	直接探针的灵敏度	zhi jie tan zhen de ling min du	2178
2453	直接注射燃烧器	zhi jie zhu she ran shao qi	0652
2454	直流电流	zhi liu dian liu	0646
2455	直流电导分析法	zhi liu dian dao fen xi fa	0649
2456	直流电弧	zhi liu dian hu	0647
2457	直流极谱法	zhi liu ji pu fa	0650
2458	直流偏压振动放电	zhi liu pian ya zhen dong fang dian	0648
2459	直线荧光	zhi xian ying guang	0653
2460	G—值	G—zhi	1008
2461	R_B—值	R_B—zhi	1994
2462	指定负载灵敏度	zhi ding fu zai ling min du	2177
2463	指示电极	zhi shi dian ji	1093
2464	指示剂[器]	zhi shi ji [qi]	1090
2465	指示剂校正	zhi shi ji jiao zheng	1092
2466	指示剂空白校正	zhi shi ji kong bai jiao zheng	1091
2467	指数衰变	zhi shu shuai bian	0850
2468	纸色层法	zhi se ceng fa	1659
2469	致电离电压	zhi dian li dian ya	1214
2470	致电离基团	zhi dian li ji tuan	1217
2471	致电离粒子	zhi dian li li zi	1212
2472	质粒基因组	zhi li ji yin zu	1781
2473	质量/电荷比	zhi liang / dian he bi	1391
2474	质量分布	zhi liang fen bu	1392
2475	质量分布比	zhi liang fen bu bi	1393
2476	(原子)质量数	(yuan zi) zhi liang shu	1394
2477	质量衰减系数	zhi liang shuai jian xi shu	1390
2478	质量作用定律	zhi liang zuo yong ding lu	1282
2479	质谱	zhi pu	1397
2480	质谱分析[学]	zhi pu fen xi [xue]	1396
2481	质谱仪	zhi pu yi	1395
2482	(正)质子,氢核	(zheng) zhi zi, qing he	1869
2483	质子化常数	zhi zi hua chang shu	1870

2484	质子平衡	zhi zi ping heng	1871
2485	质子数	zhi zi shu	1872
2486	滞后	zhi hou	1071
2487	滞流体积	zhi liu ti ji	1050
2488	置后滤波器效应	zhi hou lu bo qi xiao ying	1805
2489	中(等,度)	zhong (deng, du)	1477
2490	中(间,位)	zhong (jian, wei)	1432
2491	中(强度)电流电弧	zhong (qiang du) dian liu dian hu	1425
2492	中和[性]的	zhong he [xing] de	1553
2493	中功率	zhong gong lu	1426
2494	中间等离子体气体	zhong jian deng li zi ti qi ti	1150
2495	中间管	zhong jian guan	1151
2496	中速[能]中子	zhong su [neng] zhong zi	1149
2497	中电压火花	zhong dian ya huo hua	1427
2498	中电压火花发生器	zhong dian ya huo hua fa sheng qi	1428
2499	中子	zhong zi	1554
2500	中子俘获	zhong zi fu huo	1557
2501	中子辐射	zhong zi fu she	1563
2502	中子活化(法)	zhong zi huo hua (fa)	1555
2503	中子活化分析	zhong zi huo hua fen xi	1556
2504	中子活化仪器分析	zhong zi huo hua yi qi fen xi	1121
2505	中子密度	zhong zi mi du	1558
2506	中子通量	zhong zi tong liang	1559
2507	中子通量密度	zhong zi tong liang mi du	1560
2508	(核内)中子数	(he nei) zhong zi shu	1562
2509	中子温度	zhong zi wen du	1564
2510	终[终馏]点	zhong [zhong liu] dian	0795
2511	重力(自流)供料粉末筛子系统	zhong li (zi liu) gong liao fen mo shai zi xi tong	1004
2512	重量变化测定(法)	zhong liang bian hua ce ding (fa)	2590
2513	重量滴定	zhong liang di ding	2591
2514	重量摩尔浓度	zhong liang mo er nong du	1488
2515	重原子离子效应	zhong yuan zi li zi xiao ying	1024
2516	周期性电流	zhou qi xing dian liu	1701
2517	周期性电压	zhou qi xing dian ya	1702
2518	轴向电极	zhou xiang dian ji	0189
2519	逐级形成常数	zhu ji xing cheng chang shu	2349
2520	主[一次]电[光]源	zhu [yi ci] dian [guang] yuan	1852
2521	主[原,初级]辐射	zhu [yuan, chu ji] fu she	1850
2522	主体一客体化学	zhu ti—ke ti hua xue	1060
2523	主组[成]分,常量成分	zhu zu [cheng] fen, chang liang cheng fen	1386
2524	注入气体	zhu ru qi ti	1107
2525	注射管	zhu she guan	1109
2526	注[喷]射器	zhu [pen] she qi	1110
2527	注[喷]射温度	zhu [pen] she wen du	1108
2528	柱[塔,列]	zhu [ta, lie]	0408
2529	柱床容量	zhu chuang rong liang	0217
2530	柱床体积 [=column volume]	zhu chuang ti ji	0216
2531	柱前反应剂[器,堆]	zhu qian fan ying ji [qi, dui]	1833
2532	柱容[体]积	zhu rong [ti] ji	0412
2533	柱色谱(法)	zhu se pu (fa)	0409

2534	柱温度	zhu wen du	0411
2535	柱性能	zhu xing neng	0410
2536	转变间隔	zhuan bian jian ge	2494
2537	转动能跃迁	zhuan dong neng yue qian	2093
2538	转换[换算]因子	zhuan huan [huan suan] yin zi	0485
2539	转台	zhuan tai	2092
2540	转移等离子体	zhuan yi deng li zi ti	2491
2541	装灌溶液	zhuang guan rong ye	0881
2542	装设仪器	zhuang she yi qi	1122
2543	装柱，填充[料]	zhuang zhu, tian chong [liao]	1656
2544	锥形共振腔	zhui xing gong zhen qiang	2398
2545	准确[精]度	zhun que [jing] du	0022
2546	准直仪[管，器]平行光管	zhun zhi yi [guan,qi] ping xing guang guan	0398
2547	柱温度	zhu wen de	0411
2548	浊度测定法，散射测混法	zhuo du ce ding fa, san she ce hun fa	1547
2549	浊度滴定法	zhuo du di ding fa	1546
2550	浊度法终点检测	zhuo du fa zhong dian jian ce	1545
2551	子体产物	zi ti chan wu	0561
2552	紫外光电子光谱法(UPS)	zi wai guang dian zi guang pu fa	2537
2553	自倒转[蚀]	zi dao zhuan [shi]	2165
2554	自电极	zi dian ji	2163
2555	自动(作用)	zi dong (zuo yong)	0175,0176
2556	自动化	zi dong hua	0177,0178
2557	自动[重力(自流)]进料雾化器	zi dong [zhong li (zi liu)] jin liao wu hua qi	1003
2558	自动离子化，自动电离(作用)	zi dong li zi hua, zi dong dian li (zuo yong)	0174
2559	自发电重量分析法	zi fa dian zhong liang fen xi fa	2303
2560	自发裂变	zi fa lie bian	2305
2561	自发射跃迁概率	zi fa she yue qian gai lu	2304,2497
2562	自激电弧	zi ji dian hu	0939
2563	自屏蔽	zi ping bi	2166
2564	自然变宽	zi ran bian kuan	1529
2565	自然辐[放]射(作用)，天然放射物	zi ran fu [fang] she (zuo yong), tian ran fang she wu	1532
2566	自燃火花放电	zi ran huo hua fang dian	2164
2567	自吸收	zi xi shou	2159
2568	自吸收变宽	zi xi shou bian kuan	2160
2569	自吸收效应	zi xi shou xiao ying	2161
2570	自吸收因子	zi xi shou yin zi	2162
2571	自旋守恒定律	zi xuan shou heng ding lu	2302
2572	自氧化	zi yang hua	0180
2573	自由[游离]电子	zi you [you li] dian zi	0940
2574	自由一束缚跃迁	zi you—shu fu yue qian	0938
2575	自由振荡，自激工作方式	zi you zhen dang, zi ji gong zuo fang shi	0942
2576	自由一自由跃迁	zi you—zi you yue qian	0941
2577	总保留容积	zong bao liu rong ji	2485
2578	总分配常数	zong fen pei chang shu	1649
2579	总离子流	zong li zi liu	2483
2580	总(输出)能量峰效率	zong (shu chu) neng liang feng xiao lu	0954
2581	总吸收峰	zong xi shou feng	2482
2582	总形成常数	zong xing cheng chang shu	1650
2583	总重，负重，载荷	zong zhong, fu zhong, zai he	1342

2584	阻抗变换器	zu kang bian huan qi	1081
2585	阻尼常数	zu ni chang shu	0558
2586	阻止[滞]因素(R_f)	zu zhi [zhi] yin su	2076
2587	组分分类[分粒]	zu fen fen lei [fen li]	0449
2588	组分含量	zu fen han liang	0450
2589	组合电极	zu he dian ji	0413
2590	祖离子	zu li zi	1858
2591	钻道电极	zuan dao dian ji	0703
2592	最高峰,顶点	zui gao feng, ding dian	2368
2593	最适火花条件	zui shi huo hua tiao jian	1636
2594	最小(谱)线宽度	zui xiao (pu) xian kuan du	1465

ADDENDA

A 01	发射光谱学	fa she guang pu xue	A 09
A 02	分子发光光谱学	fen zi fa guang guang pu xue	A 08
A 03	光学声快[闸]门	guang xue sheng kuai [zha] men	A 01
A 04	黑化	hei hua	A 05
A 05	火焰光谱	huo yan guang pu xue	A 06
A 06	局部分析	ju bu fen xi	1344
A 07	内参比电极	nei can bi dian ji	A 07
A 08	入射光子	ru she guang zi	1082
A 09	X— 射线荧光光谱学	X—she xian ying guang guang pu xue	A 10
A 10	伸舌	shen she	0947
A 11	原子发射光谱学	yuan zi fa she guang pu xue	A 03
A 12	原子吸收光谱学	yuan zi xi shou guang pu xue	A 02
A 13	原子荧光光谱学	yuan zi ying guang guang pu xue	A 04

Japanese

あ行

0001	アーク	āku	0131
0002	アーク間隙	āku kangeki	0134
0003	アーク状放電	āku jō hōden	0136
0004	アーク線	āku sen	0137
0005	アーク柱	āku chū	0133
0006	アーク灯	āku tō	0135
0007	アークの電圧 － 時間関係	āku no den'atsu － jikan kankei	2572
0008	アークのよじのぼり	āku no yojinobori	0381
0009	アーク雰囲気	āku fun'iki	0132
0010	ＲＥＥＬＳ	āru－ī －ī －eru－esu	2028
0011	ＲＢＥ生物学的効果比	āru－bī －ī seibutsugaku－teki kōka－hi	2038
0012	ＲＢＳ	āru－bī －esu	2095
0013	R_B値	āru bī chi	1994
0014	アールヒード（ＲＨＥＥＤ）	āru hī do	2030
0015	ＩＳＳ	ai－esu－esu	1219
0016	ＩＰＭＡ	ai－pī －emu－ē	1218
0017	アインシュタインの遷移確率	ainshutain no sen'i kakuritsu	0729
0018	青シフト	ao shifuto	0247
0019	赤シフト	aka shifuto	2016
0020	アクチバブルな物質	akuchibaburu na busshitsu	0031
0021	アストン暗部	asuton anbu	0143
0022	アスピレータ	asupirēta	0142
0023	圧力	atsuryoku	1841
0024	圧力－電圧－電流関係	atsuryoku－den'atsu－denryū kankei	1844
0025	圧力グラジェント補正係数	atsuryoku gurajiento hosei keisū	1843
0026	圧力効果	atsuryoku kōka	1842
0027	穴あけ電極	ana ake denkyoku	0703
0028	アノーディック ストリッピング ボルタンメトリー	anōdikku sutorippingu borutanmetorī	0117
0029	アミノポリカルボン酸	amino pori karubon san	0072
0030	アルカリ滴定	arukari tekitei	0056,0057
0031	アルファ開裂	arufa kairetsu	0059
0032	アルファ係数	arufa keisū	0060
0033	アルファ線	arufa－sen	0063
0034	アルファパラメータ	arufa paramēta	0062
0035	アルファ崩壊	arufa hōkai	0061
0036	アンチコンプトン ガンマ線	anchi konputon ganma－sen	0119
0037	アンチコンプトン ガンマ線スペクトロメータ	anchi konputon ganma－sen supekutoromēta	0120
0038	安定化アーク	antei－ka āku	2318
0039	安定化リング	antei－ka ringu	2317
0040	安定度定数	anteido teisū	2315
0041	暗電流	an－denryū	0559
0042	Ｅ型遅延蛍光	ī －gata chien keikō	0822
0043	ＥＰＭＡ	ī －pī －emu－ē	0764
0044	イオン	ion	1182
0045	イオン－分子反応	ion－bunshi han'nō	1216
0046	イオン化	ion－ka	1200
0047	イオン化エネルギー	ion－ka enerugī	1206
0048	イオン化確率	ion－ka kakuritsu	1210
0049	イオン化干渉	ion－ka kanshō	1207
0050	イオン化緩衝剤	ion－ka kanshō zai	1201
0051	イオン化効率曲線	ion－ka kōritsu kyokusen	1205
0052	イオン化断面積	ion－ka dan－menseki	0526
0053	イオン活量係数	ion katsuryō keisū	1192
0054	イオン化電圧	ion－ka den'atsu	1214

0055	イオン化ポテンシャル	ion−ka potensharu	1209
0056	イオン化用針	ion−ka−yō hari	1208
0057	イオン強度	ion kyōdo	1198
0058	イオン強度調整用緩衝液	ion kyōdo chōsei−yō kanshō eki	1199
0059	イオンクロマトグラフィー	ion kuromatogurafī	1184
0060	イオン源	ion gen	1221
0061	イオン顕微鏡	ion kenbikyō	1215
0062	イオン交換	ion kōkan	1187
0063	イオン交換クロマトグラフィー	ion kōkan kuromatogurafī	1188
0064	イオン交換体	ion kōkan tai	1191
0065	イオン交換等温式	ion kōkan tō−on shiki	1189
0066	イオン交換膜	ion kōkan maku	1190
0067	イオン光路長	ion kōro−chō	1195
0068	イオンサイクロトロン共鳴質量分析計	ion saikurotoron kyōmei shitsuryō bunsekikei	1186
0069	イオン散乱分光法（ＩＳＳ）	ion−sanran bunkō−hō	1219
0070	イオン衝撃	ion shōgeki	1183
0071	イオンスペクトル	ion supekutoru	1196
0072	イオンスペクトル線	ion supekutoru sen	1197
0073	イオン生成基	ion seiseiki	1217
0074	イオン性媒体	ion−sei baitai	1194
0075	イオン線	ion sen	1193
0076	イオン［選択性］電極	ion [sentaku−sei] denkyoku	1220
0077	イオン電流	ion denryū	1185
0078	イオン特異性電極	ion tokui−sei denkyoku	1222
0079	イオンプローブ微小部分析(IPMA)	ion purō−bu bishō−bu bunseki	1218
0080	移行	ikō	1461
0081	異常グロー放電	ijōguro hōden	0001
0082	異色光度測定	ishoku kōdo sokutei	1028
0083	位相蛍光分析法	isō keikō bunseki−hō	1714
0084	位相敏感整流高調波交流ポーラログラフィー	isō binkan seiryū kōchōha kōryū pōrarogurafī	1032
0085	一官能性イオン交換体	ichi kan'nō−sei ion kōkantai	1512
0086	一次ｐＨ標準液	ichiji pī −eichi hyōjun−eki	1849
0087	一次イオン	ichiji ion	1847
0088	一次基準標準液	ichiji kijun hyōjun−eki	1851
0089	一次光源	ichiji kōgen	1852
0090	一次燃焼	ichiji nenshō	1845
0091	一次標準	ichiji hyōjun	1853
0092	一次標準物質	ichiji hyōjun busshitsu	1855
0093	一次標準溶液	ichiji hyōjun yōeki	1854
0094	一次放射光	ichiji hōsha−kō	1850
0095	一重項－一重項吸収	ichijukō − ichijukō kyūshū	2210
0096	一段階励起光源	ichidankai reiki kōgen	1617
0097	一価イオン化元素	ikka ion−ka genso	2211
0098	一貫した単位系	ikkanshita tan'i kei	0388
0099	一定荷重における感度	ittei kajū ni okeru kando	2177
0100	移動時間	idō jikan	2511
0101	移動相	idō sō	1476
0102	異方性	ihōsei	0107
0103	イムノアッセイ	imunoassei	1080
0104	イルコビッチ式の符号の規約	irukobitchi−shiki no fugō no kiyaku	1078
0105	陰イオン効果	in−ion kōka	0322
0106	陰イオン交換	in−ion kōkan	0104
0107	陰イオン交換体	in−ion kōkan−tai	0105
0108	陰イオン錯体	in−ion sakutai	0106
0109	印加電位	inka den'i	0130
0110	陰極暗部	inkyoku anbu	0312
0111	陰極降下電圧	inkyoku kōka den'atsu	0314
0112	陰極過程の（電流の）符号の規約	inkyoku katei no (denryū no) fugō no kiyaku	0317
0113	陰極グロー	inkyoku gurō	0315
0114	陰極降下領域	inkyoku kōka ryōiki	0313
0115	陰極蒸発	inkyoku jōhatsu	0321
0116	陰極スパッター	inkyoku supatta	0319
0117	陰極線ポーラログラフィー	inkyoku−sen pōrarogurafī	0318
0118	陰極層アーク	inkyoku−sō āku	0316
0119	インダクタ	indakuta	1100
0120	陰電子	in denshi	1543
0121	インピーダンス変換器	inpī dansu henkan−ki	1081
0122	隠蔽放電	inpei hōden	2190
0123	ウエストコット断面積	uesutokotto danmenseki	2592
0124	宇宙線	uchū sen	0502

0125	ウルブリヒトの球	uruburihito no kyū	2532
0126	運動量移行	undōryō ikō	1508
0127	Ａ－Ｄ変換器	e－dī henkan－ki	0078
0128	永続平衡	eizoku heikō	2153
0129	ＨＲＥＬＳ	eichi－āru－ī －eru－esu	1046
0130	泳動電流	eidō denryu	1462
0131	永年平衡	einen heikō	2153
0132	エージング	ējingu	0054
0133	エーロゾル	ēro zoru	0049
0134	エーロゾルキャリヤーガス	ēro zoru kyariyā gasu	0050
0135	液－液抽出	eki － eki chūshutsu	1332
0136	液－液分配	eki－eki bunpai	1330
0137	液－液平衡	eki－eki heikō	1331
0138	液液クロマトグラフィー	eki－eki kuromatogurafī	1329
0139	液間電位［差］	ekikan den'i［sa］	1327
0140	液固クロマトグラフィー	eki－ko kuromatogurafī	1339
0141	エキシマールミネセンス	ekisimā ruminesensu	0827
0142	液相	ekisō	1334
0143	液体吸引速度	ekitai kyūin sokudo	1990
0144	液体クロマトグラフィー	ekitai kuromatogurafī	1323
0145	液体ゲルクロマトグラフィー	ekitai geru kuromatogurafī	1325
0146	液体固定相	ekitai kotei sō	1340
0147	液体消費速度	ekitai shōhi sokudo	1991
0148	液体試料	ekitai shiryō	1335
0149	液体試料の注入	ekitai shiryō no chūnyū	1336
0150	液体シンチレータ計数器	ekitai shinchirēta keisū－ki	1337
0151	液体シンチレータ検出器	ekitai shinchirēta kenshutsu－ki	1338
0152	液体抽出	ekitai chūshutsu	1324
0153	液体容積	ekitai yōseki	1341
0154	液体レーザ	ekitai rēza	1328
0155	液滴ネブライザー	ekiteki neburaizā	0705
0156	液滴発生器	ekiteki hassei－ki	0704
0157	液絡	ekiraku	1326,1333
0158	エスカ（ＥＳＣＡ）	esuka	0767
0159	Ｘ線	ekkusu－sen	2598
0160	Ｘ線エスケープピーク	ekkusu－sen esukēpu pī ku	2604
0161	Ｘ線回析	ekkusu－sen kaiseki	2601
0162	Ｘ線強度	ekkusu－sen kyōdo	2610
0163	Ｘ線光電子	ekkusu－sen kōdenshi	2613
0164	Ｘ線光電子分光法（ＸＰＳ）	ekkusu－sen kōdenshi bunkō－hō	2612
0165	Ｘ線スペクトル	ekkusu－sen supekutoru	2614
0166	Ｘ線測定	ekkusu－sen sokutei	2611
0167	Ｘ線データの解釈	ekkusu－sen dēta no kaishaku	2600
0168	Ｘ線発光分光測定［法］	ekkusu－sen hakkō bunkō sokutei［－hō］	2602
0169	Ｘ線発光分光法	ekkusu－sen hakkō bunkō－hō	2603
0170	Ｘ線発生	ekkusu－sen hassei	2606
0171	Ｘ線分析	ekkusu－sen bunseki	2599
0172	ＸＰＳ	ekkusu－pī －esu	2612
0173	エネルギー移動	enerugī idō	0803
0174	エネルギー準位	enerugī jun'i	0799
0175	エネルギー遷移	enerugī sen'i	0804
0176	エネルギー束密度	enerugī soku mitsudo	0798
0177	エネルギー分解能	enerugī bunkai－nō	0800
0178	エネルギー分散	enerugī bunsan	0796
0179	エネルギー分散型蛍光Ｘ線分析	enerugī bunsan－gata keikō ekkusu－sen bunseki	0797
0180	エネルギー放出状態	enerugī hōshutsu jōtai	0845
0181	エネルギー目盛	enerugī memori	0801
0182	エピカドミウム中性子	epi kadomiumu chūseishi	0815
0183	エマネーション熱分析	emanēshon netsu bunseki	0787
0184	ＬＲＭＡ	eru－āru－emu－ē	1275
0185	エレクトログラフィー	erekutorogurafī	0748
0186	エレクトロスプレイ	erekutoro supurei	0774
0187	遠隔結合	enkaku ketsugō	2048
0188	塩基	enki	0207
0189	塩基性	enki－sei	0212
0190	塩橋溶液	enkyō yōeki	0267
0191	燃焼点	nenshō ten	0281
0192	塩析	enseki	2098
0193	塩析クロマトグラフィー	enseki kuromatogurafī	2099
0194	エンタルピー滴定	entarupī tekitei	0811

0195	エンタルピー滴定終点	entarupī tekitei shūten	0810
0196	円筒形	entō−kei	0557
0197	塩入	en'nyū	2097
0198	円偏光	en henkō	0376
0199	応答勾配	ōtō kōbai	2070
0200	応答時間	ōtō jikan	2071
0201	応答定数	ōtō teisū	2069
0202	オージェ効果	ōje kōka	0167
0203	オージェ収率	ōje syūritsu	0173
0204	オージェ電子	ōje denshi	0168
0205	オージェ電子収率	ōje denshi shūritsu	0170
0206	オージェ電子分光法	ōje denshi bunkō−hō	0169
0207	オージェピークエネルギー	ōje pīku enerugī	0171
0208	オージェ分光法	ōje bunkō−hō	0172
0209	オートラジオグラフ	ōto−rajiogurafu	0179
0210	オシログラフィー	oshirogurafī	1644
0211	オシロポーラログラフィー	oshiro−pōrarogurafī	1645
0212	オストワルド熟成	osutowarudo jukusei	1646
0213	汚染現象（沈殿における）	osen genshō	0455
0214	汚染（沈殿の）	osen	0454
0215	遅い中性子	osoi chūseishi	2217
0216	音衝撃波シフター系	oto shōgeki−ha shifutā kei	0030
0217	オプチカルファイバー	opuchikaru faibā	1629
0218	親イオン	oya ion	1664,1858
0219	親核種	oya kakushu	0877
0220	親物質	oya busshitsu	0876
0221	オリフィス	orifisu	1639
0222	オン オフ比	on−ofu hi	1618
0223	温度効果	ondo kōka	2402
0224	温度滴定	ondo tekitei	2448
0225	温度滴定終点	ondo tekitei shūten	2447
0226	温度プログラミングクロマトグラフィー	ondo puroguramingu kuromatogurafī	2403
	か行		
0227	カーセル	kāseru	1246
0228	カーボン カップ アトマイザ	kābon kappu atomaiza	0302
0229	カーボンフィラメントアトマイザ	kābon firamento atomaiza	0303
0230	カーボンロッド炉	kābon roddo ro	0304
0231	ガイガーミュラー計数管	gaigā−myurā keisu−kan	0980
0232	ガイガ計数域	gaigā keisū iki	0981
0233	ガイガしきい値	gaigā shikii−chi	0982
0234	開管カラム	kaikan karamu	1622
0235	開管カラムクロマトブラフィー	kaikan karamu kuromatogurafī	1621
0236	開口しぼりの径	kaikō shibori no kei	0123
0237	開始	kaishi	1106
0238	開始線	kaishi sen	1300,2335
0239	開始点	kaishi ten	2336
0240	回収係数	kaishū keisū	2008
0241	回収試験	kaishū shiken	2009
0242	回収率	kaishū ritsu	2007
0243	回収率パーセント	kaishū ritsu pāsento	1700
0244	ガイスラー管	gaisurā kan	0983
0245	回折電子	kaisetsu denshi	0635
0246	外挿開始点	gaisō kaishi ten	0863
0247	外挿飛程	gaisō hitei	0864
0248	階段セクター	kaidan sekutā	2347
0249	階段線蛍光	kaidan sen keikō	2350
0250	階段高さ	kaidan takasa	2346
0251	階段波ポーラログラフィー	kaidan−ha porarogurafī	2320
0252	回転エネルギー遷移	kaiten enerugī sen'i	2093
0253	回転円板シャッター	kaiten enban shattā	2090
0254	回転温度	kaiten ondo	2094
0255	回転鏡シャッター	kaiten−kyō shattā	2091
0256	回転プラットホーム	kaiten puratto hōmu	2092
0257	外部重原子効果	gaibu jū−genshi kōka	0852
0258	外部帯	gaibu tai	1647
0259	外部点火器によるスパーク放電	gaibu tenka−ki ni yoru supāku hōden	0853
0260	外［部］標準	gai[bu]hyōjun	0854
0261	壊変定数	kaihen teisū	0665
0262	壊変率	kaihen ritsu	0666
0263	開放端中空陰極	kaihō−tan chūkū inkyoku	1619

0264	界面	kaimen	1143
0265	解離	kairi	0671
0266	解離干渉	kairi kanshō	0673
0267	解離定数	kairi teisū	0672
0268	解離度	kairido	0573
0269	解離ポテンシャル	kairi potensharu	0674
0270	カウント	kaunto	0505
0271	化学アクチノメータ	kagaku akuchinomēta	0348
0272	化学イオン化	kagaku ion-ka	0355
0273	化学炎	kagaku en	0353
0274	化学干渉	kagaku kanshō	0354
0275	化学吸着	kagaku kyūchaku	0363
0276	化学光量計	kagaku kōryō-kei	0348
0277	化学シフト	kagaku shifuto	0357
0278	化学熟成	kagaku jukusei	0349
0279	化学線量計	kagaku senryō-kei	0350
0280	化学的収率	kagaku-teki shūritsu	0358
0281	化学的同位体交換	kagaku-teki dōitai kōkan	0356
0282	化学天びん	kagaku tenbin	0085
0283	化学電離	kagaku denri	0359
0284	化学当量	kagaku tōryō	0352
0285	化学はかり	kagaku hakari	0085
0286	化学発光	kagaku hakkō	0360
0287	化学発光指示薬	kagaku hakkō shijiyaku	0362
0288	化学発光分析	kagaku hakkō bunseki	0361
0289	化学分析のための電子分光法(ESCA)	kagaku bunseki no tameno denshi bunkō-hō	0767
0290	化学平衡	kagaku heikō	0351
0291	化学量論	kagaku ryōron	2353
0292	化学量論的終点	kagaku ryōron-teki shūten	2352
0293	鏡の配置	kagami no haichi	1467
0294	限られた量の試料	kagirareta ryō no shiryō	2106
0295	核異性状態	kaku isei jōtai	1228
0296	核異性体	kaku isei-tai	1597
0297	核異性体転位	kaku isei-tai ten'i	1229
0298	核化学	kaku kagaku	1589
0299	拡散	kakusan	0637
0300	拡散係数	kakusan keisū	0638
0301	拡散接合形半導体検出器	kakusan setsugō-gata handōtai kenshutsu-ki	0636
0302	拡散電流	kakusan denryū	0639
0303	拡散電流定数	kakusan denryū teisū	0640
0304	核子	kakushi	1607
0305	核子数	kakushi-sū	1608
0306	核種	kakushu	1609
0307	核種の質量	kakushu no shitsuryō	1610
0308	核生成	kaku seisei	1606
0309	核生成の速度	kaku seisei no sokudo	1992
0310	核転位	kaku ten'i	1605
0311	核同重体	kaku dojū-tai	1596
0312	角度ネブライザー	kakudo neburaizā	0102
0313	核燃料	kaku nenryō	1593
0314	核燃料サイクル	kaku nenryō saikuru	0950
0315	核反応	kaku han'nō	1599
0316	核飛跡	kaku hiseki	1602
0317	核飛跡検出器	kaku hiseki kenshutsu-ki	1603
0318	角分散	kaku bunsan	0101
0319	核分裂	kaku bunretsu	0887
0320	核分裂	kaku bunretsu	1591
0321	核分裂可能［な］	kaku bunretsu kanō[na]	0888
0322	核分裂鎖列収率	kaku bunretsu saretsu shūritsu	0330
0323	核分裂収率	kaku bunretsu shūritsu	0892
0324	核分裂性	kaku bunretsu-sei	0885
0325	核分裂性核種	kaku bunretsu-sei kakushu	0886
0326	核分裂生成物	kaku bunretsu seiseibutsu	0891
0327	核分裂中性子	kaku bunretsu chūseishi	0890
0328	核分裂反応	kaku bunretsu han'nō	1592
0329	核分裂片	kaku bunretsu hen	0889
0330	核変換	kaku henkan	1604
0331	核融合	kaku yūgō	1594
0332	核融合反応	kaku yūgō han'nō	1595
0333	核粒子	kaku ryūshi	1598

0334	重ね合わせ	kasane awase	2371
0335	荷重	kajū	1342
0336	ガス安定化アーク	gasu anteika āku	0977
0337	ガスクロマトグラフィー	gasu kuromatografī	0967
0338	ガス制限オリフィス	gasu seigen orifisu	0975
0339	ガス増幅	gasu zōfuku	0966
0340	ガスホールドアップ体積	gasu hōrudoappu taiseki	0970
0341	ガス流通［の］	gasu ryūtsū ［no］	0918
0342	ガスレーザ	gasu rēza	0971
0343	過制振放電	ka－seishin hōden	1651
0344	加成性	kaseisei	0041
0345	画像装置	gazō sōchi	1079
0346	数え落し	kazoe otoshi	0511
0347	カソーディック ストリッピング	kasōdikku sutorippingu	0320
0348	偏り	katayori	0224
0349	活性個体(クロマトグラフィー固定相の)	kassei kotai	0036
0350	カップ電極	kappu denkyoku	0539
0351	活量	katsuryō	0037
0352	活量係数	katsuryō keisū	0038
0353	活量係数（重量モル濃度に基づく）	katsuryō keisū	1487
0354	活量係数（モル濃度に基づく）	katsuryō keisū	1491
0355	活量（重量モル濃度に基づく）	katsuryō	1486
0356	活量（モル濃度に基づく）	katsuryō	1490
0357	荷電交換	kaden kōkan	0340
0358	価電子	ka denshi	2551
0359	荷電電流	kaden denryū	0294,0344,0696
0360	荷電粒子	kaden ryūshi	0339
0361	荷電粒子放射化	kaden ryūshi hōsha－ka	0338
0362	可動性担体電極	kadō－sei tantai denkyoku	1475
0363	加熱曲線	kanetsu kyokusen	1022
0364	加熱曲線決定	kanetsu kyokusen kettei	1021
0365	加熱速度	kanetsu sokudo	1023
0366	かぶり	kaburi	0931
0367	壁安定化アーク	kabe anteika āku	2582
0368	過飽和	kahōwa	2373
0369	過飽和溶液	kahōwa yoeki	2372
0370	から（空）	kara	0235
0371	からがけの指示値	karagake no shiji－chi	1567
0372	ガラス電極	garasu denkyoku	0992
0373	ガラス電極誤差	garasu denkyoku gosa	0993
0374	ガラス電極のゼロ点	garasu denkyoku no zero－ten	2617
0375	ガラス電極の内部参照電極	garasu denkyoku no naibu sanshō denkyoku	1161
0376	ガラス電極の内部充塡溶液	garasu denkyoku no naibu jūten yōeki	1159
0377	カラム	karamu	0408
0378	カラム温度	karamu ondo	0411
0379	カラムクロマトグラフィー	karamu kuromatografī	0409
0380	カラム性能	karamu seinō	0410
0381	カラム容積	karamu yōseki	0412
0382	カルベ型装置	karube－gata sōchi	0289
0383	カローセク ポーラログラフィー	karōseku pōrarografī	1245
0384	間隙	kangeki	0964
0385	間隙抵抗	kangeki teikō	0965
0386	緩衝	kanshō	0271
0387	干渉	kanshō	1144
0388	干渉曲線	kanshō kyokusen	1145
0389	緩衝剤	kanshō zai	0271
0390	緩衝剤添加アーク	kanshō zai tenka āku	0275
0391	緩衝剤添加法	kanshō zai tenka－hō	0273
0392	緩衝指数	kanshō shisū	0276
0393	緩衝する	kanshō－suru	0272
0394	干渉物質	kanshō busshitsu	1146
0395	緩衝容量	kanshō yōryō	0274
0396	干渉抑制機構	kanshō yokusei kikō	2046
0397	干渉抑制剤	kanshō yokusei zai	2045
0398	間接滴定	kansetsu tekitei	1095
0399	完全蒸発	kanzen jōhatsu	0417
0400	乾燥エーロゾル	kansō ēro zoru	0709
0401	観測高さ	kansoku takasa	1027
0402	観測高さ	kansoku takasa	1612
0403	観測値	kansoku chi	1611

0404	感度	kando	2072,2176
0405	感度（イオン源の）	kando	2180
0406	感度（気体導入の）	kando	2179
0407	感度（直接導入の）	kando	2178
0408	管の組み立て	kan no kumitate	2517
0409	ガンマカスケード	ganma kasukēdo	0957
0410	ガンマ線	ganma－sen	0959,0961
0411	ガンマ線スペクトル	ganma－sen supekutoru	0963
0412	ガンマ線スペクトロメータ	ganma－sen supekutoromēta	0962
0413	ガンマ線の吸収	ganma－sen no kyūshū	0016
0414	ガンマ線捕獲	ganma－sen hokaku	0960
0415	ガンマ量子	ganma ryōshi	0958
0416	管理巾	kanri haba	0463
0417	還流型ネブライザ	kanryū－gata neburaiza	2031
0418	貫流交換容量	kanryū kōkan yōryō	0264
0419	気液クロマトグラフィー	ki－eki kuromatografī	0972
0420	機械化	kikaika	1422
0421	機械化する	kikaika－suru	1423
0422	機械的シャッター	kikai－teki shattā	1420
0423	幾何学的因子	kika－gaku－teki inshi	0988
0424	幾何学的減衰	kika－gaku－teki gensui	0987
0425	希ガス	ki gasu	1566
0426	機器中性子放射化分析	kiki chūseishi hōshaka bunseki	1121
0427	機器放射化分析	kiki hōshaka bunseki	1119
0428	機構	kikō	1421
0429	気固クロマトグラフィー	ki－ko kuromatografī	0976
0430	疑似単色光光度法	giji tanshoku－kō kōdo－hō	1909
0431	希釈剤	kishaku zai	0643
0432	希釈試験	kishaku shiken	0644
0433	基準値ｐＨ標準液	kijunchi pī －eichi hyōjun－eki	2026
0434	基準電極	kijun denkyoku	2018
0435	基準ピーク	kijun pī ku	0211
0436	基準ピークに対する［相対］強度	kijun pī ku ni taisuru ［sōtai］kyōdo	1137
0437	基準物質	kijun busshitsu	2024
0438	基準物質ホルダー	kijun busshitsu horudā	2021
0439	キセノンランプ	kisenon ranpu	2597
0440	基線	kisen	0209
0441	気体温度	kitai ondo	0978
0442	気体状	kitai－jō	0969
0443	気体比例検出器	kitai hirei kenshutsu－ki	0974
0444	気体放電	kitai hōden	0968
0445	規定液	kitei eki	1586
0446	基底状態	kitei jōtai	1006
0447	規定度	kiteido	1585
0448	起電力	kidenryoku	0752
0449	揮発	kihatsu	2567
0450	逆線分散	gyaku sen bunsan	2004
0451	逆相クロマトグラフィー	gyakusō kuromatografī	2084
0452	逆抽出	gyaku chūshutsu	0192
0453	逆抽出する	gyaku chūshutsu suru	0191
0454	逆抽出物	gyaku chūshutsu butsu	0190
0455	逆直接導入バーナ	gyaku chokusetsu dōnyū bāna	2083
0456	逆滴定	gyaku tekitei	0201
0457	逆滴定する	gyaku tekitei－suru	0200
0458	逆微分電位差滴定	gyaku bibun den’isa tekitei	1177
0459	逆冷却速度曲線	gyaku reikyaku sokudo kyokusen	1176
0460	逆火	gyakka	0900
0461	キャパシティーファクター	kyapasitī fakutā	0295
0462	キャリヤーガス	kyariyā gasu	0309
0463	吸引	kyūin	0140
0464	吸引する	kyūin－suru	0139
0465	吸引速度	kyūin sokudo	0141
0466	吸引ネブライザー	kyūin neburaizā	2367
0467	吸光係数	kyūkō keisū	0021
0468	吸光度	kyūkō do	0006
0469	吸収	kyūshū	0010
0470	吸収エネルギー	kyūshū enerugī	0014
0471	吸収係数	kyūshū keisū	0011
0472	吸収遷移確率	kyūshū sen’i kakuritsu	0019
0473	吸収遷移確率	kyūshū sen’i kakuritsu	2496

0474	吸収線の半値幅	kyūshūsen no hanchi haba	1009
0475	吸収線量	kyūshū senryō	0007
0476	吸収体	kyūshū tai	0009
0477	吸収端飛躍比	kyūshū−tan hiyaku hi	0012
0478	吸収電子	kyūshū denshi	0008
0479	吸収プロファイル	kyūshū purofairu	0020
0480	吸収率	kyūshū ritsu	0015
0481	吸収路程	kyūshū rotei	0018
0482	Qスイッチ	kyū suittchi	1892
0483	Qスイッチ動作	kyū suittchi dōsa	1893
0484	吸蔵	kyūzo	1614
0485	吸着	kyūchaku	0045
0486	吸着クロマトグラフィー	kyūchaku kuromatografī	0046
0487	吸着剤	kyūchaku zai	0044
0488	吸着指示薬	kyūchaku shijiyaku	0048
0489	吸着質	kyūchaku shitsu	0043
0490	吸着電流	kyūchaku denryū	0047
0491	吸熱	kyūnetsu	0793
0492	吸熱ピーク	kyūnetsu pī ku	0794
0493	吸収端波長	kyūshū−tan hachō	0013
0494	キュリー	kyurī	0540
0495	共イオン	kyō ion	0392
0496	凝塊	gyōkai	0051
0497	凝結	gyōketsu	0382
0498	凝集	gyōshū	0053,0904
0499	凝集する	gyōshū−suru	0052
0500	共存物	kyōzon butsu	0438
0501	共沈［殿］	kyō chin[den]	0490
0502	強度	kyōdo	1129
0503	強度の校正用機器	kyōdo no kōsei−yō kiki	1132
0504	強度の橋渡し	kyōdo no hashiwatashi	1130
0505	強度の橋渡し比	kyōdo no hashiwatashi hi	1131
0506	強度変調光束	kyōdo henchō kōsoku	1133
0507	強度変調中空陰極管	kyōdo henchō chūkū inkyoku−kan	1134
0508	共鳴エネルギー	kyōmei enerugī	2064
0509	共鳴線	kyōmei sen	2065
0510	共鳴中性子	kyōmei chūseishi	2066
0511	共鳴広がり	kyōmei hirogari	2063
0512	共鳴分光器	kyōmei bunkō−ki	2067
0513	局所気体温度	kyokusho kitai ondo	1349
0514	局所原子化率	kyokusho genshika ritsu	1346
0515	局所蒸発効率	kyokusho jōhatsu kōritsu	1345
0516	局所蒸発率	kyokusho jōhatsu ritsu	1348
0517	局所脱溶媒率	kyokusho datsu yōbai ritsu	1347
0518	局所熱平衡	kyokusho netsu heikō	1350
0519	局所分析	kyokusho bunseki	1344
0520	極性	kyokusei	1788
0521	曲線修正器	kyokusen shūsei−ki	0548
0522	霧吹き	kirifuki	2309
0523	霧吹き型ネブライザ	kirifuki−gata neburaiza	1784
0524	キレート滴定	kirēto tekitei	0347
0525	記録	kiroku	2006
0526	均一磁場	kin'itsu jiba	1056
0527	均一分布係数	kin'itsu bunpu keisū	1055
0528	均質化	kinshitsu−ka	1227
0529	均質沈殿［法］	kinshitsu chinden[−hō]	1825
0530	均質膜電極	kinshitsu maku denkyoku	1057
0531	金属イオン錯体の安定度定数	kinzoku ion sakutai no anteido teisū	2316
0532	金属蛍光指示薬	kinzoku keikō shijiyaku	1437
0533	金属指示薬	kinzoku shijiyaku	1436
0534	金属蒸気アーク	kinzoku jōki āku	1438
0535	金属蒸気アークランプ	kinzoku jōki āku ranpu	1439
0536	金属フィラメント原子化装置	kinzoku firamento genshi−ka sōchi	1435
0537	クウェンチ	kuenchi	1911
0538	空間安定化	kūkan antei−ka	2259
0539	空間電荷	kūkan denka	2242
0540	空間分解技術	kūkan bunkai gijutsu	2243
0541	空間分解［能］	kūkan bunkai[nō]	2258
0542	空間分布干渉	kūkan bunpu kanshō	2257
0543	空気ピーク	kūki pīku	0055

No.	Japanese	Romanization	Ref.
0544	空隙率	kūgeki−ritsu	1165
0545	空隙体積速度	kūgeki taiseki sokudo	1166
0546	空隙容積	kūgeki yōseki	1168
0547	空洞共振器	kūdō kyōshin−ki	2068
0548	偶発同時計数	gūhatsu dōji keisū	1987
0549	クーロメトリー	kūrometori	0504
0550	クーロン滴定	kūron tekitei	0503
0551	矩形波電流	kukei−ha denryū	2313
0552	矩形波ポーラログラフィー	kukei−ha pōrarogurafī	2314
0553	屈折効果	kussetsu kōka	2032
0554	屈折率	kussetsu ritsu	2033
0555	組合せ同時測定技法	kumiawase dōji sokutei gihō	0515
0556	グラジェント層	gurajiento sō	0997
0557	グラファイトスパーク	gurafaito supāku	1001
0558	クリーンアップ	kurīn appu	0380
0559	クリーンベンチ	kurīn benchi	0377
0560	クリーンルーム	kurīn rūmu	0378
0561	クルックス暗部	kurukkusu anbu	0523
0562	グレイ	gurei	1005
0563	クレーター	kurētā	0518
0564	クレーター径	kerētā kei	0520
0565	クレーターの形	kurētā no katachi	0521
0566	クレーター深さ	kurētā fukasa	0519
0567	グロー放電	gurō hōden	0995
0568	クロノアンペロメトリー	kurono−anperometorī	0369
0569	クロノクーロメトリー	kurono−kūrometorī	0371
0570	クロノクーロメトリー定数	kurono−kūrometorī teisū	0370
0571	クロノポテンショメトリー	kurono−potenshometorī	0375
0572	クロノポテンショメトリー終点	kurono−potenshometorī shūten	0373
0573	クロノポテンショメトリー定数	kurono−potenshometorī teisū	0372
0574	クロノポテンショメトリー滴定	kurono−potenshometorī tekitei	0374
0575	グロビュールアーク	gurobyūru āku	0994
0576	クロマトグラフ	kuromatogurafu	0366
0577	クロマトグラフィー	kuromatogurafī	0368
0578	クロマトグラフィーを行う	kuromatogurafī wo okonau	0367
0579	クロマトグラム	kuromatoguramu	0365
0580	蛍光	keikō	0908
0581	蛍光Ｘ線	keikō ekkusu−sen	2605
0582	蛍光効率	keikō kōritsu	0910
0583	蛍光収率	keikō shūritsu	0915
0584	蛍光滴定	keikō tekitei	0917
0585	蛍光滴定の終点	keikō tekitei no shūten	0916
0586	蛍光の全量子収率	keikō no zen ryōshi shūritsu	2484
0587	蛍光のパワー収率	keikō no pawā shūritsu	1817
0588	蛍光分光光度計	keikō bunkō kōdo−kei	0913
0589	蛍光分析	keikō bunseki	0909
0590	蛍光量子効率	keikō ryōshi kōritsu	0912,1902
0591	蛍光量子収率	keikō ryōshi shūritsu	1905
0592	蛍光力効率	keikō ryoku kōritsu	0911
0593	傾斜充てん［法］	keisha jūten[−hō]	0998
0594	傾斜溶離［法］	keisha yōri[−hō]	0996
0595	形状に関する量	keijō ni kansuru ryō	0986
0596	計数	keisū	0505
0597	計数回路	keisū kairo	2120
0598	計数管	keisū−kan	0508
0599	計数効率	keisū kōritsu	0509,0604,0608
0600	計数精度	keisū seido	0512
0601	計数装置	keisū sōchi	2119
0602	計数値の標準偏差	keisūchi no hyōjun hensa	2323
0603	計数のジェオメトリー	keisū no jeometorī	0510
0604	計数率	keisū ritsu	0513,0514
0605	計数率計	keisū ritsu−kei	1989
0606	計装	keisō	1122
0607	ゲート型光検出器	gēto−gata hikari kenshutsu−ki	0979
0608	結果	kekka	2075
0609	結合	ketsugō	0228
0610	結合エネルギー	ketsugō enerugī	0229
0611	結合効率	ketsugō kōritsu	0516
0612	結合装置	ketsugō sōchi	0517
0613	結晶回折	kesshō kaisetsu	0529

No.	Term	Romanization	Ref.
0614	結晶回折スペクトロメータ	kesshō kaisetsu supekutorometā	0530
0615	結晶格子	kesshō kōshi	0531
0616	結晶性イオン選択性電極	kesshō−sei ion−sentakusei denkyoku	0532
0617	結晶特性［値］	kesshō tokusei [chi]	0527
0618	ゲッター作用	gettā sayō	0990
0619	ゲル浸透クロマトグラフィー	geru shintō kuromatografī	0985
0620	ゲルラッハの等強線対	gerurahha no tōkyō sentsui	0989
0621	ゲルろ過	geru roka	0984
0622	限界泳動電流	genkai eidō denryū	1296
0623	限界拡散電流	genkai kakusan denryū	1294
0624	限界吸着電流	genkai kyūchaku denryū	1291
0625	限界接触電流	genkai sesshoku denryū	1292
0626	限界電流	genkai denryū	1293
0627	限界反応電流	genkai han'nō denryū	1295
0628	原子核の崩壊	genshi kaku no hōkai	1590
0629	原子化効率	genshi−ka kōritsu	0162
0630	原子化装置	genshi−ka sōchi	0164
0631	原子化率	gensi−ka ritsu	0163
0632	原子間衝突	genshi−kan shōtotsu	1139
0633	原子吸光	genshi kyūkō	0146
0634	原子吸光分析	genshi kyūkō bunseki	0147
0635	原子蛍光	genshi keikō	0151
0636	原子蛍光分光分析	genshi keikō bunkō bunseki	0152
0637	原子質量	genshi shitsuryō	0154
0638	原子質量単位	genshi shitsuryō tan'i	0155
0639	原子衝撃	genshi shōgeki	0148
0640	原子スペクトル	genshi supekutoru	0157
0641	原子スペクトル線	genshi supekutoru sen	0158
0642	原子線	genshi sen	0153
0643	原子の同位体	genshi no dōitai	1238
0644	原子発光	genshi hakkō	0149
0645	原子発光分光分析	genshi hakkō bunkō bunseki	0150
0646	原子番号	genshi bangō	0156
0647	原子分光学	genshi bunkōgaku	0159
0648	原子分光法における基線法	genshi bunkō−hō ni okeru kisen−hō	0210
0649	検出	kenshutsu	0603
0650	検出器	kenshutsu−ki	0606
0651	検出器の有効容積	kenshutsu−ki no yūkō yoseki	2175
0652	検出器不感時間	kenshutsu−ki fukan jikan	0607
0653	検出器分解能	kenshutsu−ki bunkainō	0609
0654	検出限界	kenshutsu genkai	0605,1297
0655	検出能	kenshutsu−nō	1818
0656	原子量	genshi ryō	0160
0657	原子炉	genshi ro	1600
0658	減衰	gensui	0165
0659	減衰係数	gensui keisū	0166
0660	元素	genso	0775
0661	元素間効果	genso−kan kōka	1142
0662	減速	gensoku	1477
0663	減速材	gensoku zai	1478
0664	元素分析	genso bunseki	0776
0665	原点	genten	2333,2334
0666	検量関数	kenryō kansū	0087,0093
0667	検量線	kenryō sen	0088,0092
0668	検量線法	kenryō sen hō	0089
0669	検量法	kenryō hō	0086
0670	高圧アーク	kōatsu āku	1041
0671	高圧キセノンランプ	kōatsu kisenon ranpu	1042
0672	高圧スパーク	kōatsu supāku	1047
0673	高温中空陰極	kō−on chūkū inkyoku	1064
0674	効果	kōka	0718
0675	光学系	kōgaku kei	1635
0676	光学信号変調	kōgaku shingō henchō	1634
0677	光学的に薄いプラズマ	kōgaku−teki ni usui purazuma	1631
0678	光学的発光分光分析法	kōgaku−teki hakkō bunkō bunseki−hō	1628
0679	光学的量	kōgaku−teki ryō	1633
0680	光学フィルター	kōgaku firutā	1630
0681	光学密度	kōgaku mitsudo	1627
0682	項間交差	kōkan kōsa	1169
0683	後期沈殿	kōki chinden	1806

0684	高強度中空陰極管	kō−kyōdo chūkū inkyoku−kan	1039
0685	高繰り返しスパーク	kō−kurikaeshi supāku	1045
0686	高繰り返し予備スパーク	kō−kurikaeshi yobi supāku	1044
0687	光子	kōshi	1745
0688	光子照射によるX線発生	kōshi shōsha ni yoru ekkusu−sen hassei	2608
0689	光子束	kōshi soku	1750
0690	光子束密度	kōshi soku mitsudo	1751
0691	光子発揮スペクトル	kōshi hakki supekutoru	1747
0692	光子発光収率	kōshi hakkō shūritsu	1749
0693	光子発光分光学	kōshi hakkō bunkōgaku	1748
0694	光子放出率	kōshi hōshutsu ritsu	1749
0695	高周波昇圧トランス	kōshūha shōatsu toransu	1037
0696	高周波短絡コンデンサー	kōshūha tanraku kondensa	1036
0697	高周波滴定	kōshūha tekitei	1038
0698	高周波電導度終点	kōshūha dendōdo shūten	1033
0699	高周波電導度測定〔法〕	kōshūha dendōdo sokutei〔−hō〕	1035
0700	高周波電導度滴定	kōshūha dendōdo tekitei	1034
0701	高周波の重なり	kōshūha no kasanari	1019
0702	高周波プラズマ	kōshūha purazuma	1974
0703	高周波ポーラログラフィー	kōshūha pōrarogurafī	1975
0704	校正曲線	kōsei kyokusen	0287
0705	酵素−基質電極	kōso kishitsu denkyoku	0814
0706	光束	kōsoku	0919,1374
0707	高速原子衝撃法	kōsoku genshi shōgeki−hō	0873
0708	高速透過電子線回折（ＴＨＥＥＤ）	kōsoku tōka denshi−sen kaisetsu	2504
0709	光束の幅	kōsoku no haba	2594
0710	高速反射電子線回折(RHEED)	kōsoku hansha denshi−sen kaisetsu	2030
0711	高調波交流ポーラログラフィー	kōchoha kōryū pōrarogurafī	1031
0712	光電減弱係数	kōden genjaku keisū	1726
0713	光電効果	kōden kōka	1727
0714	光電子	kōdenshi	1729
0715	光電子収率	kōdenshi shūritsu	1732
0716	光電子スペクトル	kōdenshi supekutoru	1730
0717	光電子増倍管	kōdenshi zōbai kan	1744
0718	光電子分光法（ＰＥＳ）	kōdenshi bunkō−hō	1731
0719	光電真空管	kōden shinkū−kan	2549
0720	光電ピーク	kōden pī ku	1728,1753
0721	光度滴定	kōdo tekitei	1742
0722	光度滴定終点	kōdo tekitei shūten	1741
0723	高反射率	kō−hansha ritsu	1043
0724	高分解能エネルギー損失分光法(HRELS)	kō−bunkai−nō enerugī sonshitsu bunkō−hō	1046
0725	後方散乱	kōhō sanran	0197
0726	後方散乱電子	kōhō sanran denshi	0198
0727	後方散乱分光法（ＢＳＳ）	kōhō sanran bunkō−hō	0199
0728	効率	kōritsu	0724
0729	交流アーク	kōryū āku	0065
0730	交流電圧	kōryū den'atsu	0068
0731	交流電圧クロノポテンショメトリー	kōryū den'atsu kurono−potenshometorī	0070
0732	交流電圧振幅	kōryū den'atsu shinpuku	0069
0733	交流電圧ポーラログラフィー	kōryū den'atsu porarogurafī	0071
0734	交流［電流］クロノポテンショメトリー	kōryū ［denryu］ kurono−potenshometorī	0066
0735	交流電流振幅	kōryū denryū shinpuku	0064
0736	交流ポーラログラフィー	kōryū porarogurafī	0067
0737	光［量］子線	kō[ryō]shi−sen	1752
0738	光［量］子放射化	kō[ryō]shi hōsha−ka	1746
0739	高レベルファラデー整流	kō−reberu faradē seiryū	1040
0740	黒鉛カップアトマイザー	kokuen kappu atomaizā	0999
0741	黒鉛管炉	kokuen−kan ro	1002
0742	黒鉛棒炉	kokuen−bō ro	1000
0743	黒体	kokutai	0233
0744	黒体の分光放射輝度	kokutai no bunkō hōsha kido	2277
0745	黒体放射	kokutai hōsha	0234
0746	誤差	gosa	0820
0747	誤差パーセント	gosa pāsento	1699
0748	固体結晶格子	kotai kesshō kōshi	2223
0749	固体検出器	kotai kenshutsu−ki	2224
0750	個体支持体	kotai shijitai	2226
0751	固体試料	kotai shiryō	2220
0752	固体マトリックス電極	kotai matorikkusu denkyoku	2085
0753	固体レーザー	kotai rēzā	2225

0754	コッセル技術	kosseru gijutsu	1249
0755	固定化イオン	koteika ion	0894
0756	固定周波数	kotei shūha−sū	0893
0757	固定相	kotei sō	2339
0758	固定相画分	kotei sō kakubun	2340
0759	固定相容積	kotei sō yōseki	2341
0760	コヒーレント光源	kohīrento kōgen	0387
0761	コヒーレント電磁波放射	kohīrento denjiha hōsha	0385
0762	コヒーレント放射	kohīrento hōsha	0386
0763	こぶ回路	kobu kairo	1377
0764	固有振動数	koyū shindō−sū	0727
0765	固溶体	koyōtai	2221
0766	固溶体試料	koyōtai shiryō	2222
0767	コリメーション	korimēshon	0397
0768	コリメータ	korimēta	0398
0769	コリメータの分解能	korimēta no bunkainō	0399
0770	コレクター	korekutā	0396
0771	混合金属錯体	kongō kinzoku sakutai	1473
0772	混合錯体	kongō sakutai	1469
0773	混合指示薬	kongō shijiyaku	1471
0774	混合配位子錯体	kongō hai'ishi sakutai	1472
0775	混合溶液法	kongō yōeki−hō	1474
0776	混在物	konzai [butsu]	1087
0777	混晶	konshō	1470
0778	こん跡	konseki	2487
0779	コンデンサー電圧	kondensā den'atsu	0292
0780	コントロール滴定	kontorōru tekitei	0480
0781	コンパレータ	konparēta	0415
0782	コンプトン角	konputon kaku	0427
0783	コンプトン効果	konputon kōka	0428
0784	コンプトン散乱	konputon sanran	0430
0785	コンプトン電子	konputon denshi	0429
0786	コンプレクサン	konpurekusan	0418
0787	コンプレクソン	konpurekuson	0424
0788	コンボリューション積分直線[電位]掃引ボルタンメトリー	konboryūshon sekibun chokusen[den'i]sōin borutanmetorī	0486
	さ行		
0789	サーモグラフィ	sāmogurafī	2439
0790	サーモグラム	sāmoguramu	2438
0791	サーモフォトメトリー	sāmo−fotometorī	2451
0792	サイクリック クロノポテンショメトリー	saikurikku kurono−potenshometorī	0550
0793	サイクリック三角波ポーラログラフィー	saikurikku sankaku−ha porarogurafī	0552
0794	サイクリック三角波ボルタンメトリー	saikurikku sankaku−ha borutanmetorī	0553
0795	サイクリック電流ステップ クロノポテンショメトリー	saikurikku denryū−suteppu kurono−potenshometorī	0551
0796	サイクロトロン	saikurotoron	0554
0797	サイクロトロン共鳴質量分析計	saikurotoron kyōmei shitsuryō bunseki−kei	0555
0798	最高回折係数	saikō kaisetsu keisū	1688
0799	最終の焦点スポット径	saishū no shōten supotto kei	2533
0800	最小線幅	saishō senpuku	1465
0801	再生電流	saisei denryū	2034
0802	再沈殿	sai chinden	2049
0803	最適スパーク条件	saiteki supāku jōken	1636
0804	再点火電圧	sai−tenka den'atsu	2035
0805	サイリスター制御	sairisutā seigyo	2466
0806	サエド（ＳＡＥＤ）	saedo	2155
0807	錯陰イオン	saku in'ion	0419
0808	錯形成［反応］	sakukeisei[han'nō]	0420
0809	錯［体］生成平衡	saku[tai] seisei heikō	0421
0810	錯滴定	saku tekitei	0422
0811	錯滴定分析［法］	saku tekitei bunseki[−hō]	0423
0812	差動［の］	sadō [no]	0617
0813	差動排気	sadō haiki	0627
0814	サハーエガートの法則	saha−egāto no hōsoku	2096
0815	サブトレース分析	sabu−torēsu bunseki	2366
0816	サブミクロ分析	sabu−mikuro bunseki	2364
0817	作用ガス	sayō gasu	2596
0818	作用電極	sayō denkyoku	2595
0819	酸塩基指示薬	san enki shijiyaku	0023
0820	酸塩基相互作用	san enki sōgo sayō	0024
0821	酸塩基滴定	san enki tekitei	0025

0822	酸化還元指示薬	sanka kangen shijiyaku	1653,2012
0823	酸化還元滴定	sanka kangen tekitei	1654,2015
0824	三角波ポーラログラフィー	sankaku-ha pōrarogurafī	2512
0825	三角波ボルタンメトリー	sankaku-ha borutanmetorī	2513
0826	三重項－三重項吸収	sanjūko － sanjūko kyūshū	2515
0827	参照強度	sanshō kyōdo	2022
0828	参照元素	sanshō genso	2019
0829	参照元素法	sanshō genso-hō	2020
0830	参照線	sanshō sen	2023
0831	参照電極	sanshō denkyoku	2018
0832	参照ビーム	sanshō bī mu	2017
0833	参照溶液	sanshō yōeki	2025
0834	三スロットバーナ	san surotto bāna	2464
0835	酸性度関数	sanseido kansū	0029
0836	酸性度定数	sanseido teisū	0028
0837	酸滴定	san tekitei	0026,0027
0838	三電極構成	san-denkyoku kōsei	2462
0839	三電極プラズマ	san denkyoku purazuma	2463
0840	サンドの式	sando no shiki	2112
0841	サンプルインジェクター	sanpuru injekutā	2104
0842	残余液間電位［差］	zan-yo ekikan den'i [sa]	2055
0843	残余液絡	zan-yo ekiraku	2054
0844	残余電流	zan-yo denryū	2051
0845	散乱	sanran	2125
0846	散乱する	sanran (suru)	2124
0847	残留イオン化	zanryū ion-ka	2053
0848	残留ガス	zanryū gasu	2052
0849	シアンバンド	shian bando	0549
0850	g 値	jī chi	1008
0851	シーベルト	shīberuto	2195
0852	シールド	shīrudo	2192
0853	シールドフレーム	shīrudo furēmu	2191
0854	紫外電子分光法	shigai denshi bunkō-hō	2537
0855	時間間隔	jikan kankaku	2405
0856	時間制御スパーク	jikan seigyo supāku	0470
0857	時間的安定化	jikan-teki antei-ka	2406
0858	時間分解	jikan bunkai	2404
0859	時間分解技術	jikan bunkai gijutsu	2472
0860	時間分解分光法	jikan bunkai bunkō-hō	2471
0861	しきいエネルギー	shikii enerugī	0802
0862	しきいエネルギー	shikii enerugī	2465
0863	色素レーザー	shikiso rēzā	0712
0864	式量濃度	shikiryō nōdo	0933
0865	軸電極	jiku denkyoku	0189
0866	刺激発光遷移確率	shigeki hakkō sen'i kakuritsu	2498
0867	試験電極	shiken denkyoku	2409
0868	自己吸収	jiko kyūshū	2159
0869	自己吸収因子	jiko kyūshū inshi	2162
0870	自己吸収効果	jiko kyūshū kōka	2161
0871	自己吸収効率	jiko kyūshū kōritu	2241
0872	自己吸収広がり	jiko kyūshū hirogari	2160
0873	自己遮蔽	jiko shahei	2166
0874	自己点火スパーク放電	jiko tenka supāku hōden	2164
0875	自己反転	jiko hanten	2165
0876	示差クロマトグラフィー	shisa kuromatogurafī	0620
0877	示差クロマトグラム	shisa kuromatoguramu	0619
0878	示差検出器	shisa kenshutsu-ki	0621
0879	示差走査熱量測定	shisa sōsa netsuryō sokutei	0628
0880	示差電位差測定［法］	shisa den'isa sokutei[-hō]	0625
0881	示差電位差滴定	shisa den'isa tekitei	0624
0882	示差電流測定［法］	shisa denryū sokutei [-hō]	0618
0883	示差熱重量測定	shisa netsu jūryō sokutei	0633
0884	示差熱重量分析	shisa netsu jūryō bunseki	0632
0885	示差熱図形	shisa netsu zukei	0631
0886	示差熱分析	shisa netsu bunseki	0629
0887	示差［の］	shisa [no]	0617
0888	示差パルスポーラログラフィー	shisa parusu pōrarogurafī	0626
0889	示差膨張計	shisa bōchō-kei	0622
0890	示差ポーラログラフィー	shisa pōrarogurafī	0623
0891	示差ボルタンメトリー	shisa borutanmetorī	0634

No.	Japanese	Romanization	Ref.
0892	死止終点	shishi shūten	0563
0893	支持体	shiji−tai	2376
0894	指示値の精密さ	shiji−chi no seimitsusa	1832
0895	事実上無限厚さ	jijitsujō mugen atsusa	0721
0896	支持電解質	shiji denkaishitsu	0208
0897	支持電解質［溶液］	shiji denkaishitsu [yoeki]	2377
0898	指示電極	shiji denkyoku	1093
0899	支持プレート	shiji purēto	2378
0900	指示薬	shijiyaku	1090
0901	指示薬ブランク	shijiyaku buranku	1091
0902	指示薬補正	shijiyaku hosei	1092
0903	四重極［型］質量分析計	shijūkyoku[gata] shitsuryō bunseki−kei	1895
0904	四重電気光学効果	shijū denki kōgaku kōka	1894
0905	指数関数型崩壊	shisū kansū−gata hōkai	0850
0906	自然広がり	shizen hirogari	1529
0907	自然放射線	shizen hōshasen	1532
0908	自然放射能	shizen hōshanō	1533
0909	失活	shikkatsu	1911
0910	失活ガス	shikkatsu gasu	1910
0911	失活過程	shikkatsu katei	0562
0912	実験表面	jikken hyōmen	0849
0913	実効カドミウム カットオフ エネルギー	jikkō kadomiumu katto−ofu enerugī	0719
0914	実効線量当量	jikkō senryō tōryō	0720
0915	実効熱中性子断面積	jikkō netsu chūseishi danmenseki	0723
0916	実効無限厚さ	jikkō mugen atsusa	0721
0917	実際の吸収	jissai no kyūshū	1999
0918	実時間	jitsu jikan	2000
0919	室飽和	shitsu hōwa	0331
0920	実用的分解能	jitsuyō−teki bunkai−nō	1820
0921	実用ｐＨ尺度	jitsuyō pī−eichi shakudo	1624
0922	実用ｐＨセル	jitsuyō pī−eichi seru	1623
0923	実用ｐＨ標準液	jitsuyō pī−eichi hyōjun eki	1625
0924	実用比容積	jitsuyō hi−yōseki	1821
0925	質量／電荷比	shitsuryō/denka hi	1391
0926	質量減衰係数	shitsuryō gensui keisū	1390
0927	質量作用の法則	shitsuryō sayō no hōsoku	1282
0928	質量数	shitsuryō−sū	1394
0929	質量スペクトル	shitsuryō supekutoru	1397
0930	質量分析計	shitsuryō bunseki−kei	1395
0931	質量分析［法］	shitsuryō bunseki[−hō]	1396
0932	質量分布	shitsuryō bunpu	1392
0933	質量分布比	shitsuryō bunpu hi	1393
0934	時定数	ji teisū	2468
0935	自動イオン化	jidō ion−ka	0174
0936	自動化	jidō−ka	0176,0177
0937	自動化する	jidō−ka	0175,0178
0938	自動酸化	jidō sanka	0180
0939	磁場	jiba	1381
0940	磁場安定化アーク	jiba anteika āku	1379
0941	磁場グロー放電	jiba gurō hōden	1382
0942	自発核分裂	jihatsu kaku bunretsu	2305
0943	自発電解重量分析［法］	jihatsu denkai jūryō bunseki[−hō]	2303
0944	自発発光遷移確率	jihatsu hakkō sen'i kakuritsu	2304
0945	自発発光遷移確率	jihatsu hakkō sen'i kakuritsu	2497
0946	自発発光のアインシュタイン係数	jihatsu hakkō no ainshutain keisū	0728
0947	磁場偏向	ziba henkō	1380
0948	磁場ペンニング効果	jiba pen'ningu kōka	1383
0949	シフター電極	shifutā denkyoku	2196
0950	四分の一波長線	shibun no ichi hachō−sen	1908
0951	四分波電位	shibunpa den'i	1907
0952	シムス（ＳＩＭＳ）	shimusu	2147
0953	ジャイアントパルス	jaianto parusu	0991
0954	写真乾板の傾斜	shashin kanpan no keisha	2467
0955	写真測光法	shashin sokkō−hō	1737
0956	写真透過率	shashin tōka−ritsu	1738
0957	写真乳剤の校正	shashin nyūzai no kōsei	1734
0958	写真による強度測定	shashin ni yoru kyōdo sokutei	1735
0959	写真パラメータ	shashin paramēta	1736
0960	しゃへい指示薬	shahei shijiyaku	2136
0961	自由ー自由遷移	jiyū − jiyū sen'i	0941

0962	周期性電圧	shūki－sei den'atsu	1702
0963	周期性電流	shūki－sei denryū	1701
0964	重原子効果	jū genshi kōka	1024
0965	収縮（アークプラズマの）	shūshuku	0452
0966	収縮グロー放電	shūshuku gurō hōden	0451
0967	修正ネルンスト式	shūsei nerunsuto shiki	1481
0968	集束鏡	shūsoku－kyō	0930
0969	自由束縛遷移	jiyū sokubaku sen'i	0938
0970	集束レンズ	shūsoku renzu	0929
0971	収着	shūchaku	2238
0972	収着等温式	shūchaku tō－on shiki	2239
0973	終点	shūten	0795
0974	充塡液	jūten eki	0881
0975	充電回路	jūden kairo	0343
0976	充塡ガス	jūten gasu	0880
0977	充塡カラム	jūten karamu	1655
0978	自由電子	jiyū denshi	0940
0979	充電時定数	jūden jiteisū	0346
0980	充電抵抗	jūden teikō	0345
0981	充電電流	jūden denryū	0344
0982	充塡物	jūten butsu	1656
0983	自由動作型操作	jiyū dōsa－gata sōsa	0942
0984	自由燃焼アーク	jiyū nenshō āku	0939
0985	周波数	shūha－sū	0943
0986	周波数倍増	shūha－sū baizō	0944
0987	重量滴定	jūryō tekitei	2591
0988	重量変化測定［法］	jūryō henka sokutei[－hō]	2590
0989	重量モル濃度	jūryō moru nōdo	1488
0990	重力供給シフター系	jūryoku kyōkyū shifutā kei	1004
0991	重力供給ネブライザー	jūryoku kyōkyū neburaizā	1003
0992	主ガス	shu gasu	1856
0993	主供給電圧	shu kyokyū den'atsu	1385
0994	熟成	jukusei	0054,2086
0995	主成分	shu seibun	1386
0996	シュタルク広がり	shutaruku hirogari	2332
0997	出現エネルギー	shutsugen enerugī	0127
0998	出現温度	shutsugen ondo	0129
0999	出現電圧	shutsugen den'atsu	0128
1000	主放電	shu hōden	1846
1001	シュポルスキースペクトル	shuporusukī supekutoru	2194
1002	シミュレーション法	shimyurēshon－hō	2199
1003	樹脂マトリックス	jushi matorikkusu	2056
1004	準安定	jun－antei	1441
1005	準安定イオンピーク	jun－antei ion pī ku	1443
1006	準安定状態	jun－antei jōtai	1445
1007	準安定［な］	jun antei[na]	1440
1008	準安定ピーク	jun－antei pīku	1444
1009	準安定分解	jun－antei	1442
1010	瞬間測定	shunkan sokutei	1118
1011	瞬間電流	shunkan denryū	1116
1012	瞬間流速	shunkan ryūsoku	1117
1013	純ファラデー電流	jun－faradē denryū	1550
1014	純保持時間	jun hoji jikan	1551
1015	純保持容量	jun hoji yōryō	1552
1016	昇華	shōka	2363
1017	蒸気雲	jōki－gumo	2554
1018	蒸気ジェット	jōki jetto	2555
1019	蒸気相干渉	jōki sō kanshō	2556
1020	消蛍光分析法	shō keikō bunseki－hō	1912
1021	消光	shōkō	1911
1022	消光ガス	shōkō gasu	1910
1023	常磁性化合物	jōjisei kagōbutsu	1663
1024	照射	shōsha	1224
1025	死容積	shi yōseki	0566
1026	焦点距離	shōten kyori	0926
1027	焦電検出器	shōden kenshutsu－ki	1891
1028	焦点スポット	shōten suppoto	0927
1029	焦点スポットの直径	shōten supotto no chokkei	0928
1030	衝突	shōtotsu	0400
1031	衝突過程	shōtotsu katei	0406

1032	衝突失活	shōtotsu shikkatsu	0404
1033	衝突シフト	shōtotsu shifuto	0407
1034	衝突断面積	shōtotsu dan−menseki	0403
1035	衝突半値幅	shōtotsu hanchi haba	0405
1036	衝突広がり	shōtotsu hirogari	0401
1037	衝突連鎖	shōtotsu rensa	0402
1038	蒸発	jōhatsu	0823,2553
1039	蒸発剤	jōhatsu−zai	2569
1040	蒸発制御剤	jōhatsu seigyo−zai	0612
1041	蒸発速度	jōhatsu sokudo	2568
1042	蒸発平衡	jōhatsu heikō	0824
1043	消滅	shōmetsu	0109
1044	消滅する	shōmetsu−suru	0108
1045	消滅放射線	shōmetsu hōshasen	0110
1046	表面汚染	hyōmen osen	2385
1047	蒸留	jōryū	0676
1048	少量	shōryō	1290
1049	少量成分	shōryō seibun	1466
1050	初温度	sho ondo	1105
1051	初期電流	shoki denryū	1104
1052	助燃ガス	jonen gasu	1652
1053	ジラードーチャルマース効果	jirādo−charumāsu kōka	2392
1054	シリコンホトダイオード	shirikon hoto daiōdo	2198
1055	試料	shiryō	2100,2265
1056	試料カップ	shiryō kappu	2110
1057	試料側ビーム	shiryō−gawa bīmu	2101
1058	試料電極	shiryō denkyoku	2102
1059	試料導入温度	shiryō dōnyū ondo	1108
1060	試料導入ガス	shiryō dōnyū gasu	1107
1061	試料導入管	shiryō dōnyū−kan	1109,1110
1062	試料ボート	shiryō bōto	2109
1063	試料ホルダー	shiryō horudā	2103
1064	試料ホルダー系	shiryō horudā kei	2266
1065	試料溶液	shiryō yōeki	2107
1066	試料ループ	shiryō rūpu	2111
1067	真空スパーク	shinkū supāku	2550
1068	真空放電	shinkū hōden	0661
1069	真空ポンプ動作系	shinkū ponpu dōsa kei	1885
1070	シングルエスケープピーク	singuru eskēpu pīku	2206
1071	シングルチャネル波高分析器	singuru chaneru hakō bunseki−ki	2204
1072	人工放射能	jinkō hōsha−nō	0138
1073	深色シフト	shinshoku shifuto	0214
1074	真性検出効率	shinsei kenshutsu kōritsu	1172,1173
1075	真性光電ピーク効率	shinsei kōden pīku kōritsu	1175
1076	真性全エネルギーピーク効率	shinsei zen enerugī pīku kōritsu	1174
1077	迅速同時計数	jinsoku dōji keisū	1866
1078	シンチレーション	shinchirēshon	2130
1079	シンチレーション計数器	shinchirēshon keisū−ki	2131
1080	シンチレーション検出器	shinchirēshon kenshutsu−ki	2132
1081	シンチレーションスペクトロメータ	shinchirēshon supekutorométa	2133
1082	シンチレータ	shinchirēta	2134
1083	シンチレータ検出器	shinchirēta kenshutsu−ki	2135
1084	シンチレータ材料	shinchirēta zairyō	2129
1085	振動エネルギー遷移	shindō enerugī sen'i	2561
1086	振動温度	shindō ondo	2562
1087	浸透クロマトグラフィー	shintō kuromatogurafī	1703
1088	振動子強度	shindō−shi kyōdo	1643
1089	振動数	shindō−sū	0943
1090	振動スパーク	shindō supāku	1641
1091	振動反応	sindō han'nō	1640
1092	振動容量型電位計	shindō yoryō−gata den'i−kei	2560
1093	真の同時計数	shin no dōji keisū	2516
1094	水銀池電極	suigin−chi denkyoku	1431
1095	水晶制御発振器	suishō seigyo hasshin−ki	0528
1096	水素電極	suiso denkyoku	1067
1097	水平電極	suihei denkyoku	1059
1098	数値	sūchi	2559
1099	スカベンジャ	sukabenja	2127
1100	スカベンジング	sukabenjingu	2128
1101	スケーラ	sukēra	2119

1102	筋入り	suji－iri	2358
1103	スターン－ボルマーの法則	sutān － borumā no hōsoku	2351
1104	ステップ	suteppu	2345
1105	ステップワイズ溶離	suteppuwaizu yōri	2348
1106	ステム（ＳＴＥＭ）	sutemu	2123
1107	ストークス蛍光	sutōkusu keikō	2354
1108	ストリッピング	sutorippingu	2359
1109	ストリッピング分析	sutorippingu bunseki	2360
1110	ストロボ間隔	sutorobo kankaku	2362
1111	スパークイオン源イオン化［法］	supāku ion gen ion－ka[－hō]	2254
1112	スパーク間隙	supāku kangeki	2248
1113	スパーク繰り返し速度	supāku kurikaeshi sokudo	2253
1114	スパーククロス励起	supāku kurosu reiki	2245
1115	スパーク周波数	supāku shūha－sū	2247
1116	スパーク状放電	supāku－jō hōden	2252
1117	スパークスタンド	supāku sutando	2256
1118	スパークチャンバー	supāku chanbā	2244
1119	スパーク点火	supāku tenka	2250
1120	スパーク発生装置	supāku hassei sōchi	2249
1121	スパーク平衡	supāku heikō	2251
1122	スパーク放電	supāku hōden	2246
1123	スパイク	supaiku	2301
1124	スパイクの強度	supaiku no kyōdo	2300
1125	スパッタ収率	supatta shūritsu	2312
1126	スパッタ速度	supatta sokudo	2311
1127	スパッタリング	supattaringu	2310
1128	スパッタリングによるイオン化	supattaringu ni yoru ion－ka	1202
1129	スピン保存則	supin hozon－soku	2302
1130	スプレー放電	supurē hōden	2308
1131	スペクトル	supekutoru	2288,2298
1132	スペクトル干渉	supekutoru kanshō	2272
1133	スペクトル強度	supekutoru kyōdo	2271
1134	スペクトルグラフ	supekutoru gurafu	2289
1135	スペクトル線シフト	supekutoru－sen shifuto	2274
1136	スペクトル線の形	supekutoru－sen no katachi	2273
1137	スペクトル線の強度	supekutoru－sen no kyōdo	1136
1138	スペクトルの次数	supekutoru no jisū	1637
1139	スペクトルの周期	supekutoru no shūki	2275
1140	スペクトルバックグラウンド	supekutoru bakku guraundo	2267
1141	スペクトルバンド幅	supekutoru bando haba	2268
1142	スポット	supotto	2306
1143	スライドスパーク	suraido supāku	2212
1144	スリット	suritto	2213
1145	スリット高さ	suritto takasa	2214
1146	スリット幅	suritto haba	2215
1147	スロットバーナ	surotto bāna	2216
1148	正イオン	sei－ion	1803
1149	正確さ	seikakusa	0022
1150	制御間隙	seigyo kangeki	0464
1151	制御高圧スパーク発生装置	seigyo kōatsu supāku hassei sōchi	0472
1152	制御交流アーク	seigyo kōryu āku	0465
1153	制御スパーク間隙	seigyo supāku kangeki	0479
1154	制御雰囲気	seigyo fun'iki	0466
1155	制限されたグロー放電	seigen sareta gurō hōden	2074
1156	制限視野電子線回折（ＳＡＥＤ）	seigen shiya denshi－sen kaisetsu	2155
1157	整合コンデンサー	seigō kondensā	1399
1158	静止点	seishiten	2073
1159	静止電極ボルタンメトリー	seishi denkyoku borutanmetorī	2338
1160	正常グロー放電	seijō gurō hōden	1584
1161	正常な眼	seijō na me	1583
1162	清浄表面	seijo hyōmen	0379
1163	生成定数	seisei teisū	0934
1164	静電場質量分析計	seidenba shitsuryo bunseki－kei	2337
1165	精度	seido	1828
1166	制動放射	seidō hōsha	0266
1167	成分	seibun	0425,0448
1168	成分含量	seibun ganryo	0450
1169	成分分類	seibun bunrui	0449
1170	精密零点法電位差測定	seimitsu zero－ten－hō den'isa sokutei	1829
1171	整流した交流アーク	seiryū－shita koryū āku	2010

1172	整流した電源	seiryū−shita dengen	2011
1173	ゼーマン効果	zēman kōka	2616
1174	ゼーマンバックグランド補正	zēman bakkugurando hosei	2615
1175	積分球	sekibun kyū	1128
1176	積分吸収	sekibun kyūshū	1123
1177	積分クロマトグラフィー	sekibun kuromatografī	1124
1178	積分検出器	sekibun kenshutsu−ki	1125
1179	積分時間	sekibun jikan	1138
1180	積分反射係数	sekibun hansha keisū	1126
1181	絶縁破壊時間	zetsuen hakai jikan	0262
1182	絶縁破壊電圧	zetsuen hakai den'atsu	0263
1183	絶縁破壊の揺れ	zetsuen hakai no yure	0261
1184	接触電流	sesshoku denryū	0310
1185	絶対温度	zettai ondo	0005
1186	絶対計数	zettai keisū	0002
1187	絶対光電ピーク効率	zettai kōden pīku kōritsu	0004
1188	絶対全エネルギーピーク効率	zettai zen enerugī pīku kōritsu	0003
1189	セミミクロ	semi mikuro	2171
1190	セミミクロ試料	semi mikuro shiryō	2172
1191	セム（ＳＥＭ）	semu	2121
1192	セメンティション	sementeishon	0325
1193	全安定度定数	zen anteido teisū	0537
1194	全イオン電流	zen ion denryū	2483
1195	遷移確率	sen'i kakuritsu	2495
1196	遷移時間	sen'i jikan	2499
1197	全エネルギーピーク	zen enerugī pīku	0953
1198	全エネルギーピーク効率	zen enerugī pīku kōritsu	0954
1199	線エネルギー付与	sen enerugī fuyo	1307
1200	線エネルギー付与依存因子	sen enerugī fuyo izon inshi	1308
1201	前期イオン化	zenki ion−ka	1838
1202	全吸収ピーク	zen kyūshū pīku	2482
1203	前駆イオン	zenku ion	1835
1204	前駆物質	zenku busshitsu	1834
1205	線形	senkei	1320
1206	線型加速質量分析計	senkei kasoku shitsuryō bunseki−kei	1301
1207	線形光電吸収係数	senkei kōden kyūshū keisū	1310
1208	線形散乱係数	senkei sanran keisū	1315
1209	扇形磁場（１８０度）	senkei jiba(180 do)	1763
1210	扇形磁場（６０度）	senkei jiba (60 do)	1765
1211	扇形磁場（９０度）	senkei jiba(90 do)	1764
1212	線形電気光学効果	senkei denki kōgaku kōka	1306
1213	線形電子加速器	senkei denshi kasoku−ki	1305
1214	扇形電場	senkei denba	1914
1215	線源	sengen	2240
1216	線減衰係数	sen gensui keisū	1302
1217	先行核	senkō kaku	1834
1218	先行同位体	senkō dōitai	1836
1219	旋光［度］滴定	senkō [do] tekitei	1793
1220	線質係数	senshitsu keisū	1897
1221	線シフト	sen shifuto	1321
1222	洗浄	senjō	2138
1223	洗浄溶液	senjō yōeki	2139
1224	浅色シフト	senshoku sifuto	1070
1225	全生成定数	zen seisei teisū	0535
1226	全生成定数	zen seisei teisū	1650
1227	線束積分	sen−soku sekibun	0922
1228	線束低減	sen−soku teigen	0921
1229	線束密度	sen−soku mitsudo	0920
1230	選択係数	sentaku keisū	2158
1231	選択的蒸発	sentaku−teki jōhatsu	2157
1232	選択的溶離	sentaku−teki yōri	2156
1233	選択透過性	sentaku tōkasei	1704
1234	前端クロマトグラフィー	zantan kuromatografī	0946
1235	選定点	sentei ten	0435
1236	線幅	sen haba	1322
1237	線広がりのパラメータ	sen hirogari no paramēta	1318
1238	全プロトン化定数	zen puroton−ka teisū	0536
1239	線プロファイル関数	sen purofairu kansū	1319
1240	線分散	sen bunsan	1304
1241	全分配定数	zen bunpai teisū	1649

1242	全保持時間	zen hoji jikan	2485
1243	線流速	sen ryūsoku	1309
1244	線量	senryo	0689
1245	線量当量	senryō tōryō	0690
1246	増感ルミネセンス	zōkan ruminesensu	2181
1247	増強	zōkyō	0807
1248	増強磁場グロー放電	zōkyō jiba gurō hōden	0253
1249	増強出力グロー放電	zōkyō shutsuryoku gurō hōden	0254
1250	増強中空陰極	zōkyō chūkū inkyoku	0252
1251	増強リン光分析法	zōkyō rinkō bunseki−hō	0806
1252	総蛍光量子効率	sō keikō ryōshi kōritsu	0914
1253	相互干渉	sōgo kanshō	1526
1254	走査型質量分析計	sōsa−gata shitsuryō bunseki−kei	0714
1255	走査型電子顕微鏡（ＳＥＭ）	sōsa−gata denshi kenbikyō	2121
1256	走査型透過電子顕微鏡（ＳＴＥＭ）	sōsa−gata tōka denshi kenbikyō	2123
1257	操作ｐＨ尺度	sōsa pī−eichi shakudo	1624
1258	操作ｐＨセル	sōsa pī−eichi seru	1623
1259	操作ｐＨ標準液	sōsa pī−eichi hyōjun eki	1625
1260	走査レーザー分析	sōsa rēzā bunseki	2122
1261	相対保持	sōtai hoji	2042
1262	相対活量	sōtai katsuryō	2036
1263	相対誤差	sōtai gosa	2041
1264	相対的計数	sōtai−teki keisū	2039
1265	相対的原子量	sōtai−teki genshi ryō	2037
1266	相対標準偏差	sōtai hyōjun hensa	2043
1267	相対密度	sōtai mitsudo	2040
1268	相滴定	sō tekitei	1716
1269	相比	sō hi	1715
1270	増巾反応	zōfuku han'nō	0077
1271	増巾反応法	zōfuku han'nō−hō	1522
1272	相分離分析	sō bunri bunseki	1713
1273	層平衡	sō heikō	1283
1274	相容積比	sō yōseki hi	1717
1275	層流	sōryū	1253
1276	層流気体のさや	sōryū kitai no saya	1254
1277	層流フレーム	sōryū furēmu	1252
1278	ソーラブラインド光電子倍増管	sōra buraindo kōdenshi baizōkan	2218
1279	ゾーン	zōn	2619
1280	速中性子	soku chūseishi	0874
1281	測定間隔（電流の）	sokutei kankaku	2105
1282	測定器の指示値	sokutei−ki no shiji−chi	1120
1283	測定強度	sokutei kyōdo	1415
1284	測定蛍光励起	sokutei keikō reiki	1414
1285	測定値	sokutei−chi	1418
1286	測定リン光励起	sokutei rinkō reiki	1416
1287	測定ルミネセンス量子収率	sokutei ruminesensu ryōshi shūritsu	1417
1288	束縛	sokubaku	0228
1289	即発中性子	sokuhatsu chūseishi	1867
1290	阻止能	soshi−nō	2355
1291	測光［法］	sokko[−hō]	1743
1292	その場微小部エックス線回折	sonoba bishōbu ekkusu−sen kaisetsu	1115
1293	その場微小部分析	sonoba bishōbu bunseki	1114
1294	素粒子	soryūshi	0777
	た行		
1295	対イオン	tai ion	0507
1296	第１種イオン選択性電極	dai−isshu ion sentaku−sei denkyoku	1848
1297	ダイオードアレイ検出器	daiōdo arei kenshutsu−ki	0645
1298	大気圧マイクロ波誘導プラズマ	taikiatsu maikuro−ha yūdō purazuma	0145
1299	対数的減少	taisū−teki genshō	1351
1300	対数滴定曲線	taisū tekitei kyokusen	1353
1301	体積流速（移動相の）	taiseki ryūsoku	2581
1302	対［電］極	tai [den]kyoku	0506
1303	大電流アーク	dai denryū āku	1030
1304	ダイナミック系	dainamikku kei	0715
1305	対流	tairyū	0481
1306	対流クロノクーロメトリー	tairyū kurono−kūrometorī	0482
1307	対流クロノポテンショメトリー	tairyū kurono−potenshometorī	0483
1308	滞留時間	tairyū jikan	2050
1309	対流ボルタンメトリー	tairyū borutanmetorī	1066
1310	タウンゼンド第一係数	taunzendo daiichi keisū	2486

1311	楕円偏光	daen henkō	0778
1312	多核錯体	takaku sakutai	1797
1313	多官能性イオン交換体	ta kan'nosei ion kōkantai	1796
1314	妥協した条件	dakyōshita jōken	0426
1315	濁度滴定	dakudo tekitei	2524
1316	濁度滴定終点	dakudo tekitei shūten	2523
1317	濁度分析〔法〕	dakudo bunseki〔－hō〕	2525
1318	多原子［の］	ta－genshi[no]	1794
1319	多重掃引ポーラログラフィー	tajū－sōin pōrarogurafī	1525
1320	多重通過法	tajū tsūka－hō	1521
1321	タスト［時間］間隔	tasuto[jikan] kankaku	2400
1322	タストポーラログラフィー	tasuto pōrarogurafī	2401
1323	多スロットバーナ	ta－surotto bāna	1524
1324	多層	tasō	1520
1325	立上り時間	tachiagari jikan	2087
1326	立上り速度	tachiagari sokudo	2088
1327	脱イオン水	datsu ion sui	0576
1328	脱塩水	datsu en sui	0581
1329	脱活性化過程	datsu kasseika katei	0562
1330	脱出深さ	dasshutsu fukasa	0821
1331	脱マスキング	datsu－masukingu	0580
1332	脱マスキング機構	datsu－masukingu kikō	2046
1333	脱マスキング剤	datsu－masukingu zai	2045
1334	脱溶媒和	datsu－yōbaiwa	0602
1335	脱溶媒和された割合	datsu－yōbaiwa sareta wariai	0601
1336	多燃料フレーム	ta nenryō furēmu	0952
1337	ダブル エスケープ ピーク	daburu esukeipu pīku	0694
1338	ダブルジャンクション型参照電極の塩橋溶液	daburu jankushon－gata sanshō denkyoku no enkyō yōeki	0268
1339	ダブル電位ステップ クロノアンペロメトリー	daburu den'i suteppu kurono－anperometorī	0697
1340	ダブル電位ステップ クロノクーロメトリー	daburu den'i suteppu kurono－kūrometorī	0698
1341	ダブル トーン ポーラログラフィー	daburu tōn pōrarogurafī	0701
1342	多目的スパーク光源	tamokuteki supāku kōgen	1523
1343	単一交替	tan'itsu kōtai	1509
1344	単一パルス	tan'itsu parusu	2208
1345	単核三成分錯体	tankaku san seibun sakutai	1516
1346	単核二成分錯体	tankaku ni seibun sakutai	1515
1347	タンク回路	tanku kairo	2397
1348	タングステンーハロゲン ランプ	tangusuten － harogen ranpu	2521
1349	タングステンーフィラメントランプ	tangusuten － firamento ranpu	2520
1350	単元素ランプ	tan genso ranpu	2205
1351	単光束分光器	tan kōsoku bunkō－ki	2203
1352	単収束質量分析計	tan shūsoku shitsuryō bunseki－kei	2207
1353	単色光	tanshoku－kō	1510
1354	単色指示薬	tanshoku shijiyaku	1615
1355	淡色シフト	tanshoku sifuto	1069
1356	弾性散乱	dansei sanran	0732
1357	弾性衝突	dansei shōtotsu	0731
1358	単層	tansō	1513
1359	単掃引オシログラフ ポーラログラフィー	tan－sōin oshirogurafu pōrarogurafī	2209
1360	断続アーク	danzoku āku	1164
1361	担体	tantai	0305
1362	担体蒸留アーク	tantai jōryū āku	0306
1363	タンデム質量分析計	tandemu shitsuryō bunseki－kei	2396
1364	単電極マイクロ波プラズマ	tan－denkyoku maikuro－ha purazuma	1616
1365	単独溶液法	tandoku yōeki－hō	2183
1366	ダンピング定数	danpingu teisū	0558
1367	単分子被覆量	tanbunshi hifuku ryō	1514
1368	単方向スパーク	tan－hōkō supāku	2542
1369	単方向放電	tan－hōkō hōden	2541
1370	断面積	danmenseki	0525
1371	短絡放電	tanraku hōden	2193
1372	チェレンコフ検出器	cherenkofu kenshutsu－ki	0326
1373	チェレンコフ効果	cherenkofu kōka	0327
1374	チェレンコフ線	cherenkofu－sen	0328
1375	遅延蛍光	chien keikō	0577
1376	遅延係数	chien keisū	2076
1377	置換クロマトグラフィー	chikan kuromatogurafī	0670
1378	逐次生成定数	chikuji seisei teisū	2349
1379	蓄電アーク	chikuden āku	0439
1380	遅発中性子	chihatsu chūseishi	0578

1381	チャネリング	chaneringu	0333
1382	チャンバー型ネブライザー	chanbā－gata neburaizā	0332
1383	中間管	chūkan kan	1151
1384	中間電圧スパーク	chūkan den'atsu supāku	1427
1385	中間電圧スパーク発生器	chūkan den'atsu supāku hassei－ki	1428
1386	中間電流アーク	chūkan denryū āku	1425
1387	中間電力	chūkan denryoku	1426
1388	中間燃焼帯	chūkan nenshō tai	1141,1170
1389	中間プラズマガス	chūkan purazuma gasu	1150
1390	中空陰極光源	chūku inkyoku kōgen	1053
1391	中空陰極放電	chūku inkyoku hōden	1051
1392	中空陰極ランプ	chūku inkyoku ranpu	1052
1393	中空電極	chūkū denkyoku	1054
1394	抽出	chūshutsu	0859
1395	抽出係数	chūshutsu keisū	0860
1396	抽出剤	chūshutsu zai	0858
1397	抽出指示薬	chūshutsu shijiyaku	0862
1398	抽出する	chūshutsu－suru	0856
1399	抽出性	chūshutsu－sei	0857
1400	抽出定数	chūshutsu teisū	0861
1401	抽出物	chūshutsu butsu	0855
1402	中性	chūsei	1553
1403	中性子	chūseishi	1554
1404	中性子束	chūseishi soku	1559
1405	中性子増殖	chūseishi zōshoku	1561
1406	中性子放射化分析	chūseishi hōshaka bunseki	1556
1407	中性子放射化［法］	chūseishi hōshaka[－hō]	1555
1408	中性子捕獲	chūseishi hokaku	1557
1409	中性子密度	chūseishi mitsudo	1558
1410	中速中性子	chūsoku chuseishi	1149
1411	超音波ネブライザー	chō－onpa neburaizā	2535
1412	長管装置	chōkan sōchi	1355
1413	重畳場(電場と磁場の)	chōjōba	0524
1414	中性子温度	chūseishi ondo	1564
1415	中性子数	chūseishi－sū	1562
1416	中性子線	chūseishi sen	1563
1417	中性子束密度	chūseishi soku mitsudo	1560
1418	調整保持容量	chōsei hoji yoryō	0042
1419	調節剤	chōsetsu－zai	1482
1420	頂点	chōten	2368
1421	頂点電位	chōten den'i	2370
1422	頂点（テンサンメトリー波の）	chōten	0124
1423	頂点電流	chōten denryū	2369
1424	超熱ルミネセンス	chō netsu ruminesensu	2380
1425	超微量成分分析	chō－biryō seibun bunseki	2536
1426	超微量分析	chō－biryō bunseki	2534
1427	直接核分裂収率	chokusetsu kakubunretsu shūritsu	0651
1428	直接線蛍光	chokusetsu sen keikō	0653
1429	直接導入バーナ	chokusetsu dōnyū bāna	0652
1430	直接導入プローブ	chokusetsu dōnyū purōbu	0654
1431	直接秤量の補正	chokusetsu hyōryō no hosei	0501
1432	直線的減少	chokusen－teki genshō	1303
1433	直線滴定曲線	chokusen tekitei kyokusen	1317
1434	直線電位掃引	chokusen den'i sōin	1313
1435	直線電位掃引ボルタンメトリー	chokusen den'i sōin borutanmetorī	1316
1436	直線電位走査アノーディック ストリッピング クロノアンペロメトリー	chokusen den'i sōsa anōdikku sutorippingu kurono－anperometorī	0115
1437	直線偏光器	chokusen henkō－ki	1312
1438	直線偏光照射変調	chokusen henkō shōsha henchō	1484
1439	直流	chokuryū	0646
1440	直流アーク	chokuryū āku	0647
1441	直流電導度測定［法］	chokuryū dendōdo sokutei [－hō]	0649
1442	直流偏位振動放電	chokuryū hen'i shindō hōden	0648
1443	直流ポーラログラフィー	chokuryū pōrarogurafī	0650
1444	直列間隙	chokuretsu kangeki	2395
1445	直列点火	chokuretsu tenka	2189
1446	チョッパー	choppa	0364
1447	沈殿指示薬	chinden shijiyaku	1826
1448	沈殿する	chinden－suru	1824
1449	沈殿滴定	chinden tekitei	1827

1450	沈殿［物］	chinden[butsu]	1823
1451	対消滅減弱係数	tsui-shōmetsu genjaku keisū	1657
1452	通過時間	tsūka jikan	2500
1453	通気性膜	tsūkisei maku	0973
1454	低圧アークランプ	teiatsu āku ranpu	1361
1455	低圧スパーク	teiatsu supāku	1365
1456	低圧電気放電	teiatsu denki hōden	1363
1457	低圧放電ランプ	teiatsu hōden ranpu	1362
1458	低圧マイクロ波誘導プラズマ	teiatsu maikuro-ha yūdō purazuma	1364
1459	ＴＥＥＬＳ	tī-ī-ī-eru-esu	2501
1460	ＴＥＭモード	tī-ī-emu mōdo	2509
1461	Ｄ－Ａ変換器	dī-ē henkan-ki	0641
1462	ＤＡＴＡ	dī-ē-tī-ē	0560
1463	低エネルギー衝突	tei enerugī shōtotsu	1358
1464	定温示差熱分析	teion shisanetsu bunseki	0630
1465	定温度アーク	tei-ondo āku	0446
1466	Ｔ型管	tī-gata kan	2393
1467	抵抗	teikō	2057
1468	定常状態	teijō jōtai	2344
1469	定常的放射	teijō-teki hōsha	2342
1470	定数	teisū	0444
1471	定性分析	teisei bunseki	1896
1472	低速中性子	teisoku chūseishi	2217
1473	低速電子回折（ＬＥＥＤ）	teisoku densi kaisetsu	1359
1474	低損失	tei-sonshitsu	1360
1475	定電圧	tei-den'atsu	0447
1476	定電圧［分極］電流測定終点検出	tei-den'atsu [bunkyoku] denryū sokutei shūten kenshutsu	0221
1477	定電圧［分極］電流測定［法］	tei-den'atsu [bunkyoku] denryū sokutei[-hō]	0223
1478	定電圧［分極］電流滴定	tei-den'atsu [bunkyoku] denryū tekitei	0222
1479	定電位電解重量分析［法］	tei-den'i denkai jūryō bunseki[-hō]	0475
1480	定電位電解分離	tei-den'i denkai bunri	0476,0773
1481	定電位電量滴定	tei-den'i denryō tekitei	0473
1482	定電位電量分析［法］	tei-den'i denryō bunseki[-hō]	0474
1483	定電流	tei-denryū	0445
1484	低電流アーク	tei-denryū āku	1357
1485	定電流電位差測定［法］	tei-denryū den'isa sokutei[-hō]	0469
1486	定電流電位差滴定	tei-denryū den'isa tekitei	0468
1487	定電流電源	tei-denryū dengen	2319
1488	定電流電量分析［法］	tei-denryū denryō bunseki[-hō]	0467
1489	定電流［分極］電位差測定［法］	tei-denryū [bunkyoku] den'isa sokutei[-hō]	0232
1490	定電流［分極］電位差滴定	tei-denryū [bunkyoku] den'isa tekitei	0231
1491	定量限界	teiryō genkai	1298,1299
1492	定量示差熱分析	teiryō shisa netsu bunseki	1899
1493	定量分析	teiryō bunseki	1898
1494	テード（ＴＨＥＥＤ）	tēdo	2504
1495	テーパー型空洞共振器	tēpā-gata kūdō kyōshin-ki	2398
1496	テーリング	tēringu	2394
1497	滴下時間	tekika jikan	0708
1498	滴下水銀電極	tekika suigin denkyoku	0707
1499	滴下電極クーロメトリー	tekika denkyoku kūrometorī	0706
1500	滴定	tekitei	2474
1501	滴定曲線	tekitei kyokusen	2475
1502	滴定誤差	tekitei gosa	2476
1503	滴定剤	tekitei zai	2473
1504	滴定変換係数	tekitei henkan keisū	2479
1505	滴定分析	tekitei bunseki	2478
1506	滴定分析［法］	tekitei bunseki[-hō]	2480
1507	滴定率	tekitei ritsu	0575
1508	滴定レベル	tekitei reberu	1285
1509	出口圧力における空隙体積速度	deguchi atsuryoku ni okeru kūgeki taiseki sokudo	1167
1510	テスラコイル	tesura koiru	2408
1511	デバイーヒュッケルの式	debai-hyukkeru no shiki	0567
1512	テム（ＴＥＭ）	temu	2502
1513	デューティ サイクル	dyūti saikuru	0711
1514	デュエンーハントの短波長限界	dyuen-hanto no tan-pachō genkai	0710
1515	デリバトグラフィー	deribatografī	0599
1516	デリバトグラフ分析	deribatogurafu bunseki	0598
1517	デルナーホスキンスの式	derunā-hosukin no siki	0685
1518	デルベスカップ	derubesu kappu	0579
1519	電［気伝］導度	den[ki den] dōdo	0734

1520	電圧	den'atsu	2570
1521	電圧ー電流特性	den'atsu — denryū tokusei	2571
1522	電位	den'i	1807
1523	転位イオン	ten'i ion	2002
1524	電位規制 アノーディック ストリッピング クーロメトリー	den'i kisei anōdikku sutorippingu kūrometorī	0116
1525	電位規制ポーラログラフィー	den'i kisei pōrarogurafī	0477
1526	電位差重量滴定	den'isa jūryō tekitei	1812
1527	電位差測定〔法〕	den'isa sokutei〔—hō〕	1813
1528	電位差滴定	den'isa tekitei	1811
1529	電位差滴定終点	den'isa tekitei shūten	1809
1530	電位差法の選択係数	den'isa—ho no sentaku keisū	1810
1531	電位ステップ クロノクーロメトリー	den'i suteppu kurono—kūrometorī	1808
1532	転位分子イオン	ten'i bunshi ion	2001
1533	点火	tenka	1073
1534	電解液	denkaieki	0750
1535	電解質	denkaishitsu	0750
1536	電解重量分析	denkai jūryō bunseki	0749
1537	展開する	tenkai—suru	0610
1538	電荷移動	denka idō	0342
1539	電解分離	denkai bunri	0772
1540	点火型交流アーク	tenka—gata kōryū āku	1072
1541	点火器回路	tenka—ki kairo	1077
1542	添加剤	tenka zai	0040
1543	電荷ステップポーラログラフィー	denka suteppu pōrarogurafī	0341
1544	点火スパーク	tenka supāku	1076
1545	電荷増加ポーラログラフィー	denka zōka pōrarogurafī	1089
1546	点火できる静止間隙	tenka dekiru seishi kangeki	2514
1547	点火パルス	tenka parusu	1075
1548	添加物入りアーク	tenkabutsu iri āku	2154
1549	点火用回路	tenka—yō kairo	1074
1550	転換電子	tenkan denshi	0484
1551	電気アーク	denki āku	0733
1552	電気泳動	denki eidō	0311,0770
1553	電気泳動図	denki eidō zu	0771
1554	電気［化学的］活性物質	denki [kagaku—teki] kassei busshitsu	0741
1555	電気光学シャッター	denki kōgaku shattā	0769
1556	電気スパーク	denki supāku	0739
1557	電気的浸食	denki—teki shinshoku	0742
1558	電気的パラメータ	denki—teki paramēta	0737
1559	電気的励起	denki—teki reiki	0736
1560	電［気伝］導度測定［法］	den[ki den]dōdo sokutei[—hō]	0442
1561	電［気伝］導度滴定	den[ki den]dōdo tekitei	0441
1562	電気伝導率	denki dendō ritsu	0440
1563	電気放電	denki hōden	0735
1564	電極付空洞共振器プラズマ	denkyoku—tsuki kūdō kyōshin—ki purazuma	0746
1565	電極ー溶液界面	denkyoku yōeki kaimen	0747
1566	電源パラメータ	dengen paramēta	0738
1567	テンサメトリー	tensametorī	2407
1568	電子圧	denshi atsu	0763
1569	電子エネルギー	denshi enerugī	0756
1570	電子エネルギー損失分光	denshi enerugī sonshitsu bunkō	0757
1571	電子エネルギー転移過程	denshi enerugī ten'i katei	0758
1572	電子温度	denshi ondo	0768
1573	電子顕微鏡	denshi kenbikyō	0761
1574	電子衝撃イオン化	denshi shōgeki ion—ka	0759
1575	電子衝撃によるX線発生	denshi shōgeki ni yoru ekkusu—sen hassei	2607
1576	電子スペクトル	denshi supekutoru	0766
1577	電子線回折	denshi—sen kaisetsu	0755
1578	電子増倍	denshi zōbai	0762
1579	電子対生成	denshi—tsui seisei	1658
1580	電子の運動エネルギー	denshi no undō enerugī	0760
1581	電子プローブX線マイクロアナリシス	denshi purōbu ekkusu—sen maikuro anarishisu	0765
1582	電子プローブ マイクロアナリシス(EPMA)	denshi purōbe maikuroanarishisu	0764
1583	電磁放射	denji hōsha	0751
1584	電子捕獲	denshi hokaku	0753
1585	電子捕獲検出器	denshi hokaku kenshutsu—ki	0754
1586	転送されたプラズマ	tensō sareta purazuma	2491
1587	点対点配置	ten tai ten haichi	1787
1588	点対面配置	ten tai men haichi	1786

1589	天然存在度	ten'nen sonzai−do	1528
1590	天然同位体存在比	ten'nen dōitai sonzai−hi	1531
1591	天びん	tenbin	0203
1592	電離温度	denri ondo	1211
1593	電離係数	denri keisū	1204
1594	電離度	denrido	0574
1595	電離箱	denri−bako	1203
1596	電離放射線	denri hōshasen	1213
1597	電流	denryū	0541,0740
1598	電流が流れていないプラズマ	denryū ga nagareteinai purazuma	1575
1599	電流規制ポーラログラフィー	denryū kisei pōrarogurafī	0546
1600	電流−時間曲線	denryū − jikan kyokusen	0547
1601	電流測定終点検出［法］	denryū sokutei shūten kenshutsu[−hō]	0073
1602	電流測定［法］	denryū sokutei[−hō]	0075
1603	電流停止クロノポテンショメトリー	denryū teishi kurono−potenshometorī	0543
1604	電流滴定	denryū tekitei	0074
1605	電流の流れていないプラズマ	denryū no nagareteinai purazuma	0544
1606	電流の流れているアークプラズマ	denryū no nagareteiru āku purazuma	0542
1607	電流反転クロノポテンショメトリー	denryū hanten kurono−potenshometorī	0545
1608	電流プログラム クロノポテンショメトリー	denryū−puroguramu kurono−potenshometorī	1863
1609	電量滴定	denryō tekitei	0503
1610	電量分析［法］	denryō bunseki[−hō]	0504
1611	電力補償示差走査熱測定［法］	denryoku hoshō shisa netsu sokutei [−hō]	1816
1612	電離粒子	denri ryūshi	1212
1613	等圧重量変化測定	tōatsu jūryo henka sokutei	1226
1614	同位元素	dōi genso	1234
1615	同位相励起蛍光スペクトル	dōisō reiki keikō supekutoru	2389
1616	同位相励起りん光スペクトル	dōisō reiki rinkō supekutoru	2390
1617	同位体	dōitai	1234
1618	同位体イオン	dōitai ion	1242
1619	同位体希釈	dōitai kishaku	1235
1620	同位体希釈分析	dōitai kishaku bunseki	1236
1621	同位体交換	dōitai kōkan	1237
1622	同位体存在比	dōitai sonzai hi	1239
1623	同位体トレーサ	dōitai torēsa	1244
1624	同位体分析	dōitai bunseki	1240
1625	同位体分離	dōitai bunri	1243
1626	同位担体	dōi tantai	1241
1627	投影スクリーン	tōei sukurīn	1864
1628	等温式	tō−on shiki	1231
1629	等温遷移	tō−on sen'i	1232
1630	透過型電子顕微鏡（ＴＥＭ）	tōka−gata denshi kenbikyō	2502
1631	透過電子線エネルギー損失分光法(TEELS)	tōka denshi−sen enerugī sonshitsu bunkō−hō	2501
1632	透過比	tōka hi	2505
1633	透過率	tōka ritsu	2503
1634	透過率	tōka ritsu	2506
1635	同期回転間隙	dōki kaiten kangeki	2391
1636	等強線	tō kyō sen	1058
1637	統計的重み	tōkei−teki omomi	2343
1638	同軸型ネブライザー	dōjiku−gata neburaizā	0437
1639	同軸導波管	dōjiku dōha−kan	0383
1640	同時計数	dōji keisū	0390
1641	同時計数回路	dōji keisū kairo	0389
1642	同時計数分解時間	dōji keisū bunkai jikan	0391
1643	同時示差熱分析	dōji shisa netsu bunseki	2201
1644	同時測定法	doji sokutei−ho	2202
1645	同時分析システム	dōji bunseki shisutemu	2200
1646	同種電極	dōshu denkyoku	2163
1647	同中性子体	dō chūseishi tai	1233
1648	同調スタブ	dōchō sutabu	2522
1649	同調線発振器	dōchō−sen hasshin−ki	2519
1650	同調増幅器計測システム	dōchō zōfuku−ki keisoku shisutemu	2518
1651	動的示差熱測定	dōteki shisanetsu sokutei	0713
1652	動的熱機械測定	dōteki netsu kikai bunseki	0717
1653	動的熱重量分析	dōteki netsu jūryō bunseki	0716
1654	等電位点	tō−den'i ten	1230
1655	導電率	dōden ritsu	0440,0734
1656	導入系（試料の）	dōnyū−kei	1111
1657	当量	tōryō	0819
1658	当量点	tōryō ten	0818

1659	特性X線	tokusei ekkusu－sen	0337
1660	特性質量（原子吸光での）	tokusei shitsuryō	0335
1661	特性電位	tokusei den'i	0336
1662	特性濃度	tokusei nōdo	0334
1663	特定な干渉	tokutei na kanshō	2262
1664	閉じ込め板	tojikome ban	0443
1665	ドップラーシフト	doppurā shifuto	0688
1666	ドップラー半値巾（スペクトル線の）	doppurā hanchihaba	0687
1667	ドップラー広がり	doppurā hirogari	0686
1668	トラッピング	torappingu	2510
1669	取込み汚染	torikomi osen	0453
1670	ドリフト	dorifuto	0702
1671	トレーサー	torēsā	2490
1672	トロコイダル質量分析計（プロレイト）	torokoidaru shitsuryo bunsekikei	1865
	な行		
1673	内挿法	naisō－hō	0255
1674	内標準	nai hyōjun	1163
1675	内部吸光度	naibu kyūkōdo	1153
1676	内部参照線	naibu sanshō sen	1162
1677	内部充塡溶液	naibu jūten yōeki	1158
1678	内部振動振幅	naibu shindō shinpuku	1160
1679	内部制動放射	naibu seidō hōsha	1112
1680	内部電解	naibu denkai	1157
1681	内部電解重量分析［法］	naibu denkai jūryō bunseki［－hō］	1156
1682	内部転換	naibu tenkan	1154
1683	内部転換係数	naibu tenkan keisū	1155
1684	内［部］標準	nai[bu] hyōjun	1127
1685	内部変換	naibu henkan	1154
1686	内部領域	naibu ryōiki	1113
1687	ナノトレース	nano torēsu	1527
1688	ナノ微量成分［分析］	nano biryō seibun[bunseki]	1528
1689	ニーア・ジョンソン型	nīa－jonson－gata	1565
1690	ニードルバルブ	nīdoru barubu	1538
1691	二官能性イオン交換体	ni kan'nō－sei ion kōkantai	0226
1692	二原子［の］	ni genshi［no］	0613
1693	二個の衝突	niko no shōtotsu	0227
1694	二次イオン	niji ion	2146
1695	二次イオン化収率	niji ion－ka shūritsu	2148
1696	二次イオン質量分析法（ＳＩＭＳ）	niji ion shitsuryō bunseki－hō	2147
1697	二次蛍光	niji keikō	2145
1698	二次元クロマトグラフィー	nijigen kuromatogurafī	2530
1699	二次電子	niji denshi	2144
1700	二次電子増倍管	niji denshi zōbai－kan	2143
1701	二次燃焼	niji nenshō	2141
1702	二次微分電位差滴定	niji bibun den'isa tekitei	2152
1703	二次標準	niji hyōjun	2150
1704	二次標準溶液	niji hyōjun yōeki	2151
1705	二次放射［線］	niji hōsha[sen]	2149
1706	二次放電	niji hōden	2142
1707	二重アーク	nijū āku	0691
1708	二重収束型質量分析計	nijū shūsokugata shitsuryō bunseki－kei	0695
1709	二重同期光束分光器	nijū dōki kōsoku bunkō－ki	0700
1710	二色指示薬	nishoku shijiyaku	2529
1711	二波長分光光度計	nihachō bunkō kōdo－kei	0699
1712	乳剤校正関数	nyūzai kōsei kansū	0792
1713	乳剤校正曲線	nyūzai kōsei kyokusen	0791
1714	入射光子	nyūsha kōshi	1082
1715	入射スリット高さ	nyūsha suritto takasa	0812
1716	入射スリット巾	nyūsha suritto haba	0813
1717	入射放射線	nyūsha hōshasen	1086
1718	入力	nyūryoku	1085
1719	認証標準物質	ninshō hyōjun busshitsu	2327
1720	ネオジムガラス	neojimu garasu	1544
1721	ねじれひも分析	nejire himo bunseki	2481
1722	熱イオン化	netsu ion－ka	2423
1723	熱音響測定［法］	netsu onkyō sokutei［hō］	2429, 2455
1724	熱外中性子	netsu－gai chūseishi	0816
1725	熱化学反応	netsu kagaku han'nō	2432
1726	熱機械測定	netsu kikai sokutei	2446
1727	熱機械分析	netsu kikai bunseki	2445

1728	熱屈折率測定	netsu kussetsuritsu sokutei	2454
1729	熱源	netsu－gen	2428
1730	熱顕微測定	netsu kenbi sokutei	2449
1731	熱光学測定	netsu kōgaku sokutei	2453
1732	熱磁気測定	netsu jiki sokutei	2444
1733	熱重量測定	netsu jūryō sokutei	2442
1734	熱重量分析	netsu jūryō bunseki	2440
1735	熱重量分析図	netsu jūryō bunseki zu	2441
1736	熱熟成	netsu jukusei	2414
1737	熱蒸発分析	netsu jōhatsu bunseki	2457
1738	熱中性子	netsu chūseishi	2425
1739	熱中性子核分裂	netsu chūseishi kaku bunretsu	2422
1740	熱中性子柱	netsu chūseishi chū	2416
1741	熱的過程	netsu－teki katei	2426
1742	熱的蒸発	netsu－teki jōhatsu	2421
1743	熱的点火アーク	netsu－teki tenka āku	2424
1744	熱電気測定	netsu denki sokutei	2437
1745	熱電対列	netsu dentsui retsu	2452
1746	熱伝導率	netsu dendōritsu	2417
1747	熱伝導率検出器	netsu dendōritsu kenshutsu－ki	2418
1748	熱天秤	netsu tenbin	2431
1749	熱分解	netsu bunkai	2419
1750	熱分光測定	netsu bunkō sokutei	2456
1751	熱分析	netsu bunseki	2415
1752	熱分析技法	netsu bunseki gihō	2430
1753	熱平衡	netsu heikō	2420
1754	熱放射	netsu hōsha	2427
1755	熱膨張測定［法］	netsubōchō sokutei[－hō]	2433
1756	熱力学的定数	netsu rikigaku－teki teisū	2434
1757	熱力学的温度	netsu rikigaku－teki ondo	2436
1758	熱力学的平衡	netsu rikigaku－teki heikō	2435
1759	熱粒子測定	netsu ryūshi sokutei	2450
1760	熱量測定	netsu ryō sokutei	0288
1761	熱ルミネセンス	netsu ruminesensu	2443
1762	ネブライザー	neburaizā	1536
1763	ネブライザーフレーム系	neburaizā furēmu kei	1537
1764	ネルンスト応答	nerunsuto ōto	1549
1765	ネルンストの式	nerunsuto no shiki	1548
1766	燃焼	nenshō	0414
1767	燃焼速度	nenshō sokudo	0282
1768	燃焼電圧	nenshō den'atsu	0283
1769	燃焼度	nenshō do	0284
1770	燃焼率	nenshō ritsu	0285
1771	燃料	nenryō	0949
1772	燃料要素	nenryō yōso	0951
1773	濃縮	nōshuku	0808
1774	濃縮係数	nōshuku keisū	0809
1775	濃色シフト	nōshoku sifuto	1068
1776	濃度	nōdo	0431
1777	濃度計	nōdo kei	0583
1778	能動媒質	nōdō baishitsu	0035
1779	濃度比	nōdo hi	0436
1780	濃度分配比	nōdo bunpai hi	0433
1781	濃度平衡定数	nōdo heikō teisū	0432,0434
1782	ノルム温度	norumu ondo	1587

は行

1783	ハーターードリフィールド曲線	hātā－dorifīrudo kyokusen	1065
1784	バーン	bān	0206
1785	バイアス比例パルス増幅器	baiasu hirei parusu zōfuku－ki	0225
1786	配位子	hai'ishi	1286
1787	配位子架橋錯体	hai'ishi kakyō sakutai	1287
1788	配位子数	hai'ishi－sū	1288
1789	媒体	baitai	1424
1790	バイパスインジェクタ	baipasu injekuta	0286
1791	パイルアップ	pairuappu	1760
1792	量り取りの精密さ	hakaritori no seimitsusa	1831
1793	はかりの精密さ	hakari no seimitsusa	1830
1794	はかりのゼロ点	hakari no zero－ten	2618
1795	バキュームカップ電極	bakyūmu kappu denkyoku	2548

1796	箔検出器	haku kenshutsu－ki	0932
1797	白色化物質	hakushoku－ka busshitsu	0244
1798	薄層クロマトグラフィ	hakusō kuromatografī	2461
1799	白熱体	hakunetsu－tai	1083
1800	白熱体放射	hakunetsu－tai hōsha	1084
1801	薄膜	hakumaku	2459
1802	薄膜電極	hakumaku denkyoku	2460
1803	はくり溶液	hakuri yōeki	2361
1804	波形制御スパーク	hakei seigyo supāku	0478
1805	はけ状放電	hake－jō hōden	0270
1806	波高	hakō	2583
1807	波高選別器	hakō senbetsu－ki	1877
1808	波高増巾器	hakō zōfuku－ki	1874
1809	波高分析器	hakō bunseki－ki	1875
1810	波高分析器	hakō bunseki－ki	1882
1811	発色団内無放射遷移	hasshokudan－nai mu－hōsha sen'i	1171
1812	波数	hasū	2587
1813	波長型形状	hachō－gata keijō	2584
1814	波長分散	hachō bunsan	2585
1815	波長変調	hachō henchō	2586
1816	バックグラウンド	bakkuguraundo	0193
1817	バックグラウンド	bakkuguraundo	2546
1818	バックグラウンド質量スペクトル	bakkuguraundo shitsuryo supekutoru	0195
1819	バックグラウンド補正	bakkuguraundo hosei	0194
1820	バックグラウンド放射線	bakkuguraundo hōsha－sen	0196
1821	発光側モノクロメータ	hakkō－gawa monokuromēta	0788
1822	発光効率	hakkō kōritsu	1373
1823	発光効力	hakkō kōryoku	1372
1824	発光スペクトル	hakkō supekutoru	0790
1825	発光線の半値幅	hakkō sen no hanchi haba	1010
1826	発光線プロファイル	hakkō sen purofairu	0789
1827	発光層	hakkō sō	1375
1828	発光量	hakkō ryō	1376
1829	発色団間の無放射遷移	hasshokudan－kan no muhōsha sen'i	1140
1830	発振器	hasshin－ki	1642
1831	発生気体検出	hassei kitai kenshutsu	0726,0826
1832	発生気体分析	hassei kitai bunseki	0725,0825
1833	発熱	hatsunetsu	0846
1834	発熱反応	hatsunetsu han'nō	0848
1835	発熱ピーク	hatsunetsu pīku	0847
1836	幅	haba	2593
1837	ハメット指示薬	hametto shijiyaku	1018
1838	ハメットの酸性度関数	hametto no sanseido kansū	1017
1839	パラリシス回路	pararishisu kairo	1662
1840	バルク試料	baruku shiryō	0278
1841	バルク濃度	baruku nōdo	0277
1842	パルスエネルギー	parusu enerugī	1881
1843	パルス強度分布	parusu kyōdo bunpu	1876
1844	パルス継続時間	parusu keizoku jikan	1880
1845	パルス増幅器	parusu zōfuku－ki	1874
1846	パルス点火放電ランプ	parusu tenka hōden ranpu	1878
1847	パルス電力	parusu denryoku	1884
1848	パルス巾	parusu haba	1880
1849	パルス ポーラログラフィー	parusu pōrarogurafī	1883
1850	パルスレーザ光源	parusu rēza kōgen	1879
1851	半価層	hanka sō	1014
1852	半価層厚さ	hanka sō atsusa	1013,1015
1853	半Qスイッチ	han kyū suitchi	2173
1854	半減期	hangen ki	1011
1855	半サイクル当りのスパーク	han－saikuru atari no supāku	2255
1856	反射光強度	hansha－kō kyōdo	2027
1857	反射電子エネルギー損失分光法(REELS)	hansha denshi enerugī sonshitsu bunkō－hō	2028
1858	反射率	hansha ritsu	2029
1859	反ストークス蛍光	han sutōkusu keikō	0122
1860	半積分ポーラログラフィー	han sekibun pōrarogurafī	2170
1861	半値幅	hanchi haba	0955
1862	半値幅（ピークの）	hanchi haba	1697
1863	反跳	hanchō	2005
1864	反転分布	hanten bunpu	1798
1865	半導体	handōtai	2168

1866	半導体検出器	handōtai kenshutsu−ki	2169
1867	半導体表面	handōtai hyōmen	2167
1868	半透明鏡	han−tōmei−kyō	2174
1869	バンドスペクトル	bando supekutoru	0204
1870	バンド幅	bando haba	0205
1871	反応間隔	han'nō kankaku	1995
1872	反応層	han'nō sō	1996
1873	反応層の厚さ	han'nō sō no atsusa	2458
1874	反応速度分析	han'nō sokudo bunseki	1247
1875	反応電流	han'nō denryū	1248
1876	半波電位	han−pa den'i	1016
1877	半ピーク電位	han−pīku den'i	1012
1878	反粒子	han ryūshi	0121
1879	ＰＩＮ半導体検出器	pī−ai−enu handōtai kenshutsu−ki	1762
1880	ｐＨ	pī−eichi	1705
1881	ｐＨ指示薬	pī−eichi shijiyaku	1707
1882	ｐＨ実用尺度	pī−eichi jitsuyō shakudo	1710
1883	ｐＨ操作尺度	pī−eichi sōsa shakudo	1710
1884	ｐＨ測定	pī−eichi sokutei	1709
1885	ｐＨ滴定	pī−eichi tekitei	1712
1886	ｐＨ電極	pī−eichi denkyoku	1706
1887	ｐＨの解釈	pī−eichi no kaishaku	1708
1888	ｐＨの実際測定	pī−eichi no jissai sokutei	1819
1889	ｐＨ標準液	pī−eichi hyōjun eki	1711
1890	ＢＳＳ	bī−esu−esu	0199
1891	Ｐ型遅延蛍光	pī−gata chien keikō	1873
1892	ピーク	pīku	1682
1893	ピーク解析	pīku kaiseki	1683
1894	ピーク強度	pīiku kyōdo	1692
1895	ピーク極大	pīku kyokudai	1693
1896	ピークコンデンサー電圧	pīku kondensā den'atsu	1686
1897	ピークスパーク電流	pīku supāku denryū	1695
1898	ピーク高さ	pīku takasa	1691
1899	ピーク電流	pīku denryū	1687
1900	ピーク幅	pīku haba	1696
1901	ピークフィッティング	pīku fittingu	1690
1902	ピーク分離	pīku bunri	1694
1903	ピークベース	pīku bēsu	1685
1904	ピーク面積	pīku menseki	1684
1905	ピーク溶離容積	pīku yōri yōseki	1689
1906	ヒートフラックス示差走査熱測定［法］	hīto furakkusu shisa netsu sokutei[−hō]	1020
1907	比較溶液	hikaku yōeki	0416
1908	非加成性電流	hi−kaseisei denryū	1569
1909	光イオン化	hikari ion−ka	1739
1910	光イオン化分光法	hikari ion−ka bunkō−hō	1740
1911	光検出器	hikari kenshutsu−ki	1725
1912	光信号変調	hikari shingō henchō	1634
1913	光伝導率	hikari dendō ritsu	1626
1914	光電流	hikari denryū	1724
1915	光発揮分光法	hikari hakki bunkō−hō	1733
1916	光変調	hikari henchō	1289
1917	光放射過程	hikari hōsha katei	1941
1918	光放射強度	hikari hōsha kyōdo	1922
1919	［光］放射失活	[hikari] hōsha shikkatsu	1939
1920	［光］放射性遷移	[hikari] hōsha−sei sen'i	1943
1921	光放射能	hikari hōshanō	1917
1922	光放射プルム	hikari hōsha purumu	1924
1923	光放射粒子	hikari hōsha ryūshi	1923
1924	光放射量	hikari hōsha ryō	1926
1925	［光］放射励起	[hikari] hōsha reiki	1940
1926	光ポンピング	hikari ponpingu	1632
1927	ピクシス（ＰＩＸＥＳ）	pikushisu	1672,1767
1928	ピクセル	pikuseru	1766
1929	非結晶性電極	hi−kesshō−sei denkyoku	1574
1930	飛行時間質量分析計	hikō jikan shitsuryō bunsekikei	2469
1931	ピコトレース	piko torēsu	1759
1932	非コヒーレント光源	hi−kohīrento kōgen	1572
1933	非コヒーレント電磁放射光	hi−kohīrento denji hōsha−kō	1571
1934	非コヒーレント放射光パルス	hi−kohīrento hōsha−kō parusu	1088
1935	ピコ微量成分［分析］	piko biryōseibun [bunseki]	1759

1936	非水［溶媒］滴定	hisui [yōbai] tekitei	1570
1937	ヒステリシス	hisuterisisu	1071
1938	非スペクトル的干渉	hi-supekutoru-teki kanshō	1582
1939	比濁滴定	hidaku tekitei	1546
1940	比濁滴定終点決定	hidaku tekitei shūten kettei	1545
1941	比濁分析［法］	hidaku bunseki[-hō]	1547
1942	非弾性散乱	hi-dansei sanran	1102
1943	非弾性衝突	hi-dansei shōtotsu	1101
1944	飛程	hitei	1988
1945	非伝導性粉末	hi-dendōsei funmatsu	1573
1946	比電離	hi-denri	2263
1947	一組	hitokumi	2188
1948	一目盛りの値	hito-memori no atai	2552
1949	比燃焼度	hi-nenshōdo	2261
1950	非破壊放射化分析	hi-hakai hōshaka bunseki	1576
1951	被爆	hibaku	0851
1952	非ファラデー アドミッタンス	hi-faradē adomittansu	1578
1953	微分クロノポテンショメトリー	bibun kurono-potenshometorī	0588
1954	微分終点	bibun shūten	0590
1955	微分電位差滴定	bibun den'isa tekitei	0592
1956	微分熱重量測定［法］	bibun netsu jūryō sokutei[-hō]	0596
1957	微分熱分析	bibun netsu bunseki	0595
1958	微分［の］	bibun [no]	0587
1959	微分パルスポーラログラフィー	bibun parusu pōrarogurafī	0593
1960	微分分光光度分析法	bibun bunkō kōdo bunseki-hō	0594
1961	微分膨張計［分析法］	bibun bōcho-kei [bunseki-hō]	0589
1962	微分ポーラログラフィー	bibun pōrarogurafī	0591
1963	微分ボルタンメトリー	bibun borutanmetorī	0597
1964	非分離性バンドスペクトル	hi-bunrisei bando supekutoru	2544
1965	非分離バンドスペクトル	hi-bunri bando supekutoru	2545
1966	比放射能	hi-hōshanō	2260
1967	比保持容量	hi-hoji yōryō	2264
1968	比誘電率	hi-yūdenritsu	0614
1969	標識	hyōshiki	1250
1970	標識する	hyōshiki-suru	1251
1971	表示体積	hyōji taiseki	0600
1972	標準差し引き法	hyōjun sashihiki-ho	2331
1973	標準試料	hyōjun shiryō	0329
1974	標準的観測者	hyōjun-teki kansoku-sha	2326
1975	標準添加法	hyōjun tenka-hō	2321
1976	標準物質	hyōjun busshitsu	2330
1977	標準偏差	hyōjun hensa	2322
1978	標準溶液	hyōjun yōeki	2025,2328,2329
1979	秒速2200m中性子束密度	byōsoku 2200m chūseishi-soku mitsudo	2531
1980	標定済滴定溶液	hyōtei-zumi tekitei yōeki	2325
1981	標定済溶液	hyōtei-zumi yōeki	2324
1982	標的	hyōteki	2399
1983	表面	hyōmen	0948,2381
1984	表面修飾活性固体（クロマトグラフ固体相の）	hyōmen shūshoku kassei kotai	1480
1985	表面障壁型半導体検出器	hyōmen shōheki-gata handōtai kenshutsu-ki	2383
1986	表面スパーク	hyōmen supāku	2388
1987	表面濃度	hyōmen nōdo	2384
1988	表面反応	hyōmen han'nō	2387
1989	表面被覆率	hyōmen hifuku-ritsu	2386
1990	表面分析	hyōmen bunseki	2382
1991	ひょう(秤)量	hyōryō	0293
1992	ひょう(秤)量の手順	hyōryō no tejun	2589
1993	微量化学はかり	biryō kagaku hakari	1448
1994	微量成分	biryō seibun	1449,2489
1995	微量成分分析	biryō seibun bunseki	2488
1996	微量天びん	biryō tenbin	1447
1997	微量分析	biryō bunseki	1446
1998	比例計数管	hirei keisū-kan	1868
1999	比例パルス増幅器	hirei parusu zōfuku-ki	1314
2000	広がり	hirogari	0269
2001	広がり（フローインジクション分析での）	hirogari	0668
2002	ピンチ効果	pinchi kōka	1761
2003	ファナデーカップコレクター	fanadē kappu korekutā	0556
2004	ファノ係数	fano keisū	0866
2005	ファラデー暗部	faradē anbu	0872

2006	ファラデーカップコレクター	faradē kappu korekutā	0556,0871
2007	ファラデー整流電流	faradē seiryū denryū	0869
2008	ファラデー定数	faradē teisū	0870
2009	ファラデー電流	faradē denryū	0867
2010	ファラデー復調電流	faradē fukuchō denryū	0868
2011	フィードバック系	fīdo bakku kei	0875
2012	フィールドイオン化	fīrudo ion−ka	0878
2013	負イオン	fu−ion	1540
2014	負イオン質量スペクトル	fu−ion shitsuryō supekutoru	1541
2015	フィラメントクロマトグラフィー	firamento kuromatogurafī	0879
2016	フィルター	firutā	0883
2017	フィルムバッジ	firumu bajji	0882
2018	封入型中空陰極管	fūnyū−gata chūkū inkyoku−kan	2140
2019	フーリエ変換質量分析計	fūrie henkan shitsuryō bunseki−kei	0935
2020	フォークト関数	fōkuto kansū	2565
2021	フォークトプロファイル	fōkuto purofairu	2566
2022	深さ方向分解能	fukasa hōkō bunkai−nō	0586
2023	深さ方向分析	fukasa hōkō bunseki	0585
2024	不活性放電雰囲気	fu−kassei hōden fun'iki	1103
2025	負荷抵抗	fuka teikō	1343
2026	不感時間	fukan jikan	0564
2027	不感時間補正	fukan jikan hosei	0565
2028	吹き込み法	fukikomi−hō	0246
2029	不規則放射光パルス	fukisoku hōsha−kō parusu	1225
2030	不均質磁場	fu−kinshitsu jiba	1579
2031	不均質物質	fu−kinshitsu busshitsu	1580
2032	不均質膜電極	fu−kinshitsu maku denkyoku	1029
2033	複核錯体	fukukaku sakutai	0230
2034	複光束系	fuku kōsoku kei	0693
2035	複光束分光光度計	fuku kōsoku bunkō kōdo−kei	0692
2036	複合電極	fukugō denkyoku	0413
2037	復調ポーラログラフィー	fukuchō pōrarogurafī	0582
2038	負グロー	fu−gurō	1539
2039	符号の規約（正負の）	fugō no kiyaku (sei−fu no)	2197
2040	不斉電位	fusei den'i	0144
2041	不足当量同位体希釈分析	fusoku tōryō dōitai kishaku bunseki	2365
2042	ふた子ネブライザー	futago neburaizā	2528
2043	復極剤	fukkyoku zai	0584
2044	物質移動律速電解速度定数	busshitsu idō rissoku denkai sokudo teisū	1398
2045	物質の分散	busshitsu no bunsan	0669
2046	沸点	futten	0249
2047	沸騰温度	futtō ondo	0250
2048	物理吸着	butsuri kyūchaku	1758
2049	物理熟成	butsuri jukusei	1754
2050	物理的干渉	butsuri−teki kanshō	1756
2051	物理的平衡	butsuri−teki heikō	1755
2052	物理的捕集	butsuri−teki hoshū	1419
2053	物理量	butsuri ryō	1757
2054	不特定干渉	fu−tokutei kanshō	1581
2055	部分蒸発	bubun jōhatsu	1666
2056	部分崩壊定数	bubun hōkai teisū	1665
2057	不飽和色素スイッチ	fuhōwa shikiso suitchi	2113
2058	浮遊周波数	fuyū shūha−sū	0903
2059	フラグメントイオン	furagumento ion	0937
2060	プラズマ	purazuma	1772
2061	プラズマインピーダンス	purazuma inpīdansu	1774
2062	プラズマ温度	purazuma ondo	1779
2063	プラズマガス	purazuma gasu	1773
2064	プラズマジェット	purazuma jetto	1776
2065	プラズマ点火	purazuma tenka	1775
2066	プラズマトーチ	purazuma tōchi	1780
2067	プラズマの観測領域	purazuma no kansoku ryōiki	1613
2068	プラズマパラメータ	purazuma paramēta	1778
2069	プラズマ負荷	purazuma fuka	1777
2070	プラズモン状態	purazumon jōtai	1781
2071	ブラッグ角	buraggu kaku	0256
2072	フラックス	furakkusu	0919
2073	フラックスパータベーション	furakkusu pātabēshon	0924
2074	フラックスモニタ	furakkusu monita	0923
2075	ブラッグの式	buraggu no shiki	0257

2076	フラッシュ蛍光分析法	furasshu keikō bunseki−hō	0901
2077	フラッシュランプ	furasshu ranpu	0902
2078	プラトー	puratō	1782
2079	プラトロード	puratorōdo	1783
2080	ブランク	buranku	0235
2081	ブランク試料のバックグラウンド	buranku shiryō no bakkuguraundo	0236
2082	ブランク測定値	buranku sokutei−chi	0238
2083	ブランク滴定	buranku tekitei	0241
2084	ブランクのばらつき	buranku no baratsuki	0239
2085	プランクの法則	puranku no hōsoku	1768
2086	ブランク補正	buranku hosei	0237
2087	ブランク溶液	buranku yōeki	0240
2088	プリズムの底辺	purizumu no teihen	1857
2089	プリズムの有効底辺長	purizumu no yūkō teihen−chō	1284
2090	負領域	fu−ryōiki	1542
2091	ブレーズ角	burēzu kaku	0242
2092	ブレーズ波長	burēzu hachō	0243
2093	フレーム	furēmu	0895
2094	フレーム温度	furēmu ondo	0899
2095	フレーム形状による干渉	furēmu keijō ni yoru kanshō	0897
2096	フレームのバックグラウンド	furēmu no bakkuguraundo	0896
2097	フレーム分光分析法	furēmu bunkō bunseki−hō	0898
2098	プレカラム反応器	pure−karamu han'nō−ki	1833
2099	プレフィルター効果	pure−firutā kōka	1837
2100	不連続同時技法	furenzoku dōji gihō	0663
2101	不連続法	furenzoku−hō	0662
2102	フローインジェクション分析	furō−injekushon bunseki	0905
2103	プログラム	puroguramu	1859
2104	プログラム	puroguramu	1861
2105	プログラムする	puroguramu−suru	1862
2106	プログラムを作る	puroguramu wo tsukuru	1860
2107	ブロック	burokku	0245
2108	プロトン化定数	puroton−ka teisū	1870
2109	プロトン化平衡	puroton−ka heikō	1871
2110	フロンティング	furontingu	0947
2111	分解	bunkai	2058
2112	分解時間	bunkai jikan	2061
2113	分解時間補正	bunkai jikan hosei	2062
2114	分解能	bunkai−nō	2060
2115	分解能（１０％谷）	bunkai−nō(10% tani)	2059
2116	分解溶解（分析試料の）	bunkai yōkai	1620
2117	分岐比	bunki−hi	0260
2118	分岐崩壊	bunki hōkai	0258
2119	分岐率	bunki ritsu	0259
2120	分光化学緩衝剤	bunkō kagaku kanshō zai	2284
2121	分光化学希釈剤	bunkō kagaku kishaku zai	2286
2122	分光化学担体	bunkō kagaku tantai	2285
2123	分光化学的応用	bunkō kagaku−teki ōyo	2283
2124	分光化学的正確さ	bunkō kagaku−teki seikakusa	2281
2125	分光化学的性質	bunkō kagaku−teki seishitsu	2287
2126	分光学	bunkō−gaku	2297
2127	分光感度曲線	bunkō kando kyokusen	2280
2128	分光器	bunkō−ki	2296
2129	分光器の分解能	bunkō−ki no bunkai−nō	2293
2130	分光計	bunkō−kei	2292
2131	分光光度滴定	bunkō kōdo tekitei	2295
2132	分光写真器	bunkō shashin−ki	2289
2133	分光写真装置	bunkō shashin sōchi	2290
2134	分光写真法	bunkō shashin−hō	2291
2135	分光特性	bunkō tokusei	2269
2136	分光分析	bunkō bunseki	2282
2137	分光分析法	bunkō bunseki−hō	2294
2138	分光放射エネルギー密度	bunkō hōsha enerugī mitsudo	2278
2139	分光放射輝度	bunkō hōsha kido	2276
2140	分光放射量	bunkō hōsha ryō	2279
2141	分散	bunsan	0667,2558
2142	分子	bunshi	1507
2143	分子イオン	bunshi ion	1501
2144	分子イオン化エネルギー	bunshi ion−ka enerugī	1492
2145	分子間無放射遷移	bunshi−kan muhōsha sen'i	1152

2146	分子吸収分光学	bunshi kyūshū bunkōgaku	1495
2147	分子吸収分光分析法	bunshi kyūshū bunkō bunseki−hō	1494
2148	分子正イオン	bunshi sei−ion	1498
2149	分子的方法	bunshi−teki hōhō	1503
2150	分子発光分光学	bunshi hakkō bunkōgaku	1500
2151	分子発光分光分析法	bunshi hakkō bunkō bunseki−hō	1499
2152	分子バンド	bunshi bando	1497
2153	分子負イオン	bunshi fu−ion	1496
2154	分子分光学	bunshi bunkōgaku	1506
2155	分子分光分析法	bunshi bunkō bunseki−hō	1505
2156	分子放射光	bunshi hōsha−kō	1504
2157	分子ルミネセンス分光分析法	bunshi ruminesensu bunkō bunseki−hō	1502
2158	分析間隙	bunseki kangeki	0094
2159	分析系	bunseki kei	0099
2160	分析結果	bunseki kekka	0096
2161	分析元素	bunseki genso	0079
2162	分析試料	bunseki shiryō	0097
2163	分析的添加法	bunseki−teki tenka−hō	0084
2164	分析的放射化学	bunseki−teki hōsha kagaku	0095
2165	分析電極	bunseki denkyoku	0090
2166	分析電子顕微鏡	bunseki denshi−kenbikyō	0091
2167	分析標準品	bunseki hyōjun−hin	0098
2168	分析目的イオン	bunseki mokuteki ion	0083
2169	分析目的原子	bunseki mokuteki genshi	0082
2170	分析目的物	bunseki mokuteki butsu	0080
2171	分析目的物添加法	bunseki mokutekibutsu tenka−hō	0081
2172	ブンゼンバーナ	bunzen bāna	0279
2173	分配	bunpai	0678
2174	分配関数	bunpai kansū	1681
2175	分配クロマトグラフィー	bunpai kuromatogurafī	1678
2176	分配係数	bunpai keisū	0679,1679
2177	分配係数比	bunpai keisū hi	2184
2178	分配質	bunpai shitsu	0677
2179	分配則	bunpai soku	0683
2180	分配定数	bunpai teisū	0680,1680
2181	分配の法則	bunpai no hōsoku	1352
2182	分配比	bunpai hi	0684
2183	分配平衡	bunpai heikō	0681
2184	分配平衡定数	bunpai heikō teisū	0682
2185	分別蒸留	bunbetsu jōryū	0936
2186	粉末技術	funmatsu gijutsu	1815
2187	粉末試料	funmatsu shiryō	1814
2188	噴霧	funmu	0161
2189	噴霧室	funmu shitsu	2307
2190	分離温度	bunri ondo	2187
2191	分離過程	bunri katei	2186
2192	分離係数	bunri keisū	2185
2193	分離フレーム	bunri furēmu	2182
2194	ベイカーーサンプソンーザイデル変換	beikā−sanpuson−zaideru henkan	0202
2195	平均寿命	heikin jumyō	0187
2196	平均	heikin	0184
2197	平均活量係数	heikin katsuryō keisū	1408
2198	平均気体圧力	heikin kitai atsuryoku	0186
2199	平均空隙体積速度（キャリヤガスの）	heikin kūgeki taiseki sokudo	1409
2200	平均値	heikin−chi	1407
2201	平均配位子数	heikin hai'ishi−sū	0188
2202	平均流速	heikin ryūsoku	0185
2203	並行ー並行鏡配置	heikō−heikō−kyō haichi	1661
2204	平衡定数	heikō teisū	0817
2205	ベイツーグッゲンハイム規約	beitsu−guggenhaimu kiyaku	0213
2206	ベイパープルム	beipā purumu	2557
2207	平面陰極グロー放電光源	heimen inkyoku gurō hōden kōgen	1769
2208	平面偏光	heimen henkō	1770
2209	平面偏光した	heimen henkō shita	1771
2210	並列点火	heiretsu tenka	1660
2211	ベースライン	bēsu rain	0209
2212	ベータ開裂	bēta kairetsu	0218
2213	ベータ線	bēta−sen	0220
2214	ベータ崩壊	bēta hōkai	0219
2215	ペーパークロマトグラフィー	pēpā kuromatogurafī	1659

2216	平均イオン活量係数	heikin ion katsuryō keisū	1410
2217	平均質量飛程	heikin shitsuryō hitei	1413
2218	平均寿命	heikin jumyō	1411
2219	平均線飛程	heikin sen kitei	1412
2220	ベクレル	bekureru	0215
2221	ペス（ＰＥＳ）	pesu	1731
2222	ベッドボリューム	beddo boryūmu	0216
2223	ベッドボリューム容量	beddo boryūmu yōryō	0217
2224	ベルスマ型装置	berusuma−gata sōchi	0248
2225	偏移	hen'i	0611
2226	偏角	henkaku	0100
2227	変換	henkan	2492
2228	変換係数（エネルギー単位の）	henkan keisū	0485
2229	変換定数	henkan teisū	2493
2230	偏光	henkō	1789
2231	偏光されていない	henkō sarete inai	2543
2232	偏光子	henkōshi	1791
2233	偏光した	henkō shita	1790
2234	変色範囲	henshoku han'i	2494
2235	変調周波数	henchō shūha−sū	0945,1483
2236	変調ポーラログラフィー	henchō pōrarogurafī	1485
2237	変動係数	hendō keisū	0384
2238	ペンニングガス混合物	pen'ningu gasu kongō−butsu	1698
2239	弁別器	benbetsu−ki	0664
2240	崩壊	hōkai	0568
2241	崩壊曲線	hōkai kyokusen	0571
2242	崩壊系列	hōkai keiretsu	0569
2243	崩壊図	hōkai zu	0572
2244	崩壊定数	hōkai teisū	0570
2245	妨害線	bōgai sen	1147
2246	妨害物質	bōgai busshitsu	1148
2247	包含［物］	hōgan [butsu]	1087
2248	放射エネルギー	hōsha enerugī	1918
2249	放射エネルギー密度	hōsha enerugī mitsudo	1919
2250	放射化	hōsha−ka	0032
2251	放射化学	hōsha kagaku	1971
2252	放射化学収率	hōsha kagaku shūritsu	1970
2253	放射化学的純度	hōsha kagaku−teki jundo	1968
2254	放射化学分離	hōsha kagaku bunri	1969
2255	放射化断面積	hōsha−ka danmenseki	0034
2256	放射化分析	hōsha−ka bunseki	0033,1588
2257	放射輝度	hōsha kido	1915
2258	放射輝度温度	hōsha kido ondo	1916
2259	放射源	hōsha gen	1935
2260	放射光	hōsha−kō	1927
2261	放射光束	hōsha kōsoku	1921
2262	放射光の強度	hōsha−kō no kyōdo	1135
2263	放射再結合	hōsha sai−ketsugo	1942
2264	放射照度	hōsha shodo	1223
2265	放射性	hōshasei	1944
2266	放射性核種	hōshasei kakushu	1960,1980
2267	放射性核種純度	hōshasei kakushu jundo	1981
2268	放射性系列	hōshasei keiretsu	1961
2269	放射性原子核	hōshasei genshi kaku	1959
2270	放射性降下物	hōshasei kōkabutsu	1952
2271	放射性指示薬	hōshasei shijiyaku	1954
2272	放射性システム	hōshasei shisutemu	1601
2273	放射性試薬	hōshasei shiyaku	1982
2274	放射性同位元素	hōshasei dōigenso	1955,1976
2275	放射性トレーサ	hōshasei torēsa	1963
2276	放射性年代	hōshasei nendai	1945
2277	放射性年代決定	hōshasei nendai kettei	1948
2278	放射性廃棄物	hōshasei haikibutsu	1964
2279	放射性半減期	hōshasei hangenki	1953
2280	放射性標識	hōshasei hyōshiki	1956
2281	放射性物質	hōshasei busshitsu	1957,1962
2282	放射性平均寿命	hōshasei heikin jumyō	1958
2283	放射性崩壊	hōshasei hōkai	1949
2284	放射性［崩壊］系列	hōshasei [hōkai] keiretsu	1946
2285	放射性流出物	hōshasei ryūshutsubutsu	1950

2286	放射線	hōshasen	1927
2287	放射線エネルギー	hōshasen enerugī	1931
2288	放射線化学	hōshasen kagaku	1928
2289	放射線計数器	hōshasen keisū－ki	1929
2290	放射線源	hōsha sengen	1935
2291	放射線検出器	hōshasen kenshutsu－ki	1930
2292	放射線散乱	hōshasen sanran	2126
2293	放射線スペクトル	hōshasen supekutoru	1937,2299
2294	放射線スペクトロメータ	hōshasen supekutorometa	1936
2295	放射線パルス	hōshasen parusu	1933
2296	放射線分解	hōshasen bunkai	1977
2297	放射線量	hōshasen－ryō	1934
2298	放射滴定	hōsha tekitei	1979
2299	放射滴定終点	hōsha tekitei shūten	1978
2300	放射能	hōshanō	0037,1965
2301	放射能濃度	hōshanō nōdo	0039
2302	放射能の成長曲線	hōshanō no seichō kyokusen	1007
2303	放射分析化学	hōsha bunseki kagaku	1966
2304	放射分析化学的精製	hōsha bunsekikagaku－teki seisei	1967
2305	放射平衡	hōsha heikō	1951
2306	放射捕獲	hōsha hokaku	1938
2307	放射露光	hōsha rokō	1920
2308	放出エネルギー	hōshutsu enerugī	0730
2309	放出剤	hōshutsu－zai	2044
2310	膨張計［分析法］	bōchō－kei ［bunseki－hō］	0642
2311	放電	hōden	0655
2312	放電回路	hōden kairo	0657
2313	放電気体	hōden kitai	0659
2314	放電電流	hōden denryū	0658
2315	放電雰囲気	hōden fun'iki	0656
2316	放電ポーラログラフィー	hōden pōrarogurafī	0660
2317	飽和	hōwa	2115
2318	飽和室	hōwa shitsu	2116
2319	飽和プラトー	hōwa puratō	2117
2320	飽和溶液	hōwa yōeki	2114
2321	ポーラスカップ電極	pōrasu kappu denkyoku	1800
2322	ポーラログラフィー	pōrarogurafī	1792
2323	ホールドアップ体積	hōrudoappu taiseki	1050
2324	捕獲	hokaku	0299
2325	捕獲ガンマ線	hokaku ganma－sen	0301
2326	捕獲断面積	hokaku danmenseki	0300
2327	保持	hoji	2077
2328	保持温度	hoji ondo	2080
2329	保持時間	hoji jikan	2081
2330	保持指数	hoji shisū	2078
2331	保持担体	hoji tantai	1049
2332	保持電極	hoji denkyoku	0307
2333	保持パラメータ	hoji parameta	2079
2334	捕集	hoshū	0395
2335	捕集速度	hoshū sokudo	2547
2336	保持容積	hoji yōseki	2082
2337	補助スパーク間隙	hojo supāku kangeki	0183
2338	補助電極	hojo denkyoku	0182,2375
2339	補助放電	hojo hōden	0181,2374
2340	ホストゲスト化学	hosuto gesuto kagaku	1060
2341	ポストフィルター効果	posuto firutā kōka	1805
2342	補正強度	hosei kyōdo	0495
2343	補正蛍光減衰時間	hosei keikō gensui jikan	0491
2344	補正発光スペクトル	hosei hakkō supekutoru	0493
2345	補正保持時間	hosei hoji jikan	0499
2346	補正保持容量	hosei hoji yōryō	0500
2347	補正リン光減衰時間	hosei rinkō gensui jikan	0492
2348	補正ルミネセンス量子収率	hosei ruminesensu ryōshi shūritsu	0498
2349	補正ルミネセンス発光偏光スペクトル	hosei ruminesensu hakkō henkō supekutoru	0496
2350	補正ルミネセンス励起偏光スペクトル	hosei ruminesensu reiki henkō supekutoru	0497
2351	補正励起スペクトル	hosei reiki supekutoru	0494
2352	ポッケルスセル	pokkerusu seru	1785
2353	ホットアトム	hotto atomu	1062
2354	ホットセル	hotto seru	1063
2355	ポテンシャル	potensharu	1807

2356	ポリクロメータ	porikurométa	1795
2357	ボルタモグラム	borutamoguramu	2576
2358	ボルタンペログラム	borutanperoguramu	2577
2359	ボルタンメトリー	borutanmetorī	2575
2360	ボルタンメトリー定数	borutanmetorī teisū	2574
2361	ボルタンメトリー分析	borutanmetorī bunseki	2573
2362	ポロード	porōdo	1799
2363	ボロメータ	boro mēta	0251
2364	ポンピングエネルギー	ponpingu enerugī	1887
2365	ポンピングキャビティ	ponpingu kyabiti	1886
2366	ポンピング時間	ponpingu jikan	1889
2367	ポンピング用電力	ponpingu—yō denryoku	1890
2368	ポンピング用ランプ	ponpingu—yō ranpu	1888
2369	本来のインピーダンス	honrai no inpīdansu	1530
	ま行		
2370	マーカ	māka	1387
2371	マイクロコンピュータ	maikuro konpyūta	1450
2372	マイクロ波	maikuro—ha	1457
2373	マイクロ波プラズマ	maikuro—ha purazuma	1459
2374	マイクロ波誘導プラズマ	maikuro—ha yūdō purazuma	1458
2375	マイクロプロセッサー	maikuro purosessā	1453
2376	膜	maku	1430
2377	マクスウエルーボルツマンの法則	makusuueru—borutsuman no hōsoku	1405
2378	マグネトロン	magunetoron	1384
2379	マクロ断面積	makuro danmenseki	1378
2380	マスキング	masukingu	1388
2381	マスキング剤	masukingu zai	1389
2382	マックラファティー転位	makkurafatī ten'i	1406
2383	マッタホーヘルツォク型	mattaho—herutsoku—gata	1404
2384	マトリックス	matorikkusu	1400
2385	マトリックス効果	matorikkusu kōka	1401
2386	マトリックス調整	matorikkusu chōsei	1402
2387	マトリックス調節剤	matorikkusu chōsetsu zai	1403
2388	マルチチャネル波高分析器	maruchi chaneru hakō bunseki—ki	1519
2389	マルチチャネル分析器	maruchi chaneru bunseki—ki	1518
2390	ミー散乱	mī sanran	1460
2391	見かけのｐＨ	mikake no pī—eichi	0126
2392	見掛けの濃度	mikake no nōdo	0125
2393	ミクロクーロメトリー	mikuro kūrometorī	1451
2394	ミクロ断面積	mikuro danmenseki	1455
2395	ミクロトレース	mikuro torēsu	1456
2396	ミクロ微量［成分］	mikuro biryō[seibun]	1456
2397	ミクロホトメータ	mikurohoto mēta	1452
2398	ミクロ［量］試料	mikuro [ryō] shiryō	1454
2399	ミスト	misuto	1468
2400	未制振の振動放電	miseishin no shindō hōden	2540
2401	ミリクーロメトリー	miri kūrometorī	1463
2402	霧化	muka	1534
2403	霧化効率	muka kōritsu	1535
2404	無関係電解質	mukankei denkaishitsu	1094
2405	娘核種	musume kakushu	0561
2406	無制御高圧スパーク発生装置	mu—seigyo kōatsu supāku hassei sōchi	2539
2407	無制御交流アーク	mu—seigyo kōryū āku	2538
2408	無担体	mu—tantai	0308
2409	無電極空洞共振器プラズマ	mu—denkyoku kūdō kyōshin—ki purazuma	0745
2410	無電極放電ランプ	mu—denkyoku hōden rampu	0743
2411	無電極励起	mu—denkyoku reiki	0744
2412	無放射遷移	mu—hōsha sen'i	1932
2413	迷光	meikō	2356
2414	迷光放射	meikō hōsha	2357
2415	名目線流	meimoku sen ryū	1568
2416	目隠し	mekakushi	2137
2417	メスバウア効果	mesubaua kōka	1517
2418	メソ［量試料］微量成分分析	meso [—ryō shiryō] biryō seibun bunseki	1434
2419	メソサンプル	meso sanpuru	1433
2420	メソ微量成分分析	meso biryō seibun bunseki	1434
2421	メソ［量］	meso [—ryō]	1432
2422	メソ［量］試料	meso [—ryō] shiryō	1433
2423	メッカーバーナ	mekkā bāna	1429
2424	目盛り拡大	memori kakudāi	2118

No.	Japanese	Romanization	Ref.
2425	毛細管	mōsaikan	0298
2426	毛細管アーク	mōsaikan āku	0296
2427	毛細管電極	mōsaikan denkyoku	0297
2428	モード	mōdo	1479
2429	目視指示薬	mokushi shijiyaku	2564
2430	目視指示薬終点	mokushi shijiyaku shūten	2563
2431	モノクロメータ	monokuromēta	1511
2432	モノクロメータを通過する光束	monokuromēta wo tsūkasuru kōsoku	0925
2433	モル吸光係数	moru kyūkō keisū	1489
2434	モル濃度	moru nōdo	1493
	や行		
2435	焼き払い効果	yakiharai kōka	0280
2436	宿物質	yado busshitsu	1061
2437	有機物効果	yūkibutsu kōka	1638
2438	有効理論段数	yūkō riron dan−sū	0722
2439	有効理論段高さ	yūkō rirondan takasa	1025
2440	誘電率測定［法］	yūdenritsu sokutei[−hō]	0616
2441	誘電率滴定	yūdenritsu tekitei	0615
2442	誘導結合プラズマ	yūdō ketsugō purazuma	1099
2443	誘導コイル	yūdō koiru	1098
2444	誘導質	yūdō sitsu	1100
2445	誘導反応	yūdō han'nō	1097
2446	誘導放射能	yūdō hōshanō	1096
2447	輸送	yusō	2507
2448	輸送干渉	yusō kanshō	2508
2449	陽イオン効果	yō−ion kōka	0103
2450	陽イオン交換	yō−ion kōkan	0323
2451	陽イオン交換体	yō−ion kōkantai	0324
2452	陽イオン衝撃によるX線発生	yō−ion shōgeki ni yoru ekkusu−sen hassei	2609
2453	溶液	yōeki	2230
2454	溶液平衡	yōeki heikō	2231
2455	溶解	yōkai	0675
2456	溶解度積	yōkaido seki	2227
2457	陽極暗部	yōkyoku anbu	0112
2458	陽極間隙	yōkyoku kangeki	0113
2459	陽極グロー	yōkyoku gurō	0114
2460	陽極蒸発	yōkyoku jōhatsu	0118
2461	陽極柱	yōkyoku chū	1801
2462	陽極柱電流密度	yōkyoku chū denryū mitsudo	1802
2463	陽子	yōshi	1869
2464	陽子数	yōshi−sū	1872
2465	溶質ー溶媒相互作用	yōshitsu yōbai sōgo sayō	2228
2466	溶質の蒸発干渉	yōshitsu no jōhatsu kanshō	2229
2467	溶出液	yōshutsu eki	0779
2468	容積分布係数	yōseki bunpu keisū	2579
2469	容積膨潤比	yōseki bōjun−hi	2580
2470	容積容量	yōseki yōryō	2578
2471	ヨウ素還元滴定	yōso kangen tekitei	1180
2472	ヨウ素還元滴定分析［法］	yōso kangen tekitei bunseki [−hō]	1181
2473	ヨウ素酸化滴定	yōso sanka tekitei	1178
2474	ヨウ素酸化滴定分析［法］	yōso sanka tekitei bunseki [−hō]	1179
2475	ヨウ素滴定分析［法］	yōso tekitei bunseki [−hō]	1181
2476	陽電子	yō−denshi	1804
2477	溶媒	yōbai	2232
2478	溶媒移動距離	yōbai idō kyori	2237
2479	溶媒効果	yōbai kōka	2234
2480	溶媒先端	yōbai sentan	2236
2481	溶媒抽出	yōbai chūshutsu	2235
2482	溶媒ブランク	yōbai buranku	2233
2483	予備放電時間	yobi hōden jikan	1822
2484	溶融	yōyū	0956
2485	溶離液	yōri eki	0780
2486	溶離曲線	yōri kyokusen	0784
2487	溶離クロマトグラフィー	yōri kuromatografī	0783
2488	溶離する	yōri−suru	0781
2489	溶離バンド	yōri bando	0782
2490	溶離ピーク	yōri pīku	0785
2491	容量結合プラズマ	yōryo ketsugo purazuma	0290
2492	溶離容積	yōri yōseki	0786
2493	容量的マイクロ波プラズマ	yōryō−teki maikuro−ha purazuma	0291

2494	抑制剤	yokusei－zai	2379
2495	横方向拡散干渉	yoko hōkō kakusan kanshō	1281
2496	予混合バーナ	yo－kongō bāna	1839
2497	予備スパーク時間	yobi supāku jikan	1840
2498	予備放電曲線	yobi hōden kyokusen	2470
2499	読み	yomi	1998
2500	読み取り限度	yomitori gendo	1997
2501	読み取り限度のミリグラム換算値	yomitori gendo no miriguramu kansan－chi	1464
2502	弱い電位勾配	yowai den'i kōbai	2588
	ら行		
2503	ラザフォード後方散乱（ＲＢＳ）	razafōdo kōhō sanran	2095
2504	ラジオクロマトグラフ法	rajio kuromatogurafu－hō	1972
2505	ラジオコロイド	rajio koroido	1973
2506	ラド	rado	1913
2507	ラフィネート	rafinēto	1983
2508	ラマン光	raman－kō	1984
2509	ラマン散乱	raman sanran	1985
2510	ラマン分光学	raman bunkōgaku	1986
2511	ラムス（ＬＡＭＭＳ）	ramusu	1267
2512	ラメス（ＬＡＭＥＳ）	ramesu	1266
2513	ランプ	ranpu	1255
2514	ランプ回路	ranpu kairo	1377
2515	乱流	ranryū	2527
2516	乱流フレーム	ranryū furēmu	2526
2517	リード（ＬＥＥＤ）	rī do	1356
2518	力価	rikika	2477
2519	力価重量	rikika jūryō	0865
2520	立体角	rittai kaku	2219
2521	立体角当りの放射強度	rittai kaku atari no hōsha kyōdo	1925
2522	粒径	ryūkei	1675
2523	粒径分析	ryūkei bunseki	1676
2524	粒径分布	ryūkei bunpu	1677
2525	粒子検出器	ryūshi kenshutsu－ki	1669
2526	粒子消滅	ryūshi shōmetsu	1667
2527	粒子線	ryūshi－sen	1674
2528	粒子線照射	ryūshi－sen shōsha	1668
2529	粒子線誘起エックス線発揮分光法(PIXES)	ryūshi－sen yūki ekkusu－sen hakki bunkō－hō	1672
2530	粒子束	ryūshi soku	1670
2531	粒子束密度	ryūshi soku mitsudo	1671
2532	粒子の吸収	ryūshi no kyūshū	0017
2533	流出速度	ryūshutsu sokudo	1648
2534	粒子励起蛍光Ｘ線分析	ryūshi reiki keikō ekkusu－sen bunseki	1673
2535	流速	ryūsoku	0907
2536	流量制御ネブライザー	ryūryō seigyo neburaizā	0471
2537	流量プログラミングクロマトグラフィー	ryūryō puroguramingu kuromatografī	0906
2538	量	ryō	1900
2539	量子計数計	ryōshi keisū－kei	1901
2540	量子収率	ryoshi shūritsu	1904
2541	両性溶媒	ryōsei yōbai	0076
2542	理論段数	riron dansū	2411
2543	理論段高さ	rirondan takasa	1026
2544	理論的終点	riron－teki shūten	2410
2545	理論比容量	riron hi－yōryō	2413
2546	理論保持容積	riron hoji yōseki	2412
2547	臨界減衰放電	rinkai gensui hōden	0522
2548	りん光	rinkō	1718
2549	りん光鏡	rinkō kyō	1723
2550	りん光分光計	rinkō bunkō－kei	1721
2551	りん光分析法	rinkō bunseki－hō	1719
2552	りん光分析法	rinkō bunseki－hō	1722
2553	りん光励起－発光スペクトル	rinkō reiki － hakkō supekutoru	1720
2554	輪状陽極	rinjō yōkyoku	0111
2555	累積核分裂収率	ruiseki kaku bunretsu shūritsu	0534
2556	累積分布	ruiseki bunpu	0533
2557	累積和	ruiseki wa	0538
2558	ルミネセンス	ruminesensu	1366
2559	ルミネセンスエネルギー収量	ruminesensu enerugī shūryō	0805
2560	ルミネセンス消光	ruminesensu shōkō	1368
2561	ルミネセンスの直線偏光	ruminesensu no chokusen henkō	1311
2562	ルミネセンスパラメータ	ruminesensu paramēta	1367

2563	ルミネセンス分光学	ruminesensu bunkōgaku	1371
2564	ルミネセンス分光光度計	ruminesensu bunkō kōdo−kei	1369
2565	ルミネセンス分光光度法	ruminesensu bunkō kōdo−hō	1370
2566	ルミネセンス量子効率	ruminesensu ryōshi kōritsu	1903
2567	ルミネセンス量子収率	ruminesensu ryōshi shūritsu	1906
2568	励起	reiki	0828
2569	励起−発光スペクトル	reiki hakkō supekutoru	0829
2570	励起エネルギー	reiki enerugī	0830,0844
2571	励起温度	reiki ondo	0840
2572	励起干渉	reiki kanshō	0831
2573	励起機構	reiki kikō	0833
2574	励起蛍光スペクトル	reiki keikō supekutoru	0841
2575	励起源	reiki gen	0838
2576	励起状態	reiki jōtai	0843
2577	励起スペクトル	reiki supekutoru	0839
2578	励起測モノクロメータ	reiki−soku monokuromēta	0835
2579	励起電圧	reiki den'atsu	0836
2580	励起モード	reiki mōdo	0834
2581	冷却曲線	reikyaku kyokusen	0488
2582	冷却した中空陰極ランプ	reikyaku−shita chūkū inkyoku ranpu	0487
2583	冷却速度曲線	reikyaku sokudo kyokusen	0489
2584	冷却（放射能の）	reikyaku	1947
2585	励起領域	reiki ryōiki	0837
2586	励起リン光スペクトル	reiki rinkō supekutoru	0842
2587	励起レベル	reiki reberu	0832
2588	冷中空陰極	rei−chūkū inkyoku	0393
2589	冷中性子	rei−chūseishi	0394
2590	レイリー散乱	reirī sanran	1993
2591	レーザー	rēzā	1256
2592	レーザー活性物質	rēzā kassei busshitsu	1258
2593	レーザー鏡	rēzā kyō	1269
2594	レーザー共振器	rēzā kyōshin−ki	1276
2595	レーザー局所分析	rēzā kyokusho bunseki	1265
2596	レーザー原子化および励起	rēzā genshika oyobi reiki	1260
2597	レーザー原子化装置	rēzā genshika sōchi	1261
2598	レーザー光源	rēzā kōgen	1278
2599	レーザー作用	rēzā sayō	1257
2600	レーザー出力	rēzā shutsuryoku	1270
2601	レーザー出力エネルギー	rēzā shutsuryoku enerugī	1271
2602	レーザー照射	rēzā shōsha	1277
2603	レーザースパイク	rēzā supaiku	1279
2604	レーザー動作しきい値	rēzā dōsa shikii−chi	1280
2605	レーザーによる蒸気雲	rēzā ni yoru jōki gumo	1273
2606	レーザーによる消耗	rēzā ni yoru shōmō	1264
2607	レーザーパルス	rēzā parusu	1274
2608	レーザービーム	rēzā bīmu	1262
2609	レーザー微小部質量分析（ＬＡＭＭＳ）	rēzā bishōbu shitsuryō bunseki	1267
2610	レーザー微小部発光分光法(LAMES)	rēzā bishōbu hakkō bunkō−hō	1266
2611	レーザープルム	rēzā purumu	1272
2612	レーザー分析	rēzā bunseki	1259
2613	レーザーマイクロプローブ分析	rēzā maikuropurōbu bunseki	1268
2614	レーザーラマン微小部分析(LRMA)	rēzā raman bishōbu bunseki	1275
2615	レーザーイオン化(法)	rēzā ion−ka(−hō)	1263
2616	レドックスイオン交換体	redokkusu ion kōkantai	2013
2617	レドックスポリマー	redokkusu porimā	2014
2618	レム	remu	2047
2619	連続供給	renzoku kyōkyū	0459
2620	連続スペクトル	renzoku supekutoru	0462,2270
2621	連続［スペクトル］放射	renzoku [supekutoru] hōsha	0458
2622	連続流れ分析	renzoku nagare bunseki	0456
2623	連続波	renzoku−ha	0460
2624	連続波操作	renzoku−ha sōsa	0461
2625	連続法	renzoku−hō	0457
2626	連続放射	renzoku hōsha	1577
2627	レントゲン	rentogen	2089
2628	沪(ろ)取	roshu	0884
2629	漏出点容量	roshutsuten yōryō	0265
2630	ロマキンーシャイベの式	romakin−shaibe no shiki	1354

ADDENDA

A01 アルカリ性化する
A02 音響光学 シャッター
A03 蛍光X線分光学
A04 原子吸光 分光学
A05 原子蛍光 分光学
A06 原子発光 分光学
A07 光学的発光分光学
A08 黒化
A09 試料の重量分類
A10 内部参照電極
A11 ヒットルフ暗部
A12 フレーム分光学
A13 分子ルミネセンス分光学

arukari−sei−ka suru 0058
onkyō kōgaku shattā A01
keikō ekkusu−sen bunkōgaku A10
genshi kyūkō bunkōgaku A02
genshi keikō bunkōgaku A04
genshi hakkō bunkōgaku A03
kōgaku−teki hakkō bunkōgaku A09
kokka A05
shiryō no jūryō bunrui 2108
naibu sanshō denkyoku A07
hittorufu anbu 1048
furēmu bunkōgaku A06
bunshi ruminesensu bunkōgaku A08

Appendix: Corrections to indexes

The indexes to both editions of the Orange Book have certain typographical errors. These are listed below, and are shown as the correct form of the entry

ORANGE BOOK I

Analysis element, 16.6.1, 16.6.2.2
Atomic line, 16.7.6.1
Blackening, 16.8.3.6
Chromatograph (n.), 14.7.04
Combination electrode, 21.1.14
Compleximetry (complexometry), 8.08
Concentration, 16.6.2.1, 17.2.1, 18.4.1
Concentration ratio, 16.6.2.2
Controlled-potential polarography, 19.1.2
Convolution-integral linear-sweep voltammetry, 19/1/8
Counter electrode, 19.1.9
Derivative, 6/1/02, 19.1.1
Derivatographic analysis, T.6.2.17
Derivatography, T.6.2.17
Dropping mercury electrode, 19.1
Enzyme substrate electrode, 21.2.3.2
External standard, 16.8.4.4
Extraction constant, 12/03
Extraction indicator, 8.17.06
Fluorescent indicator, 8.17.07
Fluorimetric end-point, 8.12.05
Focal length, 16.5.1.4, 16.5.1.6
Gas-permeable membrane, 21.2.3.1
High-pressure xenon lamp, 18.3.2
Hurter and Driffield curve, 16.8.3.6
Indicator electrode, 19.1.6
Integral absorption, T.18.5.1
Integration of flux, 16.A.2.2
Intensity, 16.6.3.1
Intensity of radiation, 16.6.3.1, T.18.5.1
Intensity of spectral line, T.18.5.1
Intensity calibrating device, 16.8.4.1
Internal absorbance, T.18.5.1
Internal reference line, 16.6.3.1
Ionization energy, 16.7.4, T.18.6.1
Ionization potential, T.18.6.1
Mercury-pool electrode, 19.1.6
Mercury electrode, dropping, 19.1
Oscillopolarography, 19.5.7

Polarographic diffusion current constants, 20
Reference electrode, 19.1.9, 21.1.09
Reflection factor, 16.5.3.1
Refractive index, 16.5.2.1
Spectrograph, 16.5.1.1
Statistical weight, of particles in excitation states, T.18.6.1
Stoicheiometric end-point, 8.13
Total density (in atomic spectroscopy), T.18.6.2
Total quantum efficiency of fluorescence in spectroscopy, T.18.5.1
Transmission factor, 16.5.3.1, T.18.5.5
Transmission probability (spectroscopic), T.18.5.1
Ultra-trace analysis, 3.3
Wavelength, 16.4.2, T.18.5.1
Weighing procedures, 2.3.19

ORANGE BOOK II

Binding energy, 17.1.1
Breakdown jitters, 10.1.4.2
Concentric nebulizer, 10.3.1.1.1.2
Convolution-integral linear-sweep voltammetry, 8.2.7
Counter electrode, 8.1.1.43, 8.2.2, 10.1.3.4.1
Coupling unit, 10.1.5.1.1
Develop, 9.4.6.5
Direct-injection burner, 10.3.1.1.1.3
Discharge, DC biased oscillating, 10.1.4.1.2
Discontinuous procedures, 10.1.3.4.2
Electro-erosion, 10.1.3.4.4
Electron multiplication, 10.1.8.1.2
Energy yield of luminescence, 10.5.4.6
Excitation temperature, 10.1.2.3.2
Exoenergetic condition, 10.1.2
Fertile nuclides, 16.2.158
Flow, rate of, 8.1.1.67
Full energy peak, 16.2.184–185, 16.2.293 (not fuel...)
Gamma ray spectrum, 16.2.194
Gap resistor, 10.1.4.2 (not gas...)
Gas lasers, 10.1.7.3.2
High repetition rate sparks, 10.1.4.1.2
Indicator electrode, 8.1.1.44, 8.1.1.46, 8.2.5
Inelastic scattering, 10.1.2.5.2, 16.2.213, 16.2.372
Initial current, 10.1.4
Integral standard — delete
Internal standard, 9.4.6.17 — insert
Interstitial velocity at outlet pressure, 9.4.8.22
Ion-exchange chromatography, 9.4.4.3
Isothermic transition — delete
Isotopic carrier, 16.1.1, 16.2.61
LEED, 17.2.9 — insert
Low energy electron diffraction (LEED), 17.2.9 — insert
Meker burner, 10.3.1.1.1.13
Moderator, 16.2.249, 12.2.407
Negatron, 16.2.133 (and in text)
Nominal linear flow, 9.4.8.21, 9.4.10

Norm temperature, 10.1.2.3.2
Nuclear fuel, 16.2.180, 16.2.182–183
Nuclear fusion, 16.2.274
Peak base, 9.4.8.12, 9.4.8.14
Planck's law, 10.1.2.2.4, 10.1.2.3.2
Point-to-plane configuration, 10.1.4.9
Projection screen, 10.1.4.5
Pulsed laser sources, 10.1.7.1
Quench gas, 10.4.2.1.5
Radiation, 10.1.2.2, 10.1.8.1.1
Radioactive, 16.2.36, 16.2.340, 16.2.345–346, 16.2.417
Radioactive dating, 16.2.102
Raleigh scattering — delete
Rayleigh scattering, 10.1.2.5.2 — insert
Receiving time — delete
Resolving time, 16.2.90, 16.2.104–105, 16.2 360, 16.2.362 — insert
Rotating disc shutters, 10.1.7.2.3 (not ...acousto-optical shutters)
Rotating mirror shutters, 10.1.7.2.3
Scanning transmission electron microscopy, 17.2.6
Solvent extraction, 9.2.1
Spin conservation rule, 10.5.2.2
Surface analysis, 17, 17.2.40
Tilt of the photographic plate, 10.2.2.1.6
Transferred plasmas, 10.1.3.3
Working gas, 10.1.4.5

Most of the typographical errors in the texts have been dealt with in the comments on the entries in the English listing, but the reader should note that in the last line of the note on sensitivity (II/5, Section 2.2.10) 'analytes' should be replaced by 'analysts'.